PROTEIN POLYMORPHISM: ADAPTIVE AND TAXONOMIC SIGNIFICANCE

Editor-in-Chief
D. L. Hawksworth PhD DSc FLS FIBiol
Commonwealth Mycological Institute, Kew

Proceedings of an International
Symposium held in York

THE SYSTEMATICS ASSOCIATION
SPECIAL VOLUME NO. 24

PROTEIN POLYMORPHISM: ADAPTIVE AND TAXONOMIC SIGNIFICANCE

Edited by

G. S. OXFORD

Department of Biology, University of York, York, U.K.

and

D. ROLLINSON

Department of Zoology, British Museum (Natural History), London, U.K.

1983

Published for the

SYSTEMATICS ASSOCIATION

by

ACADEMIC PRESS

LONDON NEW YORK

PARIS SAN DIEGO SAN FRANCISCO SÃO PAULO

SYDNEY TOKYO TORONTO

ACADEMIC PRESS INC. (LONDON) LTD.
24–28 Oval Road
London NW1 7DX

U.S. Edition published by
ACADEMIC PRESS INC.
111 Fifth Avenue
New York, New York 10003

British Library Cataloguing in Publication Data

Protein polymorphism. – (The Systematics
Association special volume, ISSN 0309-2593; no. 24)
1. Chemotaxonomy – Congresses
I. Oxford, G.S. II. Rollinson, D.
III. Series
574'.021 QH83

ISBN 0-12-531780-8
LCCCN 83-70975

TYPESET BY OXFORD VERBATIM LIMITED

Contributors

F. A. Abreu-Grobois, *Department of Genetics, University College Swansea, Singleton Park, Swansea SA2 8PP, U.K.*

J. C. Avise, *Department of Molecular and Population Genetics, University of Georgia, Athens, Georgia 30602, U.S.A.*

F. J. Ayala, *Department of Genetics, University of California, Davis, California 95616, U.S.A.*

N. H. Barton, *Department of Genetics, University of Cambridge, Downing Street, Cambridge CB2 3EH, U.K.*

J. A. Beardmore, *Department of Genetics, University College Swansea, Singleton Park, Swansea SA2 8PP, U.K.*

L. Bullini, *Institute of Genetics, University of Rome, Città Universitaria, 00185 Rome, Italy.*

A. J. Cain, *Department of Zoology, University of Liverpool, Brownlow Hill, Liverpool L69 3BX, U.K.*

B. Christensen, *Zoological Laboratory, Universitatsparken 15, DK 2100 Copenhagen, Denmark.*

G. M. Davis, *Academy of Natural Sciences of Philadelphia, 19th and the Parkway, Philadelphia, Pennsylvania 19103, U.S.A.*

A. Ferguson, *Department of Zoology, The Queen's University, Belfast BT7 1NN, N. Ireland.*

C. C. Fleming, *Department of Agriculture (Northern Ireland), Agricultural Entomology Research Division, Nematology Laboratory, Felden, Mill Road, Newtownabbey BT36 7ED, N. Ireland.*

M. W. Flowerdew, *N.E.R.C. Unit of Marine Invertebrate Biology, University College of North Wales, Marine Science Laboratories, Menai Bridge LL59 5EH, U.K.*

R. A. Galleguillos, *Department of Biology and Marine Technology, Catholic University, Cafsilla 127, Talcahuano, Chile.*

I. Gibson, *School of Biological Sciences, University of East Anglia, Norwich NR4 7TJ, U.K.*

G. M. Hewitt, *School of Biological Sciences, University of East Anglia, Norwich NR4 7TJ, U.K.*

N. Khadem, *School of Biological Sciences, University of East Anglia, Norwich NR4 7TJ, U.K.*

J. Lokki, *Department of Genetics, University of Helsinki, P. Rautatiekatu 13, SF–00100 Helsinki 10, Finland.*

M. A. Miles, *Department of Medical Protozoology, London School of Hygiene and Tropical Medicine, Keppel Street, London WC1E 7HT, U.K.*

E. Nevo, *Institute of Evolution, University of Haifa, Haifa, Israel.*

A. Saura, *Department of Genetics, University of Helsinki, P. Rautatiekatu 13, SF – 00100 Helsinki 10, Finland.*

D. O. F. Skibinski, *Department of Genetics, University College Swansea, Singleton Park, Swansea SA2 8PP, U.K.*

J. P. Thorpe, *Department of Marine Biology, University of Liverpool, The Marine Biology Station, Port Erin, Isle of Man, U.K.*

S.-L. Varvio-Aho, *Department of Genetics, University of Helsinki, P. Rautatiekatu 13, SF – 00100 Helsinki 10, Finland.*

D. Walliker, *Institute of Animal Genetics, West Mains Road, Edinburgh EH9 3JN, U.K.*

R. D. Ward, *Department of Human Sciences, Loughborough University, Loughborough LE11 3TU, U.K.*

P. J. W. Young, *John Innes Institute, Colney Lane, Norwich NR4 7UH, U.K.*

Preface

This volume is derived from an international symposium organized by the Systematics Association on "Adaptation and Taxonomic Significance of Protein Variation" held at the University of York from July 13 to July 15 1982. The Systematics Association had previously dealt with some of the general principles of chemosystematics in 1967, and more recently in 1979, but wished to consider in greater detail certain issues arising from the study of protein polymorphisms in natural populations.

The enormous quantity of data generated since the introduction and application of the techniques of electrophoresis to the study of population genetics in the mid 1960's, has raised a number of intriguing questions. For example, how useful are enzyme characters, irrespective of whether they are adaptive or neutral, in the assignment of individuals to one taxonomic group or another? How do enzyme polymorphisms aid in the understanding of population structure and breeding systems? Of particular importance is the determination of the proportion of enzyme polymorphisms which are maintained by natural selection. If much of the variation is of adaptive significance, then should not evolutionary phenomena, such as convergence and parallelism, be taken into account when enzyme data are used in phylogenetic reconstruction? The structure of the meeting was designed to explore these four interconnected questions.

Emphasis was given almost entirely to the consideration of enzyme separation by electrophoresis and most of the papers were oriented towards animals. Participants came from many countries and represented many disciplines. Professor A. J. Cain, in a summary paper of the 1967 symposium, stressed that "The new taxonomy is peculiarly a field in which team work is necessary, and . . . team work by taxonomists, chemists, biochemists and anyone else with special knowledge can be stimulating, profitable and delightful". Professor R. J. Berry, at the 1979 syposium, added geneticists to the team and it is pleasing to report that they too were well represented at this meeting. The 21 papers in this volume include all but 5 of those presented at the symposium.

We acknowledge with thanks the contributions made by the Royal Society and the British Council towards the travel and expenses of speakers coming from overseas. We are also indebted to LKB, BioRad, Shandon, M.I. Scientific Ltd, and Taylor & Francis for their support.

August 1983 G.S.O.
York and London D.R.

Contents

Part 1 Enzymes as Taxonomic Characters

1 | Enzymes as Taxonomic Characters

F. J. AYALA

Department of Genetics, University of California, Davis, California 95616, U.S.A.

Abstract: Genetically determined allozyme polymorphisms are useful taxonomic characters. In many organisms, particularly where the speciation process occurs according to the model of geographic speciation, sibling and other closely related species, as well as subspecies and incipient species, can be diagnosed by means of allozyme polymorphisms. When speciation is saltational, little genetic differentiation is usually observed between closely related species; allozyme polymorphisms cannot be used for diagnostic purposes in these cases. A simple method is presented that permits the calculation of the probability of correct diagnosis of an individual on the basis of one or more enzyme polymorphisms.

INTRODUCTION

A taxonomic character is "any attribute of a member of a taxon by which it differs or may differ from a member of a different taxon" (Mayr 1969: 121). It is not my intention here to enter into a general discussion of taxonomic principles. Readers interested in the conceptual issues raised by the deceptively simple definition just given are referred to Mayr's (1969) discussion of the topic. For my purposes, the definition largely suffices, with two additional observations.

First, taxonomic characters that uniquely specify a given taxon are known as diagnostic characters (or diagnostic character "states," in the language of the phenetics school). It is my purpose to show that

Systematics Association Special Volume No. 24, 'Protein Polymorphism: Adaptive and Taxonomic Significance', edited by G. S. Oxford and D. Rollinson, 1983, Academic Press, London and New York enzymes are excellent diagnostic characters, particularly valuable for the identification of sibling, or morphologically nearly indistinguishable, species.

The second observation is that taxonomic characters are attributes of populations and not just of individuals as such. Many enzymes are genetically polymorphic (as are blood groups and other molecular attributes). Hence, individuals of the same population may differ from each other. Differences between populations and, therefore, frequency distributions are what is taxonomically significant, not just differences between individuals.

Enzyme (and protein) polymorphisms have been studied by a number of techniques: SDS electrophoresis (e.g. Snyder, 1977), two-dimensional chromatography and electrophoresis (e.g. Aquadro and Avise, 1981), immunology (e.g. Champion *et al.*, 1974), "fingerprinting" of peptides (e.g. Fletcher *et al.*, 1978), and amino acid sequencing. I shall, however, largely concentrate on the work done by the various standard methods of gel electrophoresis. These are rather simple procedures, which have been extensively used for taxonomic purposes. (For a general review of the use of biochemical methodologies in systematics, see Ferguson, 1980).

The eminent systematist Ernst Mayr wrote as recently as 1969: "A number of recent studies on gene-controlled enzymatic intrapopulation polymorphism indicate that much of the variation of the genotype is a poor source of taxonomic characters. It is thus evident that the direct study of the genetic basis of taxonomic characters is either technically impossible or else not very helpful" (Mayr, 1969: 123). These statements have been belied by an enormous body of work accumulated during the last decade and a half. Contrary to Mayr's claims, (1) gene-controlled enzyme variation has become a rich source of taxonomic characters, and (2) the direct study of the genetic basis of these taxonomic characters (enzymes) is technically feasible as well as very helpful.

The blame for Mayr's erroneous judgments should not be attributed only to him but as well to population geneticists engaged in the study of protein polymorphisms. These geneticists had uncovered large stores of enzyme variation within populations. In addition, a few comparative studies indicated that closely related species had at least some alleles in common at every enzyme locus studied. Emphasis was placed on this overlap of allelic frequencies, with the implied or explicit conclusion that enzyme polymorphisms could not be used as diagnostic characters (e.g. Prakash, 1969).

These population geneticists had failed to notice that for diagnostic purposes the significant parameter is not the frequency of alleles, but the frequency of *genotypes*. Individuals (and populations) are the subject of examination in taxonomic practice. In a paper, advisedly entitled "Allozymes as diagnostic characters of sibling species of *Drosophila*," we (Ayala and Powell, 1972) made this rather obvious point, and developed an appropriate method for using enzyme polymorphism data as taxonomic characters. I shall brieflly describe this method and enumerate the relevant assumptions.

PROBABILITY OF CORRECT DIAGNOSIS

Table I gives the allelic frequencies at each of four gene loci coding for enzymes in two sibling species, *Drosophila pseudoobscura* and *D. persimilis*. Investigators observe and record genotypes, but their results are usually presented in the form of allele frequencies. (This is done for reasons of economy of space: if the number of alleles is n, the number of possible genotypes is $n(n+1)/2$.) In outcrossing sexual organisms, genotypic frequencies can be estimated from the allelic frequencies by means of the Hardy–Weinberg formula. The proportion of overlap of the distribution of genotypic frequencies between any two populations can then be readily obtained for each locus. For example, *D. pseudoobscura* and *D. persimilis* share alleles 108 and 110 at the *Est-5* locus (see Table I). The expected frequencies of the genotypes carrying one or both alleles are:

	108/108	*108/110*	*110/110*
D. pseudoobscura	0·0121	0·0022	0·0001
D. persimilis	0·0001	0·0010	0·0025

The overlap of the genotypic distributions of the two species is, therefore, $0{\cdot}0001 + 0{\cdot}0010 + 0{\cdot}0001 = 0{\cdot}0012$.

In order to estimate the probability of assigning correctly an individual to one of the two species, one may assume that both species are equally common in a sample. The criterion, then, is simply to assign individuals with any given genotype to the species in which that

Table I. Frequencies of alleles at four loci coding for enzymes in two species of *Drosophila*.[a]

Gene	Alleles	*D. pseudoobscura*	*D. persimilis*
Est-5	92	0·01	–[b]
	96	0·08	–
	100	0·51	–
	102	0·01	–
	104	0·18	–
	106	0·08	–
	108	0·11	0·01
	110	0·01	0·05
	112	–	0·83
	114	–	0·08
	116	–	0·03
Me-1	96	0·05	0·02
	98	–	0·11
	99	–	0·78
	100	0·24	0·09
	101	0·48	–
	102	0·21	–
	103	0·02	–
Me-2	98	0·04	–
	100	0·57	–
	102	0·30	0·007
	104	0·08	0·86
	106	–	0·13
Hk-3	98	0·02	–
	100	0·20	–
	102	0·64	–
	104	0·14	0·90
	106	–	0·09

[a] Several alleles that occur with low frequencies are not included.
[b] A dash indicates that the allele has not been found in the species.

genotype has the higher frequency. Errors of attribution will be made with a frequency of half the amount of overlap of the genotypic distributions of the two species. For example, knowing their genotype at the *Est-5* locus, individuals belonging to either *D. pseudoobscura* or *D. persimilis* will be assigned to the wrong species with a frequency of $0 \cdot 0012/2 = 0 \cdot 0006$. The probability of correct diagnosis of the species relying only on the *Est-5* locus is 0·9994.

Proceeding similarly, one may calculate from Table I that the probability of erroneous species identification is 0·0060 for the *Me-1* locus, 0·0092 for *Me-2* and 0·0098 for *Hk-3*. If all four loci are jointly used, the probability of erroneous species identification of a given individual is $0 \cdot 0006 \times 0 \cdot 0060 \times 0 \cdot 0092 \times 0 \cdot 0098 = 3 \times 10^{-10}$. The diagnosis between these two sibling species can be accomplished with virtually absolute certainty on the basis of only four taxonomic characters.

The estimation of the probability of erroneous assignment is predicted on two assumptions. First, that the population is in Hardy–Weinberg equilibrium. This assumption is required when the allele frequencies, but not the genotype frequencies, are known. Populations of outcrossing sexual organisms are known to approximate fairly closely to the Hardy–Weinberg expectations and, therefore, this assumption is not likely to be a major source of error. Moreover, the practicing biologist who is collecting his own data will know the actual genotypic frequencies and these can be used for estimating the probability of erroneous attribution. But if the sample of individuals is not very large, the genotypic frequencies observed have themselves a certain probability of error (which can be calculated by standard statistical methods). Because the number of alleles at a polymorphic locus is smaller than the number of genotypes, the observed allele frequencies estimate the population frequencies with smaller probable errors than the observed genotypic frequencies estimate the actual genotypic frequencies in the population (Spiess, 1977). Thus, allele frequencies may be preferable for estimating probabilities of correct diagnosis and will, in practice, be sufficiently accurate in any case.

The second assumption made is that the two taxa are represented by equal numbers in the sample of individuals to be diagnosed. It is relatively simple to estimate the bias that this assumption will introduce as a function of the relative frequency of each taxon in the sample and of the genotypic frequencies. In practice, however, this will hardly be a worthwhile exercise. At worst, i.e. when it turns out that all individuals in the sample belong to one taxon (but we do not know it), the error of attribution will be twice as large as calculated on the assumption of equal frequency of the two taxa. For the *Est-5* example given above, the maximum probability of erroneous diagnosis would be 0·0012 rather than 0·0006.

CRITERIA FOR A DIAGNOSTIC LOCUS

Diagnostic characters are taxonomic characters that uniquely specify a given taxon. In order to classify an enzyme polymorphism as diagnostic or not, it becomes necessary to have some rule that sets forth the probability of error that we are willing to tolerate. Ayala and Powell (1972) have suggested two criteria, the second more stringent than the first: (i) when the probability of assigning an individual to the correct taxon is 99% or higher, and (ii) when that probability is 99·9% or higher. The four enzyme loci listed in Table I are diagnostic by criterion (i), but only *Est-5* is diagnostic by criterion (ii).

One related question of practical interest is the following. Assume that we are searching for enzyme loci that would be diagnostic between certain taxa. We would want to know what proportion of the loci studied turn out to be diagnostic. The four loci displayed in Table I were found in a study of 26 enzyme loci. Thus, 4/26 = 0·154 of the enzymes are diagnostic between *D. pseudoobscura* and *D. persimilis* by criterion (i), but only 1/26 = 0·038 by criterion (ii). Table II lists the proportion of diagnostic loci between any two taxa found in a survey

Table II. Percentage of loci that are diagnostic between any two *Drosophila* species.[a]

	D. pseudoobscura	*D. persimilis*	*D. miranda*	*D. azteca*
D. pseudoobscura		15·4	27·8	47·6
D. persimilis	3·8		22·2	52·4
D. miranda	16·7	16·7		38·9
D. azteca	33·3	28·6	33·3	

[a] Above the diagonal: correct diagnosis is made with a probability 0·99 or higher at each locus. Below the diagonal: the probability of correct diagnosis is 0·999 or higher at each locus.

of enzyme polymorphisms in four species: the first three are sibling species, whereas *D. azteca* is morphologically distinguishable from the other three, although closely related to them.

GENETIC DIFFERENTIATION VERSUS TAXONOMIC RESOLUTION

Numerous studies of enzyme polymorphisms have been performed during the last 15 years in a great variety of organisms. In many instances, these studies have involved comparisons of different taxa. A variety of statistical methods have been devised to quantify the amount of genetic differentiation between populations, which measure in various ways the amount of overlap between the distributions of genotypic frequencies in two populations at each locus. The average of all the loci studied provides an estimate of the amount of genetic differentiation between the populations.

A substantial amount of genetic differentiation may, however, arise either by many loci that have a moderate amount of differentiation, or by complete differentiation at only a few loci, with great similarity or identity at the other loci studied. For taxonomic purposes, these two situations are quite distinct. A locus at which complete differentiation exists between two populations can be used to diagnose the population to which an individual belongs. A locus at which only partial differentiation occurs cannot be so used. Several loci at which two populations are partially different can, of course, be jointly used for diagnostic purposes, but statistical manipulations are more complex.

ALLOZYME DIFFERENTIATION IN *DROSOPHILA WILLISTONI*

I shall now review an extensive study of enzyme polymorphism in the *Drosophila willistoni* group of species. As we shall see, not only closely related species, but even subspecies, can be diagnosed by means of enzyme differences. Data from other organisms will be reviewed later.

The *willistoni* group of *Drosophila* consists of at least 15 closely related species endemic to the tropics of the New World. Six species are siblings, morphologically nearly indistinguishable, although the species of individual males can be identified by slight but diagnostically reliable differences in their genitalia. Two sibling species, *D. insularis* and *D. pavlovskiana*, are narrow endemics; the former in some islands of the Lesser Antilles, and the latter in Guyana. Four other siblings, namely, *D. willistoni, D. equinoxialis, D. tropicalis* and *D. paulistorum* have wide and largely overlapping geographic distributions through Central America, the Caribbean and much of continental South America.

Some sibling species consist of at least two subspecies. Populations of *D. willistoni* west of the Andes near Lima, Peru, belong to the subspecies *D. w. quechua*, while east of the Andes and elsewhere the subspecies is *D. w. willistoni*. Incipient reproductive isolation in the form of partial hybrid sterility exists between these two subspecies. Laboratory crosses of *D. w. willistoni* females with *D. w. quechua* males yield fertile males and females. However, crosses between *D. w. quechua* females and *D. w. willistoni* males from continental South America east of the Andes produce fertile females but sterile males. Laboratory tests show no evidence of ethological (sexual) isolation between the subspecies.

Two subspecies are also known in *D. equinoxialis*: *D. e. caribbensis* in Central America, north of Panama, and in the Caribbean islands; and *D. e. equinoxialis* in eastern Panama and continental South America. Crosses between the two subspecies yield fertile females but sterile males, independently of the subspecies of the female parent. As in *D. willistoni*, there is no evidence of sexual isolation between the subspecies of *D. equinoxialis*.

Evolutionary divergence beyond the taxonomic category of subspecies, but without complete achievement of speciation exists in a third sibling species, *D. paulistorum*. This "species" consists of at least six semispecies, or incipient species, named Centroamerican, Transitional, Andean-Brazilian, Amazonian, Orinocan and Interior. Laboratory crosses between the semispecies generally yield fertile females but sterile males. Sexual isolation is essentially complete between some semispecies particularly when sympatric populations are tested. Gene flow among the semispecies is nevertheless possible, particularly through populations of the Transitional semispecies.

In summary, five increasingly divergent levels of evolutionary divergence can be recognized in the *D. willistoni* group:

(1) Between geographic populations of the same taxon.

(2) Between subspecies. These are allopatric populations that exhibit incipient reproductive isolation in the form of partial hybrid sterility.

(3) Between the semispecies of *D. paulistorum*. The process of speciation is being completed between the semispecies.

(4) Between sibling species. In spite of their morphological similarity, the sibling species are completely reproductively isolated.

(5) Between morphologically distinguishable species of the same group. *D. nebulosa* is a close relative of *D. willistoni* and its siblings, but can be easily distinguished from them by external morphology.

Using electrophoretic techniques, 36 gene loci coding for enzymes have been studied in each of the sibling species of the *D. willistoni* group and in the morphologically differentiated *D. nebulosa*. The genotypes of large numbers of individuals (from several hundred to several thousand) have been ascertained at each locus in each species, except for the two narrow endemics, *D. insularis* and *D. pavlovskiana*, of which only a few genomes have been sampled. Genetic similarity and differentiation between populations at various stages of evolutionary divergence are given in Table III. The statistics used are (Nei, 1972): *I* (*genetic identity*) which estimates the proportion of genes that remain identical in two populations, and *D* (*genetic distance*) which estimates the proportion of gene substitutions that have taken place in the separate evolution of two populations. *I* may range in value from zero to one; *D* may range in value from zero to infinity, because it is possible that the complete replacement of one allele (or sets of alleles) by another may have happened more than once at any gene locus (Ayala *et al.*, 1974a).

Table III. Average genetic identity, *I*, and genetic distance, *D*, between taxa of various levels of evolutionary divergence in the *Drosophila willistoni* group.[a]

Taxonomic level	*I*	*D*
Local populations	0·970 ± 0·006	0·031 ± 0·007
Subspecies	0·795 ± 0·013	0·230 ± 0·016
Semispecies	0·798 ± 0·026	0·226 ± 0·033
Sibling species	0·563 ± 0·023	0·581 ± 0·039
Nonsibling species	0·352 ± 0·023	1·056 ± 0·068

[a] The standard error is given after each mean value.

The first level of comparison is between local populations of the same taxon. These are populations that give no evidence of having developed any degree of reproductive isolation: when intercrossed they produce completely fertile and viable offspring. There is very little genetic differentiation among local populations of the *D. willistoni* group, $I = 0{\cdot}970$; $D = 0{\cdot}031$.

The second level of comparison is between subspecies. The amount of genetic differentiation is substantial. On average, $D = 0{\cdot}230$: about 23 complete gene substitutions have taken place for every 100 loci in the separate evolution of two subspecies. The genetic distance is $D = 0{\cdot}246$ between the two subspecies of *D. equinoxialis*, and $D = 0{\cdot}214$ between the two subspecies of *D. willistoni*.

The third level of comparison is between the semispecies of *D. paulistorum*. The average genetic distance between the semispecies is $D = 0{\cdot}226$, not significantly different from the value observed between subspecies.

The fourth level of comparison is between the sibling species. The mean genetic distance between them is $D = 0{\cdot}581$, i.e. on average, about 58·1 electrophoretically detectable allelic substitutions for every 100 loci have occurred in each pair of siblings since their divergence from a common ancestral population. In spite of their morphological and ecological similarity, these species are genetically quite different.

The fifth and final level of comparison is between morphologically different (but closely related) species, i.e. between *D. nebulosa* and each one of the sibling species. The average genetic distance for these comparisons is $D = 1{\cdot}056$; thus, on average, about one electrophoretically detectable allelic substitution per locus has taken place in the separate evolution of *D. nebulosa* and each of the sibling species since they diverged from the last common ancestors.

DIAGNOSTIC LOCI IN *DROSOPHILA WILLISTONI*

For the present purposes, we are primarily interested not in the average value of I (or D), but in the distribution of loci with respect to these statistics. I shall use here distributions of genetic identity, because the range of I is bounded between 0 and 1. (D may range from 0 to infinity. Because of the simple mathematical relationship between these two statistics: $D = -\log_e I$, the distribution of I becomes exactly the distribution of D if a natural logarithm scale is used for this statistic and the polarity from 0 to 1 is reversed.)

The distribution of I for comparisons between local populations of the same taxon is shown in Fig. 1. At most loci, local populations have essentially identical allelic frequencies ($I > 0 \cdot 95$). There is, however,

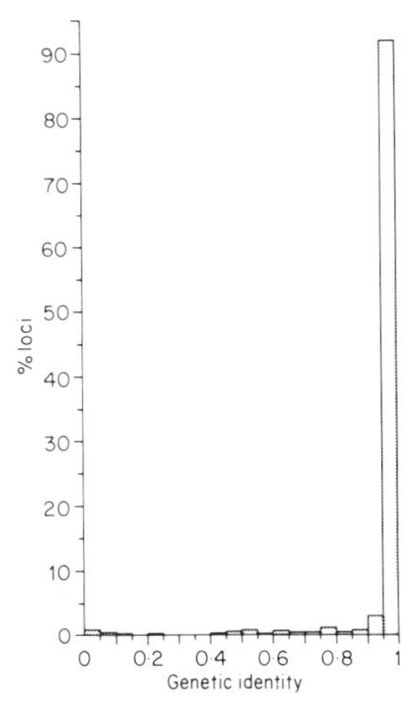

Fig. 1. Frequency distribution of genetic identity among gene loci encoding enzymes for comparisons between local populations of the same taxon of the *Drosophila willistoni* group.

an occasional locus at which some populations may differ from moderately to quite substantially. Whenever nearly complete allelic differentiation occurs, this differentiates one population from others, but not from all. One extreme example of allelic differentiation is shown in Table IV. These four Venezuelan populations of *D. equinoxialis* are quite heterogeneous at the *Ao-2* locus (Ayala *et al.*,

Table IV. Allele variation at the *Ao-2* locus in four natural populations of *Drosophila equinoxialis* from Venezuela.[a]

	Alleles					
Locality	101	102	103	104	105	106
Catatumbo	0·00	0·03	0·19	0·71	0·00	0·06
Caripito	0·00	0·40	0·05	0·48	0·00	0·05
Tucupita	0·01	0·12	0·32	0·47	0·05	0·03
El Dorado	0·12	0·82	0·01	0·01	0·00	0·00

[a] Some rare alleles are omitted.

1974b). The locus is diagnostic between El Dorado and either Catatumbo or Tucupita, but not for any of the other four possible pairwise comparisons.

The situation is more interesting for comparisons between different subspecies. Figure 2 shows that populations of different subspecies have remained essentially identical at a majority of loci, but have

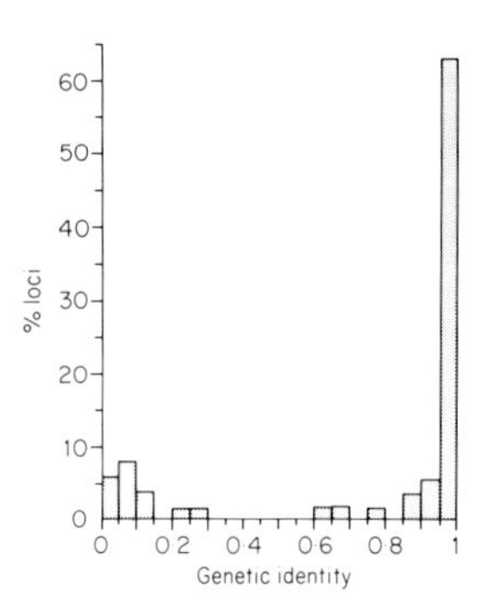

Fig. 2. Frequency distribution of genetic identity among enzyme loci for comparisons between subspecies of the *Drosophila willistoni* group.

become quite different at a few loci. Table V gives the allelic frequencies at the five loci which are most different in the two subspecies of *D. willistoni*. Four of these loci are diagnostic by the criteria given above and the fifth one (*Est-7*) very nearly so. If all five loci are jointly used, the probability of incorrect diagnosis of the subspecies of a single individual of known genotype is $3 \cdot 4 \times 10^{-14}$. Table VI displays similar information for the two subspecies of *D. equinoxialis*. In this case, one locus is diagnostic by the strictest criterion given above (probability of correct diagnosis $\geqslant 0 \cdot 999$), but four other loci are nearly diagnostic and usable for the purpose. When all five loci are jointly used, the probability of incorrect diagnosis of the subspecies of an individual is 8×10^{-11}, obviously quite satisfactory in practice.

It deserves notice that the formal description of the subspecies of *D. willistoni* and *D. equinoxialis* (Ayala, 1973) relies exclusively on enzyme differences as diagnostic characters between the subspecies. This was apparently the first instance of such practice. Another example of the use of enzyme variation in the formal description of a new subspecies is *D. pseudoobscura bogotana* (Ayala and Dobzhansky, 1974). On the basis of six enzyme loci, an individual can be assigned to this sub-

Table V. Allele frequencies at five loci coding for enzymes in two subspecies of *Drosophila willistoni*.[a]

Subspecies	Locus and alleles					Probability of correct diagnosis of the subspecies
			Xdh			
	95	*97*	*98*	*100*	*101*	
D. w. willistoni	0·000	0·007	0·114	0·468	0·322	0·99998
D. w. quechua	0·547	0·422	0·000	0·000	0·000	
			Est-2			
	98	*100*	*102*	*104*		
D. w. willistoni	0·003	0·041	0·941	0·006		0.999
D. w. quechua	0·231	0·769	0·000	0·000		
			Odh			
	96	*100*	*104*			
D. w. willistoni	0·039	0·882	0·071			0.994
D. w. quechua	0·016	0·000	0·984			
			Est-4			
	98	*100*	*102*	*104*		
D. w. willistoni	0·004	0·146	0·838	0·011		0.989
D. w. quechua	0·000	1·000	0·000	0·000		
			Est-7			
	96	*98*	*100*	*102*	*105*	
D. w. willistoni	0·025	0·147	0·563	0·211	0·049	0·974
D. w. quechua	0·874	0·000	0·153	0·000	0·000	

[a] Some rare alleles are omitted.

species or to the nominate subspecies, *D. p. pseudoobscura*, with a probability of error of 5×10^{-15}.

The distribution of *I* for comparisons between the semispecies of *D. paulistorum* is depicted in Fig. 3. Although these semispecies exhibit moderate to large sexual isolation from each other (in addition to male hybrid sterility, like the subspecies of *D. willistoni* and *D. equinoxialis*), their genetic differentiation is no greater than between subspecies (see Table III). In addition, there are fewer diagnostic loci between the semispecies, some of which cannot be reliably diagnosed by the use of enzymes (Richmond, 1972; Ayala *et al.*, 1974b).

Table VI. Allele frequencies at five loci coding for enzymes in two subspecies of *Drosophila equinoxialis*.[a]

Subspecies	Locus and alleles							Probability of correct diagnosis of the subspecies
	Mdh-2							
	94	*106*						
D. e. equinoxialis	0·994	0·000						0.9997
D. e. caribbensis	0·0006	0·989						
	Acph-1							
	88	*94*	*100*	*102*	*104*	*106*	*108*	
D. e. equinoxialis	0·000	0·013	0·172	0·000	0·811	0·000	0·014	0·986
D. e. caribbensis	0·635	0·027	0·113	0·005	0·019	0·176	0·010	
	G3pd							
	92	*96*	*100*					
D. e. equinoxialis	0·000	0·063	0·906					0·977
D. e. caribbensis	0·019	0·939	0·029					
	Hk-1							
	96	*100*						
D. e. equinoxialis	0·082	0·914						0·975
D. e. caribbensis	0·982	0·009						
	Est-4							
	98	*100*	*102*					
D. e. equinoxialis	0·150	0·769	0·081					0.967
D. e. caribbensis	0·954	0·036	0·001					

[a] Some rare alleles are omitted.

The substantial degree of genetic differentiation between the sibling species of the *D. willistoni* group (see Table III) is reflected in bimodal distributions of the individual loci. Figure 4 displays these distributions for pairwise comparisons between the four widely distributed species. About half the loci remain effectively identical between any two sibling species, and about 20–40% have become completely differentiated, with the balance representing intermediate situations. The proportion of diagnostic loci for all pairwise comparisons between the six sibling species is given in Table VII. On average, 25% of the loci studied each permit a correct decision as to which of two

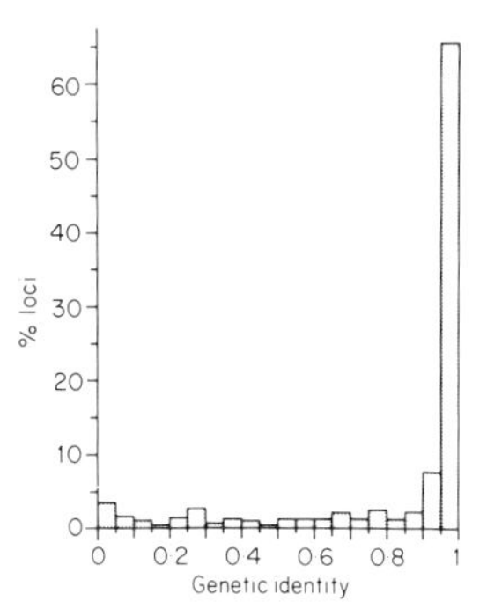

Fig. 3. Frequency distribution of genetic identity among enzyme loci for comparison between semispecies of *Drosophila paulistorum*.

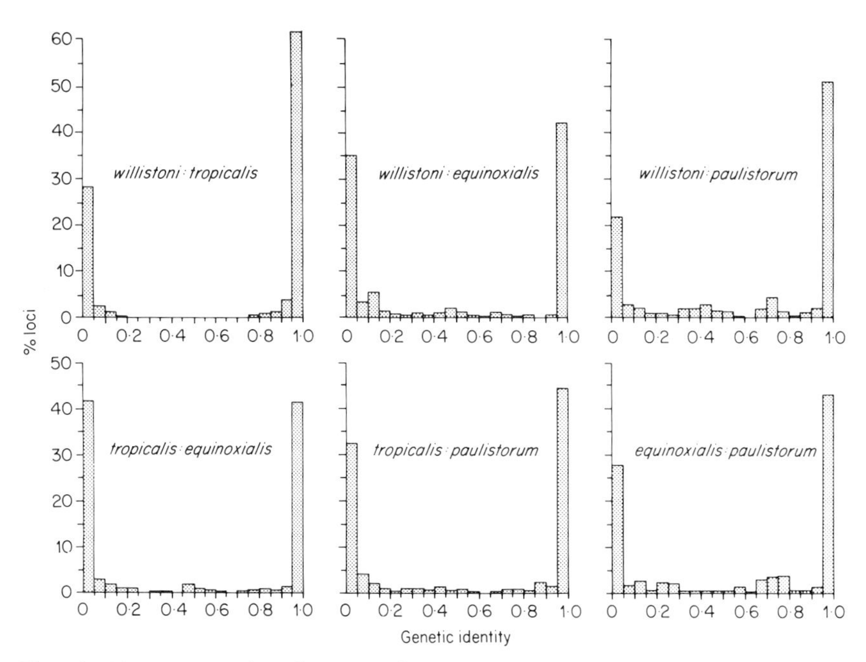

Fig. 4. Frequency distribution of genetic identity among enzyme loci for comparisons between populations of different sibling species of the *Drosophila willistoni* group. The species compared are shown in each histogram.

Table VII. Percentage of loci that are diagnostic between any two species.[a]

	D. willistoni	*D. tropicalis*	*D. equinoxialis*	*D. paulistorum*	*D. pavlovskiana*	*D. insularis*	Average
D. willistoni		17·9	21·4	25·0	25·0	32·1	24·3
D. tropicalis	10·7		21·4	35·7	28·6	28·6	26·4
D. equinoxialis	7·1	10·7		14·3	25·0	28·6	22·1
D. paulistorum	7·1	10·7	10·7		14·3	32·1	24·3
D. pavlovskiana	10·7	10·7	14·3	3·6		32·1	25·0
D. insularis	14·3	10·7	14·3	10·7	14·3		30·7
Average	10·0	10·7	11·4	8·6	10·7	12·1	

[a] Above the diagonal: correct diagnosis is made with a probability 0·99 or higher at each locus.
Below the diagonal: the probability of correct diagnosis in 0·999 or higher at each locus.

subspecies an individual belongs with a probability of 0·99 or higher; 11% of the loci make it possible to do so with a probability of 0·999 or higher for each locus. If the various diagnostic loci are jointly used, an individual can be assigned to any one of two species with a probability of error of the order or 10^{-20} or smaller.

The distribution of *I* among loci for comparisons between *D. nebulosa* and each of the four common sibling species is shown in Fig. 5. In every case, somewhat more than half the gene loci have become completely differentiated and may be used for diagnostic purposes. It should be noted, however, that this result has lesser taxonomic significance in this case than in the previous comparisons because *D. nebulosa* can be readily distinguished from the sibling species by simple examination of the external morphology.

Considering the wealth of morphologically indistinguishable taxa in the *D. willistoni* group, we may raise the question whether an individual can be correctly diagnosed on the basis of the allozymes when all taxa are simultaneously considered. Table VIII gives a general overview of the situation, based on an analysis of principal components. A given number in the body of the table (e.g. 2)

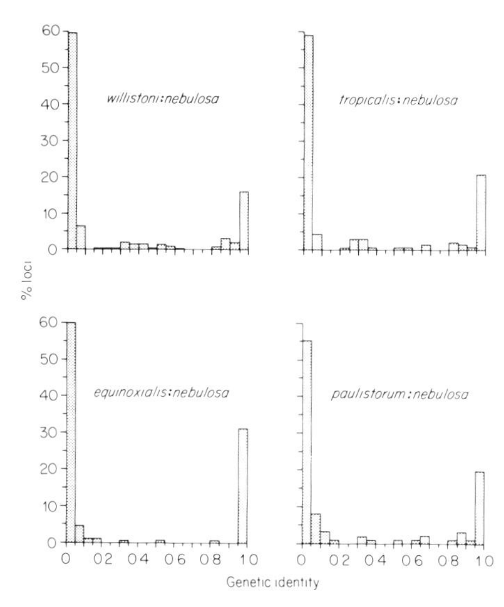

Fig. 5. Frequency distribution of genetic identity among enzyme loci for comparisons between populations of morphologically distinguishable species of the *Drosophila willistoni* group. The species compared are shown in each histogram.

Table VIII. Principal components and proportion of variance explained for each of 21 gene loci in 14 taxa of the *D. willistoni* group.

Loci	Taxa[a]														% explained of total variance
	t	*ww*	*wq*	*ee*	*ec*	AM	IN	OR	AN	CA	T	*pv*	*i*	*n*	
Lap-5	2	2	2	1	1	1	1	3	1	1	1	3	4	2	98·3
Est-2	1	2	1	2	1	1	1	1	1	1	1	3	4	3	97·0
Est-4	2	2	1	1	3	1	1	1	1	1	4	1	5	3	99·9
Est-5	2	1	1	1	3	1	1	1	1	1	1	1	2	2	99·9
Est-7	2	2	5	3	3	1	1	1	1	1	1	3	3	4	85·3
Acph-1	4	3	3	2	4	1	1	2	1	1	1	2	1	1	89·4
Ald-1	1	1	1	1	1	1	1	1	1	1	1	1	1	1	96·0
Adh	3	1	1	1	1	1	1	1	1	1	1	1	1	2	99·9
Mdh-2	3	2	2	1	4	1	1	1	1	1	1	1	3	4	86·9
α-Gpdh	1	1	1	1	1	1	1	1	1	1	1	1	1	1	98·0
Idh	1	1	1	1	1	1	1	1	1	1	1	1	1	1	99·6
Me-2	1	1	1	1	1	1	2	2	1	1	1	2	1	3	89·5
Xdh	3	1	3	1	1	1	1	1	1	1	1	1	2	2	77·3
To	1	1	1	1	1	1	1	1	1	1	1	1	2	2	92·0
Tpi-2	1	1	1	2	2	1	1	1	1	1	1	1	1	3	99·9
Pgm-1	2	1	1	2	2	1	1	1	1	1	1	1	2	3	99·9
Adk-1	1	1	1	1	1	1	1	1	1	1	1	1	2	3	99·3
Adk-2	1	1	1	2	2	1	1	1	1	1	1	1	2	2	99·5
Hk-1	2	2	2	2	3	1	1	1	1	1	1	1	3	1	90·1
Hk-2	1	1	1	1	1	1	1	1	1	1	1	1	1	1	98·0
Hk-3	1	1	1	1	1	2	2	2	1	1	1	1	3	1	99·8

[a] Symbols for the taxa: *t* = *D. tropicalis*; *ww* = *D. willistoni willistoni*; *wq* = *D. w. quechua*; *ee* = *D. equinoxialis equinoxialis*; *ec* = *D. e. caribbensis*. *D. paulistorum semispecies*: AM = Amazonian; IN = Interior; OR = Orinocan; AN = Andean; CA = Centroamerican; T = Transitional; *pv* = *D. pavlovskiana*, *i* = *D. insularis*; *n* = *D. nebulosa*.

represents a unique configuration of allelic frequencies identified as a principal component. Different principal components (e.g. 2 and 3) are orthogonal to each other, i.e. share no alleles in common. The table gives for each taxon at each locus the principal component which explains most of the variance. With the single exception of the Andean and Centroamerican semispecies of *D. paulistorum*, no two taxa share the same set of principal components. Thus, the 21 enzyme loci used for the analysis provide a distinctive pattern for every taxon, making it possible to identify to which one of 13 taxa (one of these 13 "taxa" includes the two indistinguishable semispecies just noted) an individual belongs. Needless to say, geographical location and other characteristics may help in the identification process, but the taxonomic resolving power of enzyme data is apparent. One important consequence of the widespread use of enzyme electrophoresis in population studies has been the discovery of new sibling species, particularly in organisms that do not lend themselves to traditional genetic studies (Ferguson, 1980; Bullini and Sbordoni, 1980).

ALLOZYME DIFFERENTIATION IN OTHER ORGANISMS

The degree of allozyme differentiation between taxa at various stages of evolutionary divergence observed in the *D. willistoni* group is fairly typical for those sexually reproducing organisms in which speciation occurs primarily according to the allopatric or peripatric model (Ayala, 1975). A summary is given in Table IX. The distribution of genetic identity among loci also is usually bimodal as exemplified in Fig. 6 for six sibling species of the *Anopheles maculipennis* complex (Bullini, 1980) and in Fig. 7 for nine closely related genera of fishes from California (Avise and Ayala, 1976). Typically, closely related species have I values around 0·50, around 20–30% of the loci being diagnostic between any two closely related species.

The degree of genetic differentiation between incipient or closely related species is, however, quite low in those cases where new species arise rapidly, by any of the processes encompassed under the name of "quantum" or "saltational" speciation (Ayala, 1975; White, 1978).

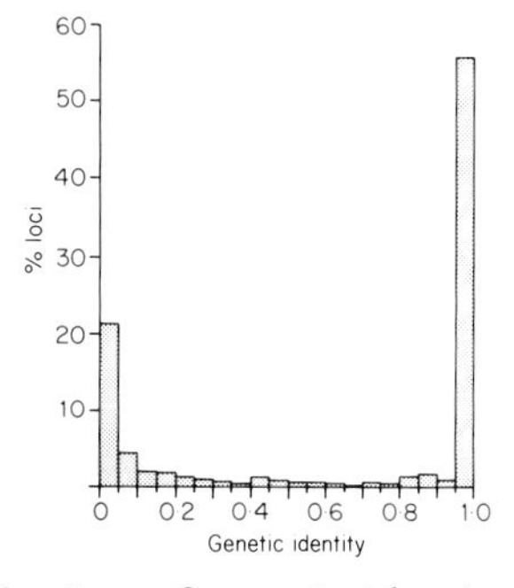

Fig. 6. Frequency distribution of genetic identity among enzyme loci for comparisons between palaearctic sibling species of the *Anopheles maculipennis* complex.

Table IX. Genetic differentiation at various stages of evolutionary divergence in several groups of organisms. The average genetic distance, D, is given first, followed by the average genetic identity, (I) in parenthesis. (After Ayala, 1982.)

Organisms	Local populations	Subspecies	Incipient species	Species and closely related genera
Drosophila	0·013 (0·987)	0·163 (0·851)	0·239 (0·788)	1·066 (0·381)
Other invertebrates	0·016 (0·985)	–	–	0·878 (0·465)
Fishes	0·020 (0·980)	0·163 (0·850)	–	0·760 (0·531)
Salamanders	0·017 (0·984)	0·181 (0·836)	–	0·742 (0·520)
Reptiles	0·053 (0·949)	0·306 (0·738)	–	0·988 (0·437)
Mammals	0·058 (0·944)	0·232 (0·793)	0·263 (0·769)	0·559 (0·620)
Plants	0·035 (0·966)	–	–	0·808 (0·510)

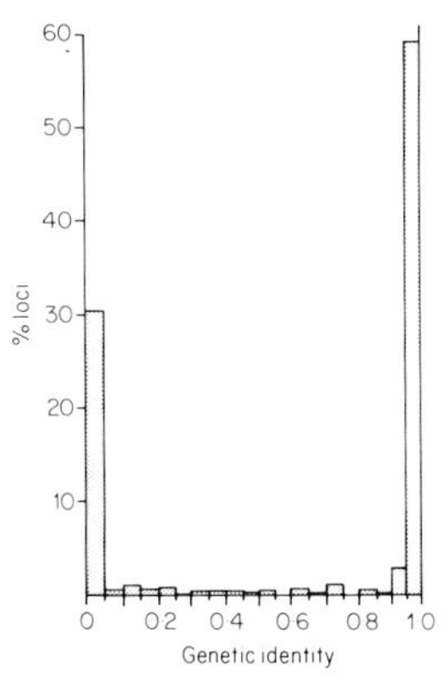

Fig. 7. Frequency distribution of genetic identity among enzyme loci for comparisons among species from nine closely related genera of California minnows.

One example is given in Table X. These species of spiny rats from Venezuela belong to the *Proechimys guairae* superspecies (Benado *et al.*, 1979). The parapatrically distributed species are morphologically virtually indistinguishable, but are reproductively isolated by Robertsonian and other chromosomal rearrangements. The diploid number of chromosomes ranges from 46 to 62; yet, with respect to allozymes, they are genetically no more different than local populations of many species and no locus is diagnostic. Other examples of little genetic differentiation between species recently arisen by saltational speciation are exhibited in Table XI. It is apparent that allozymes are not a good source of diagnostic characters in these species.

Table X. Genetic distance (above the diagonal) and genetic identity (below the diagonal) between Venezuelan species of the *Proechimys guairae* superspecies. The species are identified by their diploid (2n) chromosome number and the locality where samples were obtained.

Species	$2n = 46$ San Esteban	$2n = 48$ San Antonio	$2n = 50$ San Carlos	$2n = 62$ Barinitas
$2n = 46$		0·035	0·039	0·043
$2n = 48$	0·966		0·033	0·046
$2n = 50$	0·962	0·968		0·025
$2n = 62$	0·958	0·956	0·976	

Table XI. Genetic Identity, *I*, and genetic distance, *D*, between taxa arisen by saltational speciation.

Populations compared	*I*	*D*	Reference
Plants:			
Stephanomeria exigua versus *S. malheurensis*	0·945	0·057	Gottlieb (1975)
Rodents:			
Spalax ehrenbergi	0·978	0·022	Nevo and Shaw (1972)
Thamomys talpoides	0·925	0·078	Nevo *et al.* (1974)
Proechimys guairae complex	0·968	0·032	Benado *et al.* (1979)
Insects:			
Drosophila sylvestris versus *D. heteroneura*	0·939	0·063	Sene and Carson (1977)
Culex pipiens pipiens versus *C. p. autogenicus*	>0·950	<0·050	Bullini and Sbordoni (1980)

CONCLUDING REMARKS

In summary, allozymes are often good taxonomic characters, particularly useful to identify incipient or sibling species. Their usefulness in the case of morphologically very similar taxa is due to a number of properties, including the following: (1) Often, species or incipient species are considerably differentiated, whereas local populations of the same taxon remain practically identical. (2) Because the allozymes are immediate expressions of the genotype, they are constant and little subject to phenotypic modification by age, environmental effects, or other factors. (3) Usually not one, but several gene loci are diagnostic and each locus is an independent source of taxonomic information. (4) There is no problem with *a priori* weighting of the characters – each locus is accorded equal value; hence, several loci can be jointly used and the joint probability of incorrect diagnosis can easily be calculated as the product of the probabilities for each of the various diagnostic loci. One important restriction, however, is that allozymes can be studied only in living organisms. Moreover, in small organisms, such as many insects, the organism needs to be sacrificed in order to perform the enzyme assays. (For additional evaluation of the advantages and disadvantages of enzyme electrophoresis for taxonomic purposes, see Avise, 1974).

REFERENCES

Aquadro, D. F. and Avise, J. C. (1981). Genetic divergence between rodent species assessed by using two-dimensional electrophoresis. *Proc. Nat. Acad. Sci. U.S.A.* **78**, 3784–3788.

Avise, J. C. (1974). Systematic value of electrophoretic data. *Syst. Zool.* **23**, 465–481.

Avise, J. C. and Ayala, F. J. (1976). Genetic differentiation in speciose versus depauperate phylads: evidence from the California minnows. *Evolution* **30**, 46–58.

Ayala, F. J. (1973). Two new subspecies of the *Drosophila willistoni* group (Diptera: Drosophilidae). *Pan-Pac. Entom.* **49**, 273–279.

Ayala, F. J. (1975). Genetic differentiation during the speciation process. *Evol. Biol.* **8**, 1–78.

Ayala, F. J. (1982). "Population and Evolutionary Genetics. A Primer." Benjamin/Cummings, Menlo Park, California.

Ayala, F. J. and Dobzhansky, Th. (1974). A new subspecies of *Drosophila pseudoobscura* (Diptera: Drosophilidae). *Pan-Pac. Entom.* **50**, 211–219.

Ayala, F. J. and Powell, J. R. (1972). Allozymes as diagnostic characters of sibling species of *Drosophila*. *Proc. Nat. Acad. Sci., U.S.A.* **69**, 1094–1096.

Ayala, F. J., Tracey, M. L., Hedgecock, D. and Richmond, R. C. (1974a). Genetic differentiation during the speciation process in *Drosophila*. *Evolution* **28**, 576–592.

Ayala, F. J., Tracey, M. L., Barr, L. G., McDonald, J. F. and Pérez-Salas, S. (1974b). Genetic variation in natural populations of five *Drosophila* species and the hypothesis of the selective neutrality of protein polymorphisms. *Genetics* **77**, 343–384.

Benado, M., Aguilera, M., Reig, O. A. and Ayala, F. J. (1979). Biochemical genetics of Venezuelan spiny rats of the *Proechimys guairae* and *Proechimys trinitatis* superspecies. *Genetics* **50**, 89–97.

Bullini, L. (1980). Aspetti genetici, ecologici ed etologici del processo di speciazione negli animali. *Acc. Naz. Lincei, Contr. del Centro Linceo Interdisciplinare di scienze matematiche e loro applicazioni. VI Seminario sulla Evoluzione Biologica, Roma,* **51**, 29–59.

Bullini, L. and Sbordoni, V. (1980). Electrophoretic studies of gene-enzyme systems: microevolutionary processes and phylogenetic inference. *Boll. Zool.* **47s**, 95–112.

Champion, A. B., Prager, E. M., Wachter, D. and Wilson, A. C. (1974). Microcomplement fixation. *In* "Biochemical and Immunological Taxonomy of Animals" (C. A. Wright, ed.), pp. 397–416. Academic Press, London.

Ferguson, A. (1980). "Biochemical Systematics and Evolution." Wiley, New York.

Fletcher, T. S., Ayala, F. J., Thatcher, D. R. and Chambers, G. K. (1978). Structural analysis of the ADH^S electromorph of *Drosophila melanogaster. Proc. Nat. Acad. Sci. U.S.A.* **75**, 5609–5612.

Gottlieb, L. D. (1975). Allelic diversity in the outcrossing annual plant *Stephanomeria exigua* ssp. *carotifera* (Compositae). *Evol., Lancaster, Pa.* **29**, 213–225.

Mayr, E. (1969). "Principles of Systematic Zoology." McGraw-Hill, New York.

Nei, M. (1972). Genetic distance between populations. *Am. Nat.* **106**, 283–291.

Nevo, E. and Shaw, C. R. (1972). Genetic variation in a subterranean mammal, *Spalax ehrenbergi. Biochem. Genet.* **7**, 235–241.

Nevo, E., Kim, Y. J., Shaw, C. R. and Thaeler, C. S. (1974). Genetic variation, selection, and speciation in *Thomomys talpoides* pocket gophers. *Evol., Lancaster, Pa.* **28**, 1–23.

Prakash, S. (1969). Genic variation in a natural population of *Drosophila persimilis. Proc. Nat. Acad. Sci. U.S.A.* **62**, 778–784.

Richmond, R. (1972). Enzyme variability in the *Drosophila willistoni* group. III. Amounts of variability in the superspecies *D. paulistorum. Genetics* **70**, 87–112.

Sene, F. M. and Carson, H. L. (1977). Genetic variation in Hawaiian Drosophila. IV. Allozymic similarity between *D. sylvestris* and *D. heteroneura* from the island of Hawaii. *Genetics* **86**, 187–198.

Snyder, T. P. (1977). A new electrophoretic approach to biochemical systematics of bees. *Biochem. Syst. Evol.* **5**, 133–150.

Spiess, E. B. (1977). "Genes in Populations." Wiley, New York.

White, M. J. D. (1978). "Modes of Speciation." W. H. Freeman, San Francisco.

2 Enzyme Variation in Malaria Parasite Populations

D. WALLIKER

Institute of Animal Genetics, West Mains Road, Edinburgh EH9 3JN, U.K.

Abstract: Enzyme electrophoresis has revealed considerable genetic diversity in populations of malaria parasites. Studies on the parasites infecting rodents have shown that each species and subspecies can be distinguished by its enzymes. Within each subspecies, extensive random mating appears to occur. Hybridization studies in the laboratory have shown that subspecies of a single species from different parts of Africa can undergo cross-fertilization. Populations of the human malaria parasite *Plasmodium falciparum* are also enzymically polymorphic, but similar enzyme variants occur in parasites from different countries.

INTRODUCTION

Electrophoretic forms of enzymes have been studied extensively in malaria parasites in recent years. They have proved of particular value in (1) providing strain markers for genetic studies, (2) differentiating genetically distinct groups of parasites, and (3) investigating the breeding structure of parasite populations. Such knowledge has considerable practical importance in the planning of control measures against the disease, enabling predictions to be made concerning problems such as the spread of drug-resistance, or the capacity of parasites to express a diversity of antigens.

There are numerous species of malaria parasites infecting man and

Systematics Association Special Volume No. 24, "Protein Polymorphism: Adaptive and Taxonomic Significance", edited by G. S. Oxford and D. Rollinson, 1983, Academic Press, London and New York.

animals (see Garnham, 1966). The majority of enzyme studies have been on the species infecting rodents (Carter, 1978; Carter and Walliker, 1977b) and the human parasite *Plasmodium falciparum* (Carter and McGregor, 1973; Carter and Voller, 1975; Sanderson *et al.*, 1981; Thaithong *et al.*, 1981).

MATERIAL AND METHODS

1. *Parasite life-cycle and genetics*

The malaria parasite has a life-cycle involving two hosts, a vertebrate and a mosquito vector. In mammals, the parasite first undergoes a cycle of asexual division in cells of the liver, followed by further cycles of division in red blood cells. Gametocytes are produced in the blood, but fertilization occurs only in the midgut of the mosquito (*Anopheles* spp.) after a blood meal is taken. Zygote formation is followed by development on the midgut wall of oocysts, inside which the infective forms, the sporozoites, are formed. These migrate to the salivary glands before injection into a new mammal host.

Cytological studies have provided little information on the number of chromosomes, time of meiosis, etc. during the life-cycle. Genetic studies, however, have established that the blood forms are haploid. Crosses can be made between parasites by allowing mosquitoes to take up mixtures of gametocytes of two lines, establishing new infections from the resulting sporozoites, and examining the progeny for the presence of recombinants. Crosses involving parasite lines marked by electrophoretic forms of enzymes have shown that these characters appear to undergo a typically Mendelian type of inheritance, with recombination and segregation occurring before the emergence of parasites into the blood (Walliker *et al.*, 1975).

2. *Enzyme electrophoresis*

The stages of the parasite life-cycle most readily accessible for enzyme studies are the blood forms. The parasites are grown either in laboratory animals or in *in vitro* culture, and freed from their host-cells by methods such as saponin lysis before being examined for enzymes.

The most commonly used electrophoretic methods are starch gel and cellulose acetate electrophoresis. Six enzymes have proved to be of particular practical value, viz.: glucose phosphate isomerase (GPI, E.C.5.3.1.9) 6-phosphogluconate dehydrogenase (PGD, E.C. 1.1.1.43), lactate dehydrogenase (LDH, E.C.1.1.1.27), NADP-dependent glutamate dehydrogenase (GDH, E.C.1.4.1.2), adenosine deaminase (ADA, E.C. 3.5.4.4) and peptidase (PEP, E.C. 3.4.11 or 13). Parasite enzymes can be distinguished from those of the host by including samples of uninfected blood on each gel.

Each electrophoretic form of a given enzyme is denoted by a number, e.g. GPI-1, GPI-2, GPI-3, etc., according to its position on the gel.

ENZYMES OF RODENT MALARIA PARASITES

Malaria parasites have been isolated from rodents, principally thicket-rats (*Thamnomys rutilans* and *Grammomys surdaster*) in five African countries – Cameroun, The Central African Republic, Congo, Nigeria and Zaire (Killick-Kendrick, 1978). Four species are recognized, *Plasmodium berghei, P. yoelii, P. chabaudi* and *P. vinckei*, at least two of which occur in each country. Each species can be distinguished by its morphology and enzyme characters. Animals are frequently infected with more than one parasite species. Such mixed infections can be purified by establishing clones of blood forms from wild-caught animals in laboratory hosts.

1. *Parasites of a single region*

The parasite population which has been most extensively studied is that of the Central African Republic (Carter and Walliker, 1975; Carter, 1978). Table I shows the forms of four enzymes of cloned parasite lines derived from rodents of this region. The parasites comprise three distinct species, *P. yoelii, P. chabaudi* and *P. vinckei*, each possessing its own characteristic enzyme forms. Thus, GPI-1 and PGD-4, for example, occur only in *P. yoelii*, and GPI-4 and PGD-2 only in *P. chabaudi*. The absence of clones characterized by combinations such as GPI-1 with PGD-2 shows that no mating is taking place between the species. There is, however, considerable evidence of random mating *within* each species. This is especially clear in *P. chabaudi*,

Table I. Enzyme forms of clones of rodent malaria parasites of the Central African Republic.

Clone	GPI	PGD	LDH	GDH	
17X	1	4	1	4	
33X	2	4	1	4	*Plasmodium*
86X	1	4	1	4	*yoelii yoelii*
3AF	1	4	1	4	
6AL	4	2	2	5	
3AR	4	2	3	5	
10AS	4	2	3	5	
2BJ	4	2	4	5	*Plasmodium*
57AF	4	2	5	5	*chabaudi chabaudi*
9AJ	4	3	2	5	
14AQ	4	3	2	5	
54X	4	3	3	5	
20CE	4	3	4	5	
2CW	4	3	4	5	
3CQ	4	3	5	5	
1BS	9	5	7	6	
4BZ	9	5	7	6	*Plasmodium*
11CE	5	5	7	6	*vinckei petteri*
2CR	5	5	7	6	

where two forms of PGD (PGD-2 and -3) and four of LDH (LDH-2, -3, -4 and -5) are found. All possible combinations of these forms are seen among the clones examined, which can be assumed to result from extensive cross-fertilization of gametes of different strains.

The three species found in the Central African Republic (*P. yoelii, P. chabaudi* and *P. vinckei*) also occur together in the Congo (Carter and Walliker, 1977a) and Cameroun (Lainson, 1979). In Nigeria, only *P. yoelii* and *P. vinckei* have been found, and in Zaire only *P. berghei* and *P. vinckei* (Killick-Kendrick, 1978). Only a few isolates have been obtained from the latter two countries, however, so the existence of other species in these regions cannot be excluded.

2. *Comparison between parasites of different regions*

Parasites of a single species from different parts of Africa share forms of certain enzymes but differ in others. Table II shows the distribution of enzyme forms of *P. vinckei* and *P. yoelii* in the countries in which they occur. Variation is most pronounced in *P. vinckei*. Thus, while only one form of GDH (GDH-6) is found, in all isolates, seven forms of GPI occur, of which some (e.g. GPI-6) are found in several regions while others (e.g. GPI-7, -11, -12 and -13) are unique to one region.

Table II. Enzyme forms of *Plasmodium vinckei* and *P. yoelii* isolates from different African countries.

Species and subspecies	Country	GPI	PGD	LDH	GDH	No. of isolates examined
P. vinckei vinckei	Zaire	7	6	6	6	2
P. vinckei petteri	Central African Republic	9	5	7	6	3
		5	5	7	6	2
P. vinckei lentum	Congo	6	5	7	6	3
		11	5	9	6	1
P. vinckei brucechwatti	Nigeria	6	6	9	6	2
P. vinckei subsp.	Cameroun	5	5	7	6	1
		6	5	11	6	1
		12	5	9	6	1
		13	5	7	6	1
		6	6	11	6	1
P. yoelii yoelii	Central African Republic	1	4	1	4	17
		2	4	1	4	6
		10	4	1	4	1
P. yoelii killicki	Congo	1	4	1	1	2
P. yoelii nigeriensis	Nigeria	2	4	1	2	1
P. yoelii subsp.	Cameroun	1	4	1	4	1

The parasites in Zaire, the Central African Republic, Congo and Nigeria are considered to be sufficiently distinct to be designated as subspecies (Table II). In Cameroun, one isolate possesses enzyme forms identical to those of two Central African Republic isolates, while the other isolates possess some forms unique to Cameroun. The taxonomic status of these organisms has not yet been resolved.

P. yoelii populations in different countries are also divided into subspecies on the basis of their enzymes (Table II). The variation is less marked than between the *P. vinckei* subspecies.

3. *Hybridization between parasites of different regions*

A question arising from the enzyme studies is whether parasites of each subspecies are reproductively as well as geographically isolated. This subject has been investigated by attempting deliberate crosses between geographically remote isolates in the laboratory.

Most of this work has been carried out with *P. yoelii*. Knowles *et al.* (1981) attempted to cross a line of *P. y. yoelii* (Central African Republic) with one of *P. y. nigeriensii* (Nigeria), the two lines differing by three enzyme markers as well as by sensitivity to the drug pyrimethamine. Mosquitoes were fed on a mixture of gametocytes of the two lines, and the resulting sporozoites used to infect mice. The blood forms developing in these animals were then cloned, and each clone examined for the parent-line markers. It was expected that parent-type clones would be present, resulting from self-fertilization events in the mosquitoes; the presence of recombinants would indicate that cross-fertilization between the lines had occurred. Thirty-seven clones were obtained, of which 17 showed parental-type character combinations, while the remainder were recombinants. Recombination between each marker was detected. It could be concluded, therefore, that the two lines had undergone cross-fertilization; the proportions of parental and recombinant forms suggested that there was no preference for self-fertilization.

Similar crosses have been successfully made between the other *P. yoelii* subspecies, and between certain *P. chabaudi* subspecies (Lainson, unpublished).

ENZYMES OF HUMAN MALARIA PARASITES

Malaria in man occurs in most tropical and many subtropical countries. Four species are recognized, *Plasmodium vivax, P. malariae, P. ovale* and *P. falciparum* which are distinguishable by blood form morphology and patterns of infection. Experimental studies on these parasites have been limited by a lack of suitable methods for maintaining them in the laboratory. However, culture methods now exist for the blood form of *P. falciparum* (Trager and Jensen, 1976), and these have been exploited for studies on enzyme variation in this species.

The countries from which most *P. falciparum* isolates have been examined are The Gambia, Tanzania and Thailand (Carter and McGregor, 1973; Carter and Voller, 1975; Thaithong *et al.*, 1981). Six enzymes have been studied. Variant forms of each have been found but, with certain exceptions, similar forms occur in each country. PGD and GDH exhibit the same form in almost all isolates, the exceptions being a few rare variants in some African parasites. Variant forms of GPI, LDH, PEP and ADA are more common, and Table III

Table III. Enzyme forms of *Plasmodium falciparum* from The Gambia, Tanzania and Thailand.

	GPI				ADA			
	Number examined	1	2	1&2	Number examined	1	2	1&2
Gambia	170	64%	10%	26%	53	92%	4%	4%
Tanzania	21	43%	29%	29%	8	13%	13%	74%
Thailand	176	62%	22%	16%	135	94%	3%	3%
	LDH				**PEP**			
	Number examined	1	2	1&2	Number examined	1	2	3
Gambia	164	40%	17%	43%	52	100%	–	–
Tanzania	26	73%	8%	19%	8	–	75%	25%
Thailand	143	100%	–	–	64	100%	–	–

shows the distribution of these in the countries examined. For GPI and ADA, two forms are found, and it is noteworthy that each occurs at a very similar frequency in The Gambia and Thailand, the countries from which most samples have been obtained. The only enzymes which appear to show some regional variation are LDH and PEP. LDH-1 occurs in all regions, but LDH-2 has been found so far only in African isolates. PEP variants appear to differ in their frequency between East and West Africa.

In comparison to the rodent malaria parasites, therefore, *P. falciparum* appears to exhibit less enzyme variation when parasites of different countries are compared. Many more isolates need to be examined, however, to obtain a complete picture of the variation in this species.

CONCLUSIONS

Electrophoretic forms of enzymes have proved of considerable value in differentiating malaria parasites. Among the rodent malaria species, it has been possible to distinguish populations which are reproductively isolated from one another and, hence, to make clear demarcations between the species present in a given region. The variations found between parasites of different regions have provided evidence that geographical isolation has been sufficient to allow the accumulation of genetic differences between the populations, resulting in the formation of distinct subspecies. Hybridization studies show that the differences are not large enough to prevent interbreeding between parasites of different regions, at least under laboratory conditions. With regard to the malaria parasite of man, *Plasmodium falciparum*, there are differences in the frequencies of certain enzyme forms in different parts of the world, but these differences are less striking than those distinguishing the rodent malaria subspecies. *P. falciparum* is, thus, genetically rather similar in different countries, and probably comprises a potentially interbreeding population.

ACKNOWLEDGEMENTS

The work reviewed in this paper was supported by grants from the U.K. Medical Research Council, the Wellcome Trust and the World Health Organization.

REFERENCES

Carter, R. (1978). Studies on enzyme variation in the murine malaria parasites *Plasmodium berghei, P.yoelii, P.vinckei* and *P.chabaudi* by starch gel electrophoresis. *Parasitology* **76**, 241–267.

Carter, R. and McGregor, I. A. (1973). Enzyme variation in *Plasmodium falciparum* in The Gambia. *Trans. R. Soc. trop. Med. Hyg.* **67**, 830–837.

Carter, R. and Voller, A. (1975). The distribution of enzyme variation in populations of *Plasmodium falciparum* in Africa. *Trans. R. Soc. trop. Med. Hyg.* **69**, 371–376.

Carter, R. and Walliker, D. (1975). New observations on the malaria parasites of rodents of the Central African Republic; *Plasmodium vinckei petteri* subsp.nov and *Plasmodium chabaudi* Landau, 1965. *Ann. trop. Med. Parasit.* **69**, 187–196.

Carter, R. and Walliker, D. (1977a). Malaria parasites of rodents of the Congo (Brazzaville): *Plasmodium chabaudi adami* subsp.nov. and *Plasmodium vinckei lentum* Landau, Michel, Adam and Boulard, 1970. *Annals Parasit. hum. comp.* **51**, 637–646.

Carter, R. and Walliker, D. (1977b). Biochemical markers for strain differentiation in malaria parasites. *Bull. Wld. Hlth. Org.* **55**, 339–345.

Garnham, P. C. C. (1966). "Malaria Parasites and Other Haemosporidia". Blackwell Scientific Publications, Oxford.

Killick-Kendrick, R. (1978). Taxonomy, zoogeography and evolution. *In* "Rodent Malaria" (Killick-Kendrick, R. and Peters, W., eds), pp. 1–53. Academic Press, London.

Knowles, G., Sanderson, A. and Walliker, D. (1981). *Plasmodium yoelii*: genetic analysis of crosses between two rodent malaria subspecies. *Expl. Parasit.* **52**, 243–247.

Lainson, F. A. (1979). Enzyme variation in rodent malaria parasites isolated in Africa-Cameroun. *Parasitology* **79**, xxviii.

Sanderson, A., Walliker, D. and Molez, J-F. (1981). Enzyme typing of *Plasmodium falciparum* from some African and other old world countries. *Trans. R. Soc. trop. Med. Hyg.* **75**, 263–267.

Thaithong, S., Sueblinwong, T. and Beale, G. H. (1981). Enzyme typing of some isolates of *Plasmodium falciparum* from Thailand. *Trans. R. Soc. trop. Med. Hyg.* **75**, 268–270.

Trager, W. and Jensen, J. B. (1976). Human malaria parasites in continuous culture. *Science, N.Y.* **193**, 673–675.

Walliker, D., Carter, R. and Morgan, S. (1975). Genetic studies on *Plasmodium chabaudi*: recombination between enzyme markers. *Parasitology,* **70**, 19–24.

3 *Trypanosoma* and *Leishmania*: the Contribution of Enzyme Studies to Epidemiology and Taxonomy

M. A. MILES

Department of Medical Protozoology, London School of Hygiene and Tropical Medicine, Keppel Street, London WC1E 7HT, U.K.

Abstract: Three principal *Trypanosoma cruzi* zymodemes (Z1, Z2 and Z3), have different geographical distributions related to transmission cycles and linked circumstantially to the forms of chronic Chagas' disease in man. Heterozygous enzyme patterns, which are frequent in more temperate localities, suggest that *T. cruzi* is diploid, but as yet there is no evidence from allozyme frequencies of current gene exchange between the zymodemes. Subunit structures inferred from comparative estimates of *T. cruzi* enzyme molecular weights are compatible with interpretations of electrophoretic data.

Enzyme profiles of *T. brucei* point to the domestic pig as a reservoir host of Gambian sleeping sickness and separate *T. b. gambiense* from *T. b. rhodesiense*/*T. b. brucei*: heterozygous enzyme patterns and allozyme frequencies suggest diploidy and random mating within these two groups. In contrast, there is homogeneity between New and Old World isolates of both *T. vivax* and *T. evansi*. Forest and savannah zymodemes of *T. congolense* have been described.

Three agents of human cutaneous leishmaniasis in the Amazon basin, *Leishmania mexicana amazonensis*, *L. braziliensis braziliensis* and *L. b. guyanensis*, have been distinguished but *L. b. guyanensis* and Central American *L. b. panamensis* are extremely similar. A major agent of New World visceral leishmaniasis appears to be *L. donovani infantum* of the Mediterranean area. *L. major* and *L. tropica* have been validated as distinct Old World taxa.

Systematics Association Special Volume No. 24, "Protein Polymorphism: Adaptive and Taxonomic Significance", edited by G. S. Oxford and D. Rollinson, 1983, Academic Press, London and New York.

INTRODUCTION

The explosion of research progress generated by the use of enzymes as genetic markers for a wide variety of living organisms (Markert, 1975), subsequently produced a secondary eruption in the field of parasitology. Realization spread from the early work on trypanosomes and *Leishmania* that this novel approach would have tremendous impact on the understanding of the epidemiology and taxonomy of parasitic protozoa and helminths (Bagster and Parr, 1973; Kilgour and Godfrey, 1973, Gardener *et al.*, 1974). Enzyme studies as diverse as those employing the exquisite properties of restriction endonucleases (Borst *et al.*, 1981; Majiwa *et al.*, 1982) or comparisons of metabolism (Kilgour, 1980a) might be encompassed by the title of this chapter. There is no doubt, however, that the technique foremost in this context is that of enzyme electrophoresis. Letch (1979) has used this method to prove that trypanosome infections in two fish species were caused by the same organism, and the taxonomy of Old and New World bat trypanosomes has been re-evaluated, partially on the basis of isozyme data (Baker *et al.*, 1978, Baker and Miles, 1979). Nevertheless, within the genera *Trypanosoma* and *Leishmania*, enzyme electrophoresis has been used almost exclusively for taxa that infect, or are suspected of infecting, man and his domestic animals (Godfrey, 1978, 1979). There have been major contributions to the epidemiology and taxonomy of South American trypanosomiasis, African trypanosomiasis and cutaneous and visceral leishmaniasis. These will be presented here but, because of the limitations of space, interested readers are referred to the literature cited for detailed descriptions of isozyme patterns and the concomitant epidemiologies.

SOUTH AMERICAN TRYPANOSOMIASIS (Chagas' disease)

Throughout the Americas, *Trypanosoma cruzi*, the aetiological agent of Chagas' disease in man, is reported from an enormous number of mammal species and the majority of blood-sucking triatomine bug species (Hemiptera, Reduviidae). The human infection is a primary cause of heart disease in South America. Diverse behaviour in experimental studies, differences in response to chemotherapy and especially the enigmatous geographical distribution of chronic sequelae to infection (megaoesophagus and megacolon) have suggested heterogeneity within the species *T. cruzi* (Miles, 1979). Biological characters have failed to produce intrinsic parameters for the reliable identification of strains. In applying enzyme electrophoresis to investigate the epidemiology and taxonomy of *T. cruzi*, three questions have been asked: first, can *T. cruzi* be considered as a single entity, secondly, if there is heterogeneity within the species, does this relate to particular hosts and vectors or silvatic and domestic cycles of transmission and, thirdly, are regional differences in pathology strain-dependent?

The first epidemiological study was undertaken in a rural Brazilian town, and *T. cruzi* isolates were collected from both a domestic cycle of transmission, involving man and his domestic animals, and a silvatic cycle of transmission involving the principal feral host, the opossum *Didelphis albiventris* (Miles *et al.*, 1977). Domestic *T. cruzi* stocks from man, dogs, cats and rodents were distinguished from those from opossums and a silvatic vector species by isozyme electrophoresis. These differences were radical and consistent, the domestic and silvatic strains, named zymodemes 2 and 1 respectively, were separable by ten of 18 enzymes (Miles *et al.*, 1980). In a further locality where silvatic *T. cruzi* was alleged to have been introduced into houses, under conditions of severe drought, with invading peridomestic rodents and triatomine bugs, the silvatic zymodeme was found in man and his domestic animals (Barrett *et al.*, 1980). Yet another distinct zymodeme, Z3, was isolated from armadillos and very occasionally from man. In the Amazon basin, where silvatic transmission is prolific but there are no domestic vectors, the origin of sporadic acute cases in man could be traced to silvatic foci of either zymodemes 1 or 3 in suburban communities (Miles *et al.*, 1978, 1981a). It was quite clear from investigations in these three regions that there was considerable heterogeneity within what was considered to be a single trypanosome species. Three major zymodemes were identifiable and these were indeed associated with particular mammals and vectors and could be correlated with types of transmission cycle. This strengthened the concept of considering the regional epidemiology of Chagas' disease in terms of separate, overlapping or entirely enzootic transmission cycles (Miles, 1979). As far as the enigmatous distribution of chagasic syndromes was concerned, the absence of Z2 from both the Amazon basin and Venezuela (mega syndromes absent) and the common occurrence of this zymodeme in central and southern Brazil (mega syndromes present) suggested, although the evidence was purely circumstantial, that different infecting agents might be responsible for regional variation in the form of chronic Chagas' disease (Miles *et al.*, 1981b).

The term (zymo)deme (not greeted with universal approval amongst geneticists), was defined as a trypanosome population that differs from others of the same species or subspecies in a specified property (isozyme) or set of properties (set of isozymes) (W.H.O., 1978). With respect to *T. cruzi*, the term has been ascribed to populations that are radically dissimilar and distinguishable by a set of isozyme differences rather than by a single isozyme difference (Miles *et al.*, 1981a). Numerical taxonomy has been used to assess the degree of difference between the *T. cruzi* zymodemes. Ready and Miles (1980) used paired comparisons with absence/presence of isozyme bands as characters to calculate similarity quotients. Tibayrenc and Miles (1983) calculated genetic distance according to Nei from allozyme frequencies (Nei, 1975; Tibayrenc, 1980). Either method confirms the high degree of divergence between the three *T. cruzi* zymodemes Z1, Z2 and Z3, with, for example, genetic distances calculated from 12 loci of 1·36 for Z1/Z2 and Z1/Z3. Had these genetic distances been backed up by a significant framework of other biological characters, the three zymodemes would have been described as discrete taxa. As it is not definitely known, however, whether *T. cruzi* is an asexual or sexual organism, it is conceivable that genetic exchange might occur between the zymodemes in certain localities. Recently, typical heterozygous enzyme patterns have emerged in *T. cruzi* isolates, including clones, from Bolivia (Tibayrenc and Miles, 1983), Paraguay, the extreme south of Brazil and Chile (unpublished). It appears that there may have been a contribution from the three *T. cruzi* zymodemes nominated Z1, Z2 and Z3 to the genetic constitution of these heterozygous strains but a parental relationship is not apparent. Allozyme frequencies give no evidence of Hardy–Weinberg equilibrium and current gene exchange. Furthermore, the heterozygosity seems to be fixed, with five out of 12 enzymes consistently giving heterozygous patterns (Tibayrenc *et al.*, 1981; Tibayrenc and Desjeux, 1983; Tibayrenc and Miles, 1983). Because of the fixed nature of the heterozygosity, it is unlikely that those stocks have arisen simply by mutation. It is most probable that all the *T. cruzi* zymodemes are diploid and that diploid heterozygotes have been isolated, by the dissemination of particular clones in domestic transmission cycles, the so-called Founder Effect, or by selection. Interestingly, the heterozygous *T. cruzi* strains have been collected from regions that suffer wide variation in temperatures and it is well known that heterozygosity may confer physiological advantages in such conditions (Ferguson, 1980). There is independent evidence suggesting that *T. cruzi* may be diploid (Lanar *et al.*, 1981) with ten chromosome units (Solari, 1980). A tetraploid hybrid may be postulated as a less likely explanation for the fixed heterozygosity in *T. cruzi* – and more bizarre explanations cannot be excluded (Tibayrenc and Miles, 1983). If *T. cruzi* proves to be a sexual organism, consideration of *T. cruzi* as a polytypic species or as a species complex will largely depend on the presence or absence of significant genetic exchange between the zymodemes.

To confirm the validity of interpretations of isozyme data Jeremiah *et al.* (1982) have attempted to obtain independent evidence of the subunit structures of *T. cruzi* enzymes. Molecular weight estimates of *T. cruzi* enzymes have been used to infer subunit structures by comparisons with equivalent mammalian enzymes, although, in some cases, such as with the peptidases (Letch and Gibson, 1981), precise equivalence has been difficult to assess. The enzymes glucosephosphate isomerase (GPI, E.C.5.3.1.9) and isocitrate dehydrogenase (ICD, E.C.1.1.1.42) were considered to be dimeric on the basis of molecular weight estimates and phosphoglucomutase (PGM, E.C.2.7.5.1) was though to be monomeric. These results are in accord with the heterozygous isozyme patterns observed. Carvalho (1981) has failed to generate recombinant patterns from mixed zymodeme extracts by dissociation and reassociation *in vitro*, although mouse-human liver hybrids were obtained. Cyclical transmission of mixed zymodemes has also failed to produce detectable recombination (Miles, 1982a). Nevertheless, in both of the latter experimental approaches, conditions necessary to produce recombination are difficult to predict. A more complete understanding of the distribution and relationships between the *T. cruzi* zymodemes is dependent on further investigations.

AFRICAN TRYPANOSOMIASIS

Human sleeping sickness has been a scourge of the African continent since ancient times, and although major epidemics are now less common, it is still an important public health problem. Trypanosomiasis of livestock, especially cattle, horses and pigs continues to be a severe blight on the economic development of the continent, with vast tracts of land denied use. Nevertheless, the ecology of African trypanosomiasis is intricate and Ormerod (1979) has even suggested that there are beneficial consequences in the limitation of overgrazing and the spread of desertification.

Since the discovery of African trypanosomiasis, much attention has been devoted to determining the role of animals as reservoir hosts of human disease. The disease is traditionally divided into two forms, East African or acute rhodesian sleeping sickness and West African or chronic gambian sleeping sickness. In East Africa, the importance of wild and domestic animals as reservoir hosts, implicit in the epidemiology of the disease, was shown by the inoculation of volunteers (Heisch *et al.*, 1958) and subsequently confirmed by Gibson *et al.* (1980) on the basis of isozyme comparisons. In West Africa, the epidemiology of gambian sleeping sickness was allegedly explicable by the presence of human carriers. Nevertheless, trypanosome stocks isolated from human cases readily infect a wide range of domestic and wild animals and these animals become infective to tsetse flies. Crucial evidence of an animal reservoir clearly depends, however, on the demonstration that natural animal infections are transmissible to man, and this was lacking – apart from the experiment of Denecke (1941) who, in an endemic area, succeeded in transferring an infection from a dog to himself. Strong circumstantial evidence that both dogs and pigs were important reservoir hosts of gambian sleeping sickness was obtained by Mehlitz (1977a, b). Forty-nine stocks of trypanosomes were isolated in Liberia from 117 domestic pigs and 106 dogs by inoculation of blood into the multimammate rat, *Mastomys natalensis*, which is particularly susceptible to isolates from human cases of gambian sleeping sickness. Several stocks from pigs and dogs were found to be resistant to human plasma, a test which has been taken to give a broad indication of man-infectivity. Electrophoresis of 11 enzymes showed that isozyme patterns of some stocks from pigs did correspond with those stocks from human cases, in the presence of a characteristic slow alanine aminotransferase (ALAT, E.C.2.6.1.2) described as a marker for the agent of gambian sleeping sickness. Subsequently, it was found that this ALAT marker was not invariably diagnostic for *T. b. gambiense* but isolates from man and domestic pigs in the same village in Ivory Coast were indistinguishable by isozymes, demonstrating the possible role of pigs as reservoir hosts of sleeping sickness in West Africa (Godfrey and Kilgour, 1976; Gibson *et al.*, 1978a, 1979, 1980; Mehlitz *et al.*, 1982).

Gibson *et al.* (1980) have made an extensive study of 160 *Trypanozoon* stocks with diverse origins, using 12 enzymes. Comparisons with trypanosomes of the subgenera *Duttonella* and *Nannomonas* supported the separation of the three subgenera. Within the subgenus *Trypanozoon*, however, isozyme data suggested that the existing nomenclature was inappropriate, and Gibson *et al.* (1980) proposed that the three subspecies *T. b. gambiense, T. b. rhodesiense, T. b. brucei* and the species *T. evansi* be united under the single species *T. brucei*. They further suggested a working nomenclature for recognizable infraspecific groups and tentatively defined six such groups. Although there was an overall degree of homogeneity of *Trypanozoon* stocks, with little or no variation for six of the 12 enzymes, there was a wide range of minor variation. Using the term zymodeme to apply to trypanosome stocks with any single unique isozyme character, in contrast with the designation of major infraspecific groups in *T. cruzi*, 59 zymodemes were described and the relationships between them were explored by cluster analysis, which revealed a dichotomy between West and East African stocks.

In clones, some isozyme patterns of the enzymes ICD and PGM were those expected for heterozygotes assuming that ICD was dimeric and PGM monomeric (Gibson *et al.*, 1980). Furthermore, the multibanded patterns in the Lake Victoria area suggested recombination between West and East African forms. Tait (1980) studied the electrophoretic variation of eight enzymes in 17 stocks from Lugala, Uganda, and also observed hybrid banding patterns that would be expected in heterozygotes. The allozyme frequencies were calculated and compared with those anticipated if the population was in Hardy–Weinberg equilibrium. There was a remarkably close agreement, between observed and predicted proportions of genotypes. Fifty out of a possible 90 recombinants occurred; this was taken to strongly suggest diploidy and random mating. Preliminary work with mixed zymodeme cultures had indicated that such recombination may possibly occur *in vitro* – although recombinants could not be isolated by cloning (Tait, pers. comm.). Although genetic distances for the *T. brucei* zymodemes are much lower than those separating the principal *T. cruzi* zymodemes, Tait (pers. comm.) suspects that there are separate *T. b. gambiense* and *T. b. rhodesiense*/*T. b. brucei* populations. The potential occurrence of mating in *T. brucei* obviously affects the use of the term zymodeme as, in populations undergoing random mating, zymodemes may correspond to the level of individual variation. The use of this term for trypanosomes undergoing random mating, for trypanosome stocks differing in single isozymes or for major subspecific groups discernable on isozyme data, clearly requires clarification (Miles *et al.*, 1981a).

T. evansi is widely distributed outside the African continent including South America, India and parts of the Far East, where there are no tsetse fly vectors. It is an important pathogen of horses, cattle, camels and other domestic stock, and also occurs in wild hosts such as the capybara, *Hydrochoerus hydrochaeris*, in South America (Morales *et al.*, 1976). *T. evansi* is transmitted mechanically by blood sucking flies, for example Tabanidae and also, in South America, by the vampire bat, *Desmodus rotundus*. It is thought to have been carried from the African tsetse belts across the Sahara and then to India and Asia by camel caravans and introduced into South America by the export of Arabian horses during the Spanish conquest (see Hoare, 1972 and Ramirez *et al.*, 1979 for an epidemiological review). *T. evansi* is morphologically indistinguishable from *T. brucei* but may have entirely lost the capacity to undergo cyclical development in the tsetse fly and this, together with the geographical distribution, is the main basis for considering it to be a distinct species. Gibson *et al.* (1980) were also unable to separate *T. evansi* from West African *T. brucei* by isozyme data and concluded that the two species, as mentioned above, should be united under *T. brucei*. The most interesting aspects of isozyme studies of *T. evansi* are the lack of extensive variation and the fact that stocks from South America and West Africa were electrophoretically indistinguishable, which supports the African origin of *T. evansi* (Gibson, 1980; Gibson *et al.*, 1978b, 1980). This lack of variation amongst stocks of *T. evansi* is precisely what one would expect if *T. evansi* has spread recently and is indeed denied the opportunity to undergo mating, by restricted development in the vector.

The epidemiology of the bovine trypanosome *T. vivax* (subgenus *Duttonella*) is comparable to that of *T. evansi* in that it has also spread beyond the tsetse belts to the West Indies, Central and South America and Mauritius. It is also thought to be transmitted entirely mechanically by biting flies. We might expect then to find limited isozyme variation and comparable patterns in Africa and the New World. Kilgour and Godfrey (1977) described two stable Nigerian *T. vivax* zymodemes but based on two enzymes only; Murray (1982) has found limited variation and just as with *T. evansi*, stocks of African and South American origin were indistinguishable.

Bearing these observations in mind, it is interesting to turn to *T. congolense* (subgenus *Nannomonas*) which is restricted to the tsetse fly belts of tropical Africa and develops in both proboscis and mid-gut of the vector. Young (1980) could easily separate *T. congolense* stocks from representatives of the subgenera *Trypanozoon* and *Duttonella* by eight of 14 enzymes and recorded an astonishing 75 zymodemes amongst 78 stocks examined. Computer-assisted numerical taxonomy revealed a major dichotomy between savannah stocks and riverine or forest stocks, and there was some suggestion that the savannah tsetse species *Glossina morsitans* was a more effective vector of the corresponding zymodeme. Although the potential occurrence of mating has not been invoked or tested for *T. congolense*, it seems rational that the observed extensive variation may be explicable in these terms. These results once more highlight the need to reconsider the definition of the term zymodeme.

With both *T. brucei* and *T. congolense*, some enzymes showed little or no variation whilst others were highly variable, suggesting that a system of weighting of enzymes could be introduced for numerical taxonomy.

LEISHMANIASIS

The taxonomy of *Leishmania* has been built up on epidemiological differences, especially the clinical picture in man (Bray, 1974). This has been expanded to include a series of other characters, particularly for New World *Leishmania*, that include experimental and natural development in sandflies, behaviour in inoculated hamsters, growth rates *in vitro*, morphology, serotype and latterly DNA buoyancy and isozyme patterns (Williams and Coelho, 1978; Chance, 1979; Lainson and Shaw, 1979). The need for a range of parameters and the importance of intrinsic parameters, such as isozymes, that do not depend on a series of interactions with a vector or mammal host, can be illustrated by evidence that at least two aetiological agents may cause visceral leishmaniasis in man (Peters *et al.*, 1981; Schnur *et al.*, 1981), and by the well known genetic basis of susceptibility in experimental animals (Howard *et al.*, 1980; Plant *et al.*, 1982), which might be analogous to racial differences in the progression of human muco-cutaneous disease (Walton and Valverde, 1979). In the most recent classification, the *Leishmania* are divided into three sections: the Hypopylaria consisting of two Old World lizard species; the Peripylaria with two Old World lizard species and the New World *L. braziliensis* complex, and the Suprapylaria covering the New World *L. mexicana* and *L. hertigi* complexes, the New and Old World *L. donovani* complex and the Old World *L. tropica* and *L. major* complexes. Division into the three sections is based, respectively, on whether development in sandflies is in the posterior intestine, posterior and anterior intestine or anterior intestine alone (Lainson and Shaw, 1979; Lainson, 1981).

The isozymes of *Leishmaniae* in the Amazon basin have been studied in some detail, and have been most important in elucidating the complexities of local epidemiologies. Three aetiological agents of human cutaneous leishmaniasis – *L. mexicana amazonensis, L. braziliensis braziliensis* and *L. braziliensis guyanensis* – were separated by isozyme patterns. Ten of 14 enzymes distinguished *L. m. amazonensis* from the two *L. braziliensis* subspecies, confirming the major differences between the *L. mexicana* and *L. braziliensis* complexes, whilst the somewhat controversial splitting of *L. b. braziliensis* and *L. b. guyanensis*, on epidemiological grounds or poorly defined growth characteristics, accorded with slight mobility differences of four enzymes (Miles *et al.*, 1981c). Thirty Brazilian stocks of *L. m. amazonensis* from widely separated areas were extremely homogeneous, varying on two of 18 enzymes only and independently of whether isolates were from simple cutaneous or grave, incurable, diffuse cutaneous leishmaniasis (Miles *et al.*, 1979). This latter observation supported the idea that the progression to diffuse leishmaniasis was not strain-dependent but reflected immunological incompetence of the host. Until recently, the reservoir hosts of *L. b. guyanensis* were unknown. Isozyme data have played a role in demonstrating a silvatic cycle involving sloths (*Choloepus didactylus*) and anteaters (*Tamandua tetradactyla*) (Lainson *et al.*, 1981a, Gentile *et al.*, 1981) and a suburban cycle involving the opossum *Didelphis marsupialis* (Arias and Naiff, 1981). In addition, an unsuspected, extremely close relationship, in agreement with some epidemiological similarities, emerged from comparisons of *L. b. guyanensis* and the Central American *L. b. panamensis*, which were indistinguishable by ten enzymes (Miles *et al.*, 1981c). A major point of contention concerning New World leishmaniasis is whether neotropical visceral leishmaniasis is imported from the Old World, or due to indigenous strains, or both (Killick-Kendrick *et al.*, 1980). It was thought to be unlikely, despite the similar domestic epidemiology involving the dog as a reservoir host, that parasites imported from the Old World could attain such a wide distribution in South America and be readily transmitted by the principal vector *Lutzomyia longipalpis* (Lainson and Shaw, 1979). Nevertheless, Killick-Kendrick *et al.* (1980) have shown that laboratory-bred *Lu. longipalpis* is readily susceptible, to at least the initial stages of infection with a Mediterranean strain of the visceralizing *Leishmania, L. donovani infantum*. This suggests that *L. d. infantum* could have been introduced to South America by the importation of infected dogs with the conquistadors. Isozyme data have confirmed this theory as *Leishmania* stocks isolated from visceral cases in Brazil were indistinguishable by ten enzymes from *L. d. infantum* (Lainson *et al.*, 1981b). It seems probable, therefore, that *L. d. infantum* is an agent of visceral leishmaniasis in the New World.

Isozyme studies have also improved understanding of New World animal *Leishmaniae* not, as yet, known to infect man, with the distinction of *L. hertigi hertigi* and two groups of *L. h. deanei* of the porcupine (Miles *et al.*, 1979) and the discovery of a new *Leishmania* from the armadillo, *Dasypus novemcinctus* (Lainson *et al.*, 1982). The ability to culture flagellates directly from wild-caught, infected sandflies (Arias, unpublished) promises rapid new progress in unravelling the extraordinary complexities of neotropical forest leishmaniasis where there may be transmission of many species of *Leishmania, Endotrypanum* and *Trypanosoma* involving numerous sandfly vector species in the same small area of forest (Lainson *et al.*, 1981c). The status of the other New World species, not mentioned so far, that is: *L. b. peruviana, L. m. mexicana, L. m. pifanoi, L. m. aristedesi* and *L. m. enriettii* has not been investigated fully using isozymes, but preliminary results for those that have been examined (Lainson *et al.*, in press) support the basis for their integrity (Lainson and Shaw, 1979).

In the Old World, investigations have not, as yet, uncovered such a profusion of different leishmanias as in the New World. Enzymes have been important in pin-pointing reservoir hosts in various regions (Dedet *et al.*, 1979; Mutinga *et al.*, 1980; Peters *et al.*, 1977) and have supported the separation of the two principal Old World types of cutaneous leishmaniasis, the zoonotic rural form due to *L. major* and the urban form (oriental sore) due to *L. tropica* (Al-Taqi and Evans, 1978; Rassam *et al.*, 1979; Aljeboori and Evans, 1980b). In addition, *L. aethiopica* which causes cutaneous leishmaniasis and is also associated with diffuse cutaneous leishmaniasis in Ethiopia, was readily identifiable (Chance *et al.*, 1978) – although its status as a full species rather than as *L. tropica aethiopica* is still not wholly accepted (Lainson, 1981). Studies of visceral leishmaniasis in the Old World indicated the relatively minor nature of the differences between *L. d. donovani* and *L. d. infantum* (Lainson *et al.*, 1981b; Peters, *et al.*, 1981; Schnur *et al.*, 1981) as well as uncovering *L. tropica* as a potential visceralizing organism in man, which is perhaps not unexpected in view of the variety of behaviour of *Leishmania* taxa in experimental and wild animals (Lainson and Shaw, 1979; Aljeboori and Evans, 1980a).

A numerical taxonomic approach has been applied to the classification of Old World *Leishmaniae* using simple indices of similarity or genetic distances calculated from allozyme frequencies. Similar dendrograms were produced with both approaches, clearly delimiting *L. major, L. tropica, L. aethiopica* and *L. donovani* (Lanotte *et al.*, 1981). Heterozygous enzyme patterns appear to be rare amongst the *Leishmaniae*, although three possible heterozygous stocks of *L. d. infantum, L. tarentolae* and *L. aethiopica* have been reported(Maazoun *et al.*, 1981). There is no direct evidence of any genetic recombination in the *Leishmaniae*. More general relationships between Old and New World cutaneous leishmanias remain to be explored and, whilst intercontinental movement of parasites is usually considered in the Old–New World direction, the ancient nature of South American mammals and ecosystems suggest that a two-way interchange cannot be ruled out.

GENETICS OF INSECT VECTORS

Closely related vector species may present a confusing array of morphological characters, isozyme electrophoresis can be used to define interbreeding populations, to study vector capacity and to attempt to understand the mechanism of insecticide resistance which, in mosquito vectors of malaria, has been linked to prominent esterase isozymes. Studies are beginning on triatomine vectors of Chagas' disease (Dórea *et al.*, 1982), sandflies (Miles and Ward, 1978; Ward *et al.*, 1981a, b) and tsetse flies (Letch, pers. comm.). There is no doubt that this will contribute significantly to future research on the vector component of the epidemiologies of trypanosomiasis and leishmaniasis.

SIMPLIFICATION OF METHODS

Most of the work to which I have referred has been based on starch-gel electrophoresis, polyacrylamide-gel electrophoresis or occasionally isoelectric focusing. The improvement in cellulose acetate electrophoretic procedures heralds the wide availability of isozyme characterization to simply equipped laboratories worldwide (Kreutzer and Christensen, 1980; Lanham *et al.*, 1981; Kreutzer and Sousa, 1981). There is the prospect that, with provision of extracts of relevant standard organisms, improvement of stability of development reagents, and the further simplification of apparatus, a simple field kit may be produced to identify infecting agents in endemic areas, at least to species level. The development of field kits for isozyme electrophoresis is discussed in detail by Lanham (1982).

A CAUTIONARY NOTE

There is a real danger that the transposition of the relatively simple technique of enzyme electrophoresis to the field of parasitology may produce a miscellaneous assemblage of observations without due reference to significant epidemiological questions, adequate controls or continuity with future work. The results obtained with enzyme electrophoresis are only as good as the preceeding epidemiological and routine laboratory methods allow. A frightening series of errors have come to light that testify to the capacity of research workers to mix up or mis-identify laboratory stocks of organisms (Miles, 1982b). Mixed populations also occur in nature (Scott, 1981). The intricacies of local epidemiologies may be daunting, suggesting that isozyme characterization is best applied extensively in particular localities before absolute confidence can be placed in broader comparisons. In characterizing stocks, it is sound practice to question stated labels and origins, cryopreserve materials from early passages, keep precise records, prepare Giemsa-stained films of each population harvested and to use several enzymes repeatedly, under control conditions with consistent marker stocks. It is also advisable to use the same phase of the life-cycle of the organism, as electrophoretic mobilities may vary between, for example, blood stream and culture forms of *T. brucei* (Kilgour, 1980b).

CONCLUSION

It is quite evident from the foregoing account that enzyme studies have made a most significant contribution to the epidemiology and taxonomy of *Trypanosoma* and *Leishmania* and clearly there will be more important progress in this field. An additional exciting prospect is that with the development of fusion protocols for diverse cell types using polyethylene glycol (Crane and Dvorak, 1980) a system, analogous to the somatic hybrid approach to mapping mammalian genes, will be developed using protozoa carrying drug resistance or other selective markers. In contrast, some of the usefulness of enzymes may be superceded by monoclonal antibodies or by recombinant DNA techniques which may provide reagents of exquisite specificities for analysing the variation of natural populations (Pratt and David, 1981).

When I entered the field of medical research, I gave little credence to the development of research in classical evolutionary biology. The revolutionary effect of the use of enzymes as genetic markers belies this misguided opinion. The scope and epidemiological relevance of the investigations that I have described are small tribute to the imaginative workers who produced the fruitful combination of isozyme studies and numerical taxonomy.

ACKNOWLEDGEMENTS

I sincerely thank the Wellcome Trust for financial support.

REFERENCES

Aljeboori, T. I. and Evans, D. A. (1980a). *Leishmania* spp. in Iraq. Electrophoretic isoenzyme patterns. I. Visceral leishmaniasis. *Trans. R. Soc. trop. Med. Hyg.* **74**, 169–177.

Aljeboori, T. I. and Evans, D. A. (1980b). *Leishmania* spp. in Iraq. Electrophoretic isoenzyme patterns. II. Cutaneous leishmaniasis. *Trans. R. Soc. trop. Med. Hyg.* **74**, 178–184.

Al-Taqi, M. and Evans, D. A. (1978). Characterization of *Leishmania* spp. from Kuwait by isoenzyme electrophoresis. *Trans. R. Soc. trop. Med. Hyg.* **72**, 56–65.

Arias, J. R. and Naiff, R. D. (1981). The principal reservoir host of cutaneous leishmaniasis in the urban areas of Manaus, Central Amazon of Brazil. *Mem. Inst. Oswaldo Cruz* **76**, 279–286.

Bagster, I. A. and Parr, C. W. (1973). Trypanosome identification by electrophoresis of soluble enzymes. *Nature, Lond.* **244**, 364–366.

Baker, J. R. and Miles, M. A. (1979). *Trypanosoma (Schizotrypanum) dionisii breve* n. subsp. from Chiroptera. *Syst. Parasitol.* **1**, 61–65.

Baker, J. R., Miles, M. A. , Godfrey, D. G. and Barrett, T. V. (1978). Biochemical characterization of some species of *Trypanosoma (Schizotrypanum)* from bats (Microchiroptera). *Am. J. trop. Med. Hyg.* **27**, 483–491.

Barrett, T. V., Hoff, R. H., Mott, K. E., Miles, M. A., Godfrey, D. G., Teixeira, R., Almeida de Souza, J. A. and Sherlock, I. A. (1980). Epidemiological aspects of three *Trypanosoma cruzi* zymodemes in Bahia State, Brazil. *Trans. R. Soc. trop. Med. Hyg.* **74**, 84–90.

Borst, P. Fase-Fowler, F. and Gibson, W. C. (1981). Quantitation of genetic differences between *Trypanosoma brucei gambiense, rhodesiense* and *brucei* by restriction enzyme analysis of kinetoplast DNA. *Molec. Biochem. Parasitol.* **3**, 117–131.

Bray, R. S. (1974). *Leishmania. Ann. Rev. Microbiol.* **28**, 189–217.

Carvalho, N. (1981). "Investigation of the recombination potential of *Trypanosoma cruzi* using hybridization of zymodemes". M.Sc. thesis, University of London.

Chance, M. L. (1979). The identification of *Leishmania. Symp. Brit. Soc. Parasitol.* **17**, 55–74.

Chance, M. L., Schnur, L. F., Thomas, S. C. and Peters, W. (1978). The biochemical and serological taxonomy of *Leishmania* from the Aethiopian zoogeographical region of Africa. *Ann. trop. Med. Parasit.* **72**, 533–542.

Crane, M. St. J. and Dvorak, J. A. (1980). Vertebrate cells express protozoan antigen after hybridization. *Science, N.Y.* **208**, 194–196.

Dedet, J. P., Derouin, F., Hubert, B., Schnur, L. F. and Chance, M. L. (1979). Isolation of *Leishmania major* from *Mastomys erythroleucus* and *Tatera gambiana* in Senegal. *Ann. trop. Med. Parasit.* **73**, 433–437.

Denecke, K. (1941). Menschenpathogene Trypanosomen des Hundes auf Fernando Poo. Ein Betrag zur Epidemiologie der Schlafkrankheit. *Arch. Hyg. Bakteriol.* **126**, 38–42.

Dórea, R. C., Póvoa, M. M., Miles, M. A. and Souza, A. A. de. Electroforèse de enzímas para estudos de triatomineos com referência especial a sub-populações de *Panstrongylus megistus*. *Rev. Bras. Biol.*, **42**, 521–526.

Ferguson, A. (1980). "Biochemical Systematics and Evolution". Blackie, Glasgow and London.

Gardener, P. J., Chance, M. L. and Peters, W. (1974). Biochemical taxonomy of *Leishmania*. II. Electrophoretic variation of malate dehydrogenase. *Ann. trop. Med. Parasit.* **68**, 317–325.

Gentile, B., Le Pont, F., Pajot, F. X. and Besnard, R. (1981). Dermal leishmaniasis in French Guyana: the sloth (*Choloepus didactylus*) as a reservoir host. *Trans. R. Soc. trop. Med. Hyg.* **75**, 612–613.

Gibson, W. (1980). *Trypanosoma evansi*: similarity of stocks from Central America and West Africa. *Trans. R. Soc. trop. Med. Hyg.* **74**, 112.

Gibson, W., Mehlitz, D., Lanham, S. M. and Godfrey, D. G. (1978a). The identification of *Trypanosoma brucei gambiense* in Liberian pigs and dogs by isoenzymes and by resistance to human plasma. *Tropenmed. Parasit.* **29**, 335–345.

Gibson, W. C., Parr, C. W., Swindlehurst, C. A. and Welch, S. G. (1978b). A comparison of the isoenzymes, soluble proteins, polypeptides and free amino acids from ten isolates of *Trypanosoma evansi*. *Comp. Biochem. Physiol.* **60B**, 137–142.

Gibson, W., Mehlitz, D., Zillmann, U. and Godfrey, D. G. (1979). The search for reservoir hosts of *Trypanosoma* (*Trypanozoon*) *brucei gambiense* in West Africa. *Trans. R. Soc. trop. Med. Hyg.* **73**, 134–135.

Gibson, W. C., Marshall, T. F. de C. and Godfrey, D. G. (1980). Numerical analysis of enzyme polymorphism: A new approach to the epidemiology and taxonomy of trypanosomes of the subgenus *Trypanozoon*. *Adv. Parasitol.* **18**, 175–246.

Godfrey, D. G. (1978). Identification of economically important parasites. *Nature, Lond.* **273**, 600–604.

Godfrey, D. G. (1979). The zymodemes of trypanosomes. *Symp. Brit. Soc. Parasitol.* **17**, 31–53.

Godfrey, D. G. and Kilgour, V. (1976). Enzyme electrophoresis in characterizing the causative organism of Gambian trypanosomiasis. *Trans. R. Soc. trop. Med. Hyg.* **70**, 219–224.

Heisch, R. B., McMahon, J. P. and Manson-Bahr, P. E. C. (1958). The isolation of *Trypanosoma rhodesiense* from a bushbuck. *Br. med. J.* **2**, 1203–1204.

Hoare, C. A. (1972). "The Trypanosomes of Mammals". Blackwell, Oxford.

Howard, J. G., Hale, C. and Liew, F. Y. (1980). Genetically determined susceptibility to *Leishmania tropica* infection is expressed by haematopoietic donor cells in mouse radiation chimaeras. *Nature, Lond.* **288**, 161–162.

Jeremiah, S. J., Povey, S. and Miles, M. A. (1982). Molecular size of enyzmes in *Trypanosoma cruzi* considered in relationship to the genetic interpretation of isozyme patterns. *Molec. Biochem. Parasitol.*, **6**, 297–302.

Kilgour, V. (1980a). *Trypanosoma*: Intricacies of biochemistry, morphology and environment. *J. Int. Biochem.* **12**, 325–332.

Kilgour, V. (1980b). The electrophoretic mobilities and activities of eleven enzymes of bloodstream and culture forms of *Trypanosoma brucei* compared. *Molec. Biochem. Parasitol.* **2**, 51–62.

Kilgour, V. and Godfrey, D. G. (1973). Species-characteristic isoenzymes of two aminotransferases in trypanosomes. *Nature, Lond.* **244**, 69–70.

Kilgour, V. and Godfrey, D. G. (1977). The persistence in the field of two characteristic isoenzyme patterns in Nigerian *Trypanosoma vivax*. *Ann. trop. Med. Parasit.* **71**, 387–389.

Killick-Kendrick, R., Molyneux, D. H., Rioux, J. A., Lanotte, G. and Leaney, A. J. (1980). Possible origins of *Leishmania chagasi*. *Ann. trop. Med. Parasit.* **74**, 563–565.

Kreutzer, R. D. and Christensen, H. A. (1980). Characterization of *Leishmania* spp. by isozyme electrophoresis. *Am. J. trop. Med. Hyg.* **29**, 199–208.

Kreutzer, R. D. and Sousa, O. E. (1981). Biochemical characterization of *Trypanosoma* spp. by isozyme electrophoresis. *Am. J. trop. Med. Hyg.* **30**, 308–317.

Lainson, R. (1981). Epidemiologia e ecologia de leishmaniose tegumentar na Amazônia. *Hiléia Médica, Belém.* **3**, 35–40.

Lainson, R. and Shaw. J. J. (1979). The role of animals in the epidemiology of South American leishmaniasis. *In* "Biology of the Kinetoplastida" (W. H. R. Lumsden and D. A. Evans, eds), Vol. 2, pp. 1–116. Academic Press, London.

Lainson, R. Shaw, J. J. and Póvoa, M. (1981a). The importance of edentates (sloths and anteaters) as primary reservoirs of *Leishmania braziliensis guyanensis*, causative agent of "pian bois" in north Brazil. *Trans. R. Soc. trop. Med. Hyg.* **75**, 611–612.

Lainson, R., Miles, M. A. and Shaw, J. J. (1981b). On the identification of viscerotropic leishmanias. *Ann. trop. Med. Parasit.* **75**, 251–253.

Lainson, R., Shaw, J. J., Ready, P. D., Miles, M. A. and Póvoa, M. (1981c). Leishmaniasis in Brazil: XVI. Isolation and identification of *Leishmania* species from sandflies, wild mammals and man in north Pará State, with particular reference to *L. braziliensis guyanensis* causative agent of "pian-bois". *Trans. R. Soc. trop. Med. Hyg.* **75**, 530–536.

Lainson, R., Shaw, J. J., Miles, M. A. and Póvoa, M. Leishmaniasis in Brazil: XVII. Enzymic characterization of a *Leishmania* from the armadillo, *Dasypus novemcinctus* (Edentata), from Pará State. *Trans. R. Soc. trop. Med. Hyg.*, **76**, 810–811.

Lanar, D. E., Levy, L. S. and Manning, J. E. (1981). Complexity and content of the DNA and RNA in *Trypanosoma cruzi*. *Molec. Biochem. Parasitol.* **3**, 327–341.

Lanham, S. M. (1982). Kits for isoenzyme characterization of Leishmania isolates in the field. *In* "Biochemical Characterization of *Leishmania*". UNDP World Bank, Geneva.

Lanham, S. M., Grendon, J. M., Miles, M. A., Póvoa, M. M. and Almeida, A. A. A. de (1981). A comparison of electrophoretic methods for isoenzyme characterization of trypanosomatids. I: Standard stocks of *Trypanosoma cruzi* zymodemes from northeast Brazil. *Trans. R. Soc. trop. Med. Hyg.* **75**, 742–750.

Lanotte, G., Rioux, J. A., Maazoun, R., Pasteur, N., Pratlong, F. and Lepart, J. (1981). Application de la méthode numérique à la taxonomie du genre *Leishmania* Ross, 1903 – A propos de 146 souches originaires de l'Ancien Monde. Utilisation des allozymes. Corollaires épidémiologiques et phylétiques. *Annls Parasit. hum. comp.* **56**, 575–592.

Letch, C. A. (1979). Host restriction, morphology and isoenzymes among trypanosomes of some British freshwater fishes. *Parasitology* **79**, 107–117.

Letch, C. A. and Gibson, W. C. (1981). *Trypanosoma brucei*: The peptidases of bloodstream trypanosomes. *Expl. Parasit.* **52**, 86–90.

Maazoun, R., Lanotte, G., Rioux, J. A., Pasteur, N., Killick-Kendrick, R., Pratlong, F. (1981). Signification du polymorphisme chez les Leishmanies. A propos de trois souches hétérozygotes de *Leishmania infantum* Nicolle, 1908, *Leishmania tarentolae* Wenyon, 1921, et *Leishmania aethiopica* Bray, Ashford et Bray, 1973. *Annls. Parasit. hum. comp.* **56**, 467–475.

Majiwa, P. A. O., Young, J. R., Englund, P. T., Shapiro, S. J. and Williams, R. O. (1982). Two distinct forms of surface antigen gene rearrangement in *Trypanosoma brucei*. *Nature, Lond.* **297**, 514–516.

Markert, C. L. (1975). (Ed.) "Isozymes". Academic Press, New York.

Mehlitz, D. (1977a). The behaviour in the blood incubation infectivity test of four *Trypanozoon* strains isolated from pigs in Liberia. *Trans. R. Soc. trop. Med. Hyg.* **71**, 86.

Mehlitz, D. (1977b). Untersuchunger uber das Vorkommen von Trypanosomeninfektionen bei einheimischen Schweinen und Hunden in Liberia und zum Verhalten von isolierten Trypanozoon-stammen im "Blood-incubation-Infectivity-Test". *Tropenmed. Parasit.* **28**, 278–279.

Mehlitz, D., Zillman, U., Scott, C. M. and Godfrey, D. G. (1982). Epidemiological studies on the animal reservoir of *gambiense* sleeping sickness. Part IV. Characterization of *Trypanozoon* stocks by isoenzymes and sensitivity to human serum. *Tropenmed. Parasit.* **33**, 113–118.

Miles, M. A. (1979). Transmission cycles and heterogeneity of *Trypanosoma cruzi*. *In* "Biology of the Kinetoplastida". (W. H. R. Lumsden and D. A. Evans, eds), Vol. 2, pp. 117–196. Academic Press, London.

Miles M. A. (1982a). *Trypanosoma cruzi*: epidemiology *In* "Perspectives in Trypanosomiasis Research" (J. Baker, ed.), pp. 1–15. Wiley, Chichester.

Miles, M. A. (1982b). *Leishmania*: Culture and biochemical comparisons – some difficulties. *In* "Biochemical Characterization of *Leishmania*". UNDP World Bank, Geneva.

Miles, M. A. and Ward, R. D. (1978). Preliminary isoenzyme studies on phlebotomine sandflies (Diptera: Psychodidae). *Ann. trop. Med. Parasit.* **72**, 398–400.

Miles, M. A., Toyé, P. J., Oswald, S. C. and Godfrey, D. G. (1977). The identification by isoenzyme patterns of two distinct strain-groups of *Trypanosoma cruzi* circulating independently in a rural area of Brazil. *Trans. R. Soc. trop. Med. Hyg.* **71**, 217–225.

Miles, M. A., Souza, A., Póvoa, M., Shaw, J. J., Lainson, R. and Toyé, P. J. (1978). Isozymic heterogeneity of *Trypanosoma cruzi* in the first autochthonous patients with Chagas' disease in Amazonian Brazil. *Nature, Lond.* **272**, 819–821.

Miles, M. A., Póvoa, M. M., Souza, A. A. de, Lainson, R. and Shaw, J. J. (1979). Some methods for the enzymic characterization of Latin-American *Leishmania* with particular reference to *Leishmania mexicana amazonensis* and subspecies of *Leishmania hertigi*. *Trans. R. Soc. trop. Med. Hyg.* **74**, 243–252.

Miles, M. A., Lanham, S. M., Souza, A. A. de and Póvoa, M. (1980). Further enzymic characters of *Trypanosoma cruzi* and their evaluation for strain identification. *Trans. R. Soc. trop. Med. Hyg.* **74**, 221–237.

Miles, M. A., Póvoa, M. M., Souza, A. A. de, Lainson, R., Shaw, J. J. and Ketteridge, D. S. (1981a). Chagas' disease in the Amazon Basin: II. The distribution of *Trypanosoma cruzi* zymodemes 1 and 3 in Pará State, north Brazil. *Trans. R. Soc. trop. Med. Hyg.* **75**, 667–674.

Miles, M. A., Cedillos, R. A., Póvoa, M. M., Souza, A. A. de, Prata, A. and Macedo, V. (1981b). Do radically dissimilar *Trypanosoma cruzi* strains (zymodemes) cause Venezuelan and Brazilian forms of Chagas' disease? *Lancet*, June 20, 1338–1340.

Miles, M. A., Lainson, R., Shaw, J. J., Póvoa, M. and Souza, A. A. de (1981c). Leishmaniasis in Brazil: XV. Biochemical distinction of *Leishmania mexicana amazonensis, L. braziliensis braziliensis* and *L. braziliensis guyanensis* – aetiological agents of cutaneous leishmaniasis in the Amazon Basin of Brazil. *Trans. R. Soc. trop. Med. Hyg.* **75**, 524–529.

Morales, G. A., Wells, E. A. and Angel, D. (1976). The capybara (*Hydrochoerus hydrochaeris*) as a reservoir host for *Trypanosoma evansi*. *J. Wildl. Dis.* **12**, 572–574.

Murray, A. K. (1982). Characterization of stocks of *Trypanosoma vivax*. I. Isoenzyme studies. *Ann. trop. Med. Parasit.* **76**, 275–282.

Mutinga, M. J., Ngoka, J. M., Schnur, L. F. and Chance, M. L. (1980). The isolation and identification of leishmanial parasites from domestic dogs in the Machakos district of Kenya, and the possible role of dogs as reservoirs of kala-azar in East Africa. *Ann. trop. Med. Parasit.* **74**, 139–144.

Nei, M. (1975). "Molecular Population Genetics and Evolution". North Holland Publishing, Amsterdam.

Ormerod, W. E. (1979). Human and animal trypanosomiasis as world public health problems. *Pharmac. Therap.* **6**, 1–40.

Peters, W., Chance, M. L., Mutinga, M. J., Ngoka, J. M. and Schnur, L. F. (1977). The identification of human and animal isolates of *Leishmania* from Kenya. *Ann. trop. Med. Parasit.* **71**, 501–502.

Peters, W., Chance, M. L., Chowdhury, A. B., Dastivar, B. G., Nandy, A., Kalra, J. L., Sanyal, R. K., Sharma, M. I. D., Srivastava, L. and Schnur, L. F. (1981). The identity of some stocks of *Leishmania* isolated in India. *Ann. trop. Med. Parasit.* **75**, 247–249.

Plant, J. E., Blackwell, J. M., O'Brien, A. D., Bradley, D. J. and Glynn, A. A. (1982). Are the Lsh and Ity disease resistance genes at one locus on mouse chromosome 1? *Nature, Lond.* **297**, 510–511.

Pratt, D. M. and David, J. R. (1981). Monoclonal antibodies that distinguish between New World species of *Leishmania. Nature, Lond.* **291**, 581–583.

Ramirez, L. E., Wells, E. A. and Betancourt, A. (1979). "La tripanosomiasis en los animales domesticos en Colombia. Revision bibliografica. *Trypanosoma evansi*". Centro International de Agricultura Tropical, Ser. 09FG/1, Cali, Colombia.

Rassam, M. B., Al-Mudhaffar, S. A. and Chance, M. L. (1979). Isoenzyme characterization of *Leishmania* species from Iraq. *Ann. trop. Med. Parasit.* **73**, 527–534.

Ready, P. D. and Miles, M. A. (1980). Delimitation of *Trypanosoma cruzi* zymodemes by numerical taxonomy. *Trans. R. Soc. trop. Med. Hyg.* **74**, 238–242.

Schnur, L. F., Chance, M. L., Ebert, F., Thomas, S. C. and Peters, W. (1981). The biochemical and serological taxonomy of visceralizing *Leishmania. Ann. trop. Med. Parasit.* **75**, 131–144.

Scott, C. M. (1981). Mixed populations of *Trypanosoma brucei* in a naturally infected pig. *Tropenmed. Parasit.* **32**, 221–222.

Solari, A. J. (1980). The 3-dimensional fine structure of the mitotic spindle in *Trypanosoma cruzi. Chromosoma* **78**, 239–255.

Tait, A. (1980). Evidence for diploidy and mating in trypanosomes. *Nature, Lond.* **287**, 536–538.

Tibayrenc, M. (1980). Application of the calculations of genetic distance for flagellate systematics. *Cah. O.R.S.T.O.M., sér. Ent. méd. et Parasitol.* **18**, 301–302.

Tibayrenc, M. and Desjeux, P. The presence in Bolivia of two distinct zymodemes of *Trypanosoma cruzi*, circulating sympatrically in a domestic transmission cycle. *Trans. R. Soc. trop. Med. Hyg.*, **77**, 73–75.

Tibayrenc, M. and Miles, M. A. A genetic comparison between Brazilian and Bolivian zymodemes of *Trypanosoma cruzi. Trans. R. Soc. trop. Med.* Hyg., **77**, 76–83.

Tibayrenc, M., Carioue, M. L., Solignac, M. and Carlier, Y. (1981). Arguments génétiques contre l'existence d'une sexualité actuelle chez *T. cruzi*. Implications taxonomiques. *C. r. hebd. Séanc. Acad. Sci., Paris* **293**, 207–209.

Walton, B. C. and Valverde, L. (1979). Racial differences in espundia. *Ann. trop. Med. Parasit.* **73**, 23–29.

Ward, R. D., Pasteur, N. and Rioux, J. A. (1981a). Electrophoretic studies on genetic polymorphism and differentiation of phlebotomine sandflies (Diptera: Psychodidae) from France and Tunisia. *Ann. trop. Med. Parasit.* **75**, 235–245.

Ward, R. D., Bettini, S., Maroli, M., McGarry, J. W. and Draper, A. (1981b). Phosphoglucomutase polymorphism in *Phlebotomus perfiliewi perfiliewi* Parrot (Diptera: Psychodidae) from central and northern Italy. *Ann trop. Med. Parasit.* **75**, 653–661.

World Health Organization (1978). Proposals for the nomenclature of salivarian trypanosomes and for the maintenance of reference collections. *Bull. W. H. O.* **56**, 467–480.

Williams, P. and Coelho, M. V. (1978). Taxonomy and transmission of *Leishmania*. *Adv. Parasitol.* **16**, 1–42.

Young, C. (1980). "Zymodemes of *Trypanosoma congolense* and a preliminary assessment of their epidemiological significance". Ph.D. thesis, University of London.

4 | Protein Variation in Protozoa

I. GIBSON and N. KHADEM

School of Biological Sciences, University of East Anglia, Norwich NR4 7TJ, U.K.

Abstract: The use of electrophoresis to study the proteins of the free-living ciliated protozoa, *Paramecium aurelia, Tetrahymena pyriformis* and *Paramecium caudatum* has supplemented the mating type test. This paper reviews the use of both of these methods to classify the breeding groups of these organisms. *P. caudatum* has not received the same attention as the other two but here also electrophoresis has aided the mating type test in defining breeding groups.

The electrophoretic technique also holds promise for study of gene frequencies and their adaptive significance in the natural habitats of ciliates.

INTRODUCTION

The best known proteins in protozoa are probably the immobilization surface antigens of *Paramecium aurelia*. A combination of biochemical and genetical studies has revealed a complex network of control over the expression of certain nuclear genes, involving cytoplasmic "conditions" and the prevailing influence of the environment. By changing the latter, e.g. by temperature shifts, "new" genes are expressed and "old" genes switched off. The details of the control process remain to be understood but they seem to involve the elaboration of new messenger-RNA molecules and, therefore, suggest that control of gene expression operates at the level of gene transcription into RNA,

Systematics Association Special Volume No. 24, "Protein Polymorphism: Adaptive and Taxonomic Significance", edited by G. S. Oxford and D. Rollinson, 1983, Academic Press, London and New York.

or in the processing of the RNA (Preer *et al.*, 1981). The new phenotype, i.e. antigen, is stable and transmitted through cell generations, both asexual and sexual, until the growth conditions change. What is impressive is the rapidity of cellular phenotypic change. In relation to the subject matter of this symposium, however, comparison between such proteins in terms of the different breeding groups or species was seen as a useful means of identification. Indeed, cross-reactions between different breeding groups (historically in the literature referred to as varieties or syngens), did occur (Finger, 1974). Furthermore since anti-sera were available, studies on natural isolates from ponds were carried out in an attempt to understand the genetic structure and breeding strategies of this group of organisms (Pringle, 1956; Pringle and Beale, 1960).

In the late sixties, however, following the apparent successes in other organisms, workers set out to investigate protozoa using modern-day biochemical techniques to compare nucleic acids and proteins. The background of genetical work and carefully defined collections of certain ciliates gave them confidence that interesting results might appear. The aims were basically simple. It was hoped to supplement the large body of work which had defined breeding groups by mating tests using genetic exchange as the major criterion, serotyping, temperature tolerance and geographical distribution (Sonneborn, 1957). Hopefully, it might be possible to rationalize the "species" problem in ciliates, to extend the approach to other protozoa, e.g. trypanosomes, where pure culture techniques would be of great advantage and also, of course, help us to understand biochemical evolution, its rate, the selective forces, if any, involved and the apparent conservatism of morphological form within a particular group of protozoa. Already in the antigenic studies, there were indications that rapid changes could occur under environmental pressures. A group of ciliates like *Tetrahymena pyriformis* could also, for all its morphological similarity, live at amazingly different temperature extremes. The hope was to understand this in biochemical terms with the ready-made breeding groups, defined by mating tests.

Other papers in this symposium will deal with the parasitic protozoa and the usefulness of biochemical studies on their proteins but here we wish to concentrate on the extensive investigations made on *Paramecium aurelia, Tetrahymena pyriformis* and, more recently, by Gibson and Khadem on *Paramecium caudatum*.

COMPARISONS OF BREEDING GROUPS

The breeding groups of *Paramecium aurelia*, called syngens*, were initially defined as consisting of isolates which exchanged genes amongst themselves, i.e. they mated and undertook genetic recombination via chromosomal and extra-chromosomal mechanisms. Such exchange between the syngens was seriously restricted or non-existent (Sonneborn, 1957). Technical considerations were important in the judgment not to name the group by species names, since tester mating cultures from all the groups would need to be available before designating an unknown to a group. However, following the work published in the early 1970s and, more recently, on the isozymes of *Paramecium aurelia*, the syngens were elevated to species status (Sonneborn, 1974). With nine enzymes controlled by nine separate and apparently unlinked loci, it became possible to separate all the fourteen breeding groups (Tait, 1970a, 1978; Allen and Gibson, 1971, 1975; Allen *et al*., 1973). In Fig. 1, we show some of the separations possible with the esterases of *P. aurelia* syngens. None of the enzymes varied, however, when any two of the breeding groups were compared. Similarly, in *Tetrahymena pyriformis*, another ciliate, syngens were distinguished and given species names (Borden *et al*., 1973b; Nanney and McCoy, 1976). Recently, however, a more guarded approach has been taken with the *Tetrahymena* data for reasons we will discuss later. The problems of species identification in *Tetrahymena pyriformis* remain to be solved, although the literature still uses species names and syngen numbers (Nanney *et al*., 1980).

In *Paramecium caudatum*, four breeding groups were obtained from Professor Hiwatashi and isozyme studies initiated. These four groups, unlike experiences with the other two ciliates, proved to be exceptional in that studies with seven enzymes (used to differentiate the other ciliates) showed an amazing uniformity of enzyme type. Five stocks of syngen 1, six stocks of syngen 3, 11 of syngen 12 and two stocks of syngen 13 had similar patterns of enzymatic activity on starch gels, e.g. the esterases and phosphoglucose isomerase. Variations of type did occur but these were within stocks of the same syngen and this was confirmed by mating and breeding analysis with the

* We will use syngen, breeding or mating group synonymously in this paper for all the protozoa.

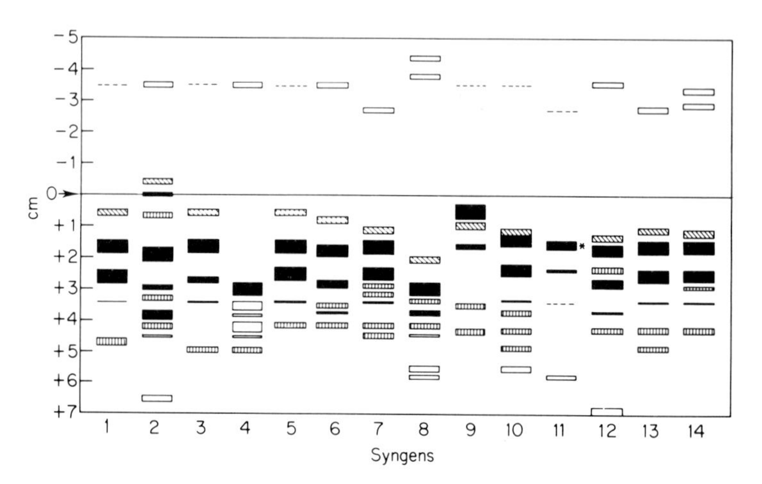

Fig. 1. Diagrams of the esterases in axenic stocks of the 14 syngens of *P. aurelia* in electrostarch. Types of esterases are represented by the following symbols: ▬ A type; ▭ B type; ▭ C type; ▭ D type.

There is considerable variation between stocks of syngen 2 in the mobilities of the A estrases. The * represents a hidden B esterase in syngen 11. Dashed lines represent esterases of low activity. Distances in migration are marked off in centimetres from the origin (o) on the margin in the diagram.

variants. A similar observation had been made with the enzyme lactate dehydrogenase in 44 stocks from the same four syngens where the only variant was within one syngen (Agatsuma and Tsukii, 1980). The syngens, true to definition, did not form mating pairs. Further studies on natural isolates showed that there were major differences with the same enzymes and along with other evidence could be used to justify classifying some of these isolates as new syngens (Fig. 2). However, at least four of the syngens, all isolated in Japan, are difficult to differentiate, which is different from the picture with any four syngens of *Paramecium aurelia* and *Tetrahymena pyriformis*. Representatives from these four syngens would have been placed in the same breeding group if isozyme analysis was the sole criterion.

Ideally, observers had been looking for restricted variations within breeding groups and great variations between. *Tetrahymena* and *Paramecium aurelia* seemed to fit this requirement but with *Paramecium caudatum*, we need to qualify this statement. The results indicated

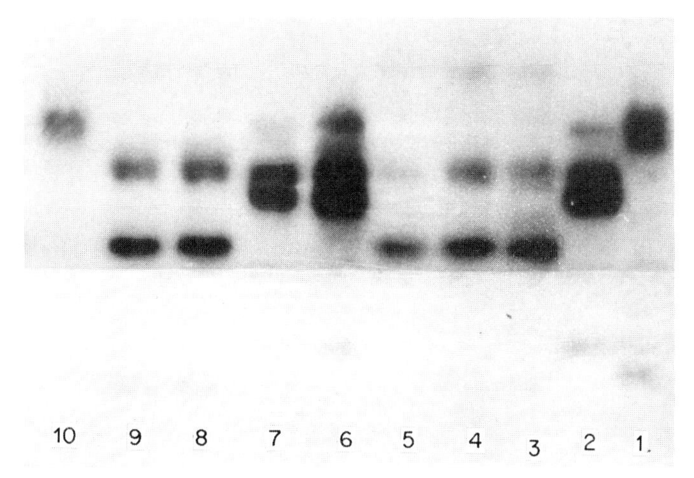

Fig. 2. Variations in α naphthyl butyrate esterase between 85 stocks of *P. caudatum* isolated from nature. (See Table I for description of the sources of these isolates.)

1, 10	=	Stocks No. 1–No 45, Mar. 1–Mar. 10, Syngen, 13.
2, 6, 7,	=	Stocks Cal 1–Cal 10.
3, 4, 5, 8, 9	=	LLSCO1–LLSCO10, ALSCO1–ALSCO10, No. 9, 35, 37.

more intra-syngenic than inter-syngenic variation. The qualification involves the possibility that if more syngens of *P. caudatum* were available, and it is documented there are sixteen, then the results so far may not be typical of this ciliate.

Crucial, however, to the validity of the inter-syngenic comparisons is the extent of intra-syngenic variation. Is it more or less than that between syngens and are the variants different in any way from the inter-syngenic? If they were, it would make syngenic identification easier.

COMPARISONS OF ORGANISMS WITHIN THE SAME BREEDING GROUP

In most breeding groups of *T. pyriformis* and *P. aurelia*, only 1% of the natural isolates were variants for any of the enzymes initially studied.

In *Paramecium*, seven of the nine enzymes had variant forms each controlled by a separate locus (Tait, 1968, 1970b; Allen and Golembiewski, 1972; Cavill and Gibson, 1972). The variants were, however, unique to that breeding group. In *Paramecium caudatum*, low numbers of isolates show variations within a breeding group and these also are controlled by genes. Certain groups of this ciliate have been isolated from the same pond and defined by breeding analysis so that we can conclude they are in the same breeding group. A group of 45 isolates from a Norfolk pond had three members with variations for butyrl esterase, another five (including only one of the previous three) varied for phosphogluco-isomerase. This indicates that different conclusions about the extent of variation between breeding groups depend on the enzyme under study. Some enzymes vary more than others. Tait (1978) showed that the same enzyme, phosphogluco-isomerase, showed great variation within a species of *P. aurelia* and indeed some of these variants were found to be indistinguishable from variants in other species. Other enzymes, like succinic dehydrogenase, showed no variation either within or between species. Malic enzyme does not vary in *P. caudatum* but shows polymorphism in *Tetrahymena* and *Leishmania* (Borden *et al.*, 1977). Clearly the choice of enzyme in any comparison is critical and can affect the results whether for inter- or intra-breeding group analysis.

Another complication which arose when variations within breeding groups were analysed was that some groups showed more than others. Species 1 of *Tetrahymena* and Species 2 of *P. aurelia* had more variants than others (Tait, 1970a; Allen and Gibson, 1971; Allen and Weremiuk, 1971; Allen *et al.*, 1971; Allen *et al.*, 1973; Allen and Gibson, 1975). Six out of ten enzymes showed variants in Species 2 of *P. aurelia* but only three in Species 1 and two in Species 4, 8 and 9. We have found syngen 3 of *P. caudatum* to show more variants than the other three (see Fig. 3). In Fig. 3, the intra-syngenic variations in syngen 3 are shown and also this is contrasted with the similarity of the other syngens.

Finally, the actual extent of variation within any group may depend on the electrophoretic techniques used to seek it, the conditions of analysis of the gels, the levels of enzymes in the cells, the uniformity of growth conditions, and stage in the cell cycle of the cells being examined. We know, for example, that bacteria may inhibit the formation of *Paramecium* esterases or stimulate the activity of others. Acetate in the growth medium affects the level of activity of an

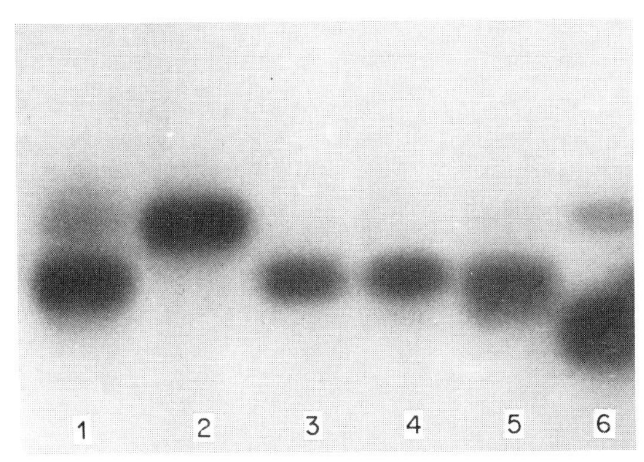

Fig. 3. Intra-syngenic variation and inter-syngenic similarities for phosphogluco-isomerase 1. KOK (Syn 3); 2. KT2 (Syn 3); 3. ISn (Syn 1); 4. Hj-6 (Syn 13); 5. Hj-1 (Syn 12); 6. YT3 (Syn 3).

esterase in *Paramecium* and this sensitivity is itself under genetic control (Allen and Nerad, 1978; Gibson and Cavill, 1973; Rowe *et al.*, 1971).

A major problem in looking again at *Tetrahymena* enzymes has been finding conditions and procedures for assessing the same enzyme in all the species. In fact, recently, Nanney *et al.* (1980) have only felt confident in comparing three enzymes out of their 16 species as a result of such considerations. Similarly, in *P. caudatum*, we found that certain enzymes, e.g. hexokinase, which showed intra-syngenic variations in the named syngens, could not be detected in the natural isolates. Tait (1970a) recorded a similar problem in *P. aurelia* with a mitochondrial enzyme.

Using heat effects on extracts greater variations were discovered in certain enzymes which had the same mobility in starch gels (Allen and Gibson, 1975). This was a lesson learned from studies on the surface antigens where cross reactions with sera were so strong that allelic products could not be distinguished. The same was true with non-allelic products. However, when more sophisticated studies were made on purified antigens both allelic and non-allelic proteins showed peptide map differences. It appears that the similarities were more apparent than real (Finger, 1974).

Finally, on the question of the extent of intra-syngenic variants, it has been pointed out that it is necessary to screen many isolates within a group before the true amount can be assessed. In this respect, it is interesting that as our number of natural isolates of *P. caudatum* increased, variants appeared in some of the enzymes which looked invariant within a group. Only superoxide dismutase and malic enzyme remained invariant in 85 isolates but the percentage of varieties of phosphogluco-isomerase and butyrl esterase increased. Clearly, then, besides a large number of isolates a survey of a larger number of loci becomes necessary. This leads us back to the difficulties in detecting certain enzymes. We have found it impossible to detect in *P. caudatum* isocitrate dehydrogenase, glutamate dehydrogenase, alcohol dehydrogenase and xanthine dehydrogenase, although further work should continue.

With these provisos on the problems of assessing variations, however, we can return to the use of protein comparisons to define the breeding groups of protozoa.

COMPARISON OF BREEDING GROUPS AGAIN

The original observations with *Tetrahymena pyriformis* and *P. aurelia* were not complicated by large amounts of intra-syngenic variation. In fact, it was apparent that there was more variation between the breeding groups of *Tetrahymena* than those of *Paramecium* (Adams and Allen, 1975; Borden *et al.*, 1973a). The extent of the differences in the latter were equated with sibling species of *Drosophila* and the former with non-sibling species. In an attempt to lay down rules whereby an unknown could be placed in a group, Nanney and McCoy (1976), studied eight enzymes of *Tetrahymena* within its breeding groups as defined by mating tests, and suggested 67% similarity in enzyme banding was a cut-off point. Anything below this justified calling the two being compared, separate species. Adams and Allen (1975) were confident enough of the data in *Paramecium and Tetrahymena* to construct dendrograms and concluded there were four groupings of the fourteen breeding sets in *P. aurelia*. These contradicted the groupings as defined by Sonneborn (1957) based on serotypes and the system of mating type inheritance.

Many groupings, when enzymes are compared, have a higher coefficient of identity than 67% and some have lower. In these cases, it is important to re-emphasize the problems that can arise with particular enzymes. If there is little variation then first of all non-mating protozoa would be classified as within the same syngen, since enzyme analysis is all that is available. If there is great variation with an enzyme within a group and this variation is of the same nature as between groups (Tait, 1978) then, without other criteria, the unknowns would be placed within the same group. Species 1 and 5 of *P. aurelia* may be separable with the phosphogluco-isomerase enzyme studies but variants in Species 1 resemble those in 9 and 12. Here, then, the 67% cut-off point would be irrelevant.

Studies on natural isolates of *P. caudatum* show how we might use mating type tests and isozyme studies effectively in protozoa. The variations in the known syngens are restricted to the butyrl esterases and phosphogluco-isomerases. The variants in the latter, unlike those in *P. aurelia* and those in the former, are specific to the syngen and, therefore, open up the possibility of allowing syngen identification. However, mating type tests were crucial in defining the syngens of *P. caudatum*.

It was interesting to see what happened with the 85 isolates from nature and other sources (see Table I) when we examined seven enzymes. The coefficients of identity showed clearly almost all the stocks from the same pond were similar in their enzyme patterns. Rare variants were seen in some ponds for the enzymes, esterases, superoxide dismutase and phosphogluco-isomerase, but, nevertheless, high levels of 100% identity of enzymes in stocks from the same pond were found (see Table II). These results have been compared with the variations within syngens where 3/10 possible pairwise combinations

Table I. Stocks collected from nature and their geographical origins.

Stocks	Geographical origin
No. 1–No. 14	Hethersett, Norfolk
No. 15–No. 24	U.E.A. Broad, Norfolk
No. 25–No. 45	Marlingford, Norfolk
Mar. 1–Mar. 10	Margate, Kent
ALSCO1–ALSCO10	Auchencrieff Loch, Scotland
LLSCO1–LLSCO10	Lochmaben Loch, Scotland
Cal 1–Cal 10	Manhattan Beach, California
2C (Syn 13)	Devon, England
Hj-6 (Syn 13)	Japan

Table II. The coefficient of identity between stocks collected from nature.

Stocks	No. 1–No. 45[a]	Mar. 1–Mar. 10	LLSCO1–LLSCO10	ALSCO1–ALSCO10	Cal 1–Cal 10	2C	Hj-6
No. 1–No. 45[a]	100	72	36	36	27	100	70
Mar. 1–Mar. 10		100	33	33	25	72	60
LLSCO1–LLSCO10			100	100	44	36	22
ALSCO1–ALSCO10				100	44	36	22
Cal 1–Cal 10					100	27	40
2C (Syn 13)						100	70
Hj-6 (Syn 13)							100

[a] Stock No. 1–No. 45 were considered as samples from a single geographical site, because of the close location of the three ponds from where the samples were collected and also the most common patterns of enzyme activity in these collections were identical.

Coefficient of identity (Borden *et al.*, 1973a) represents number of similar electrophoretic forms which cannot be distinguished between two stocks. The number of identical bands is divided by the total number of band comparisons made for any two stocks. Absolute identity is 100%. These are based on comparisons of seven loci.

of stocks in syngen 1, 3/15 in syngen 3, 26/55 in syngen 12, and 1/1 in syngen 13 had 100% identity of enzymes. However, when inter-syngenic comparisons were made for the same seven enzymes, 10/10 pairwise combinations had a 100% degree of similarity. Even absolute similarity then with the stocks from nature does not mean they are in the same syngen. We are left then with the mating type test to confirm or otherwise the similarity of the stocks as to their breeding groups. This has been done with each pond sample and it can be concluded they are in the same breeding group.

This leaves us with the comparisons between ponds from different geographical areas. Here, again, the coefficient of identity has been used (Borden *et al.*, 1973b). The results are shown in Table II and include comparisons with stocks from a known syngen (13). The results show the unknowns can be placed in one of three groups.

(i) Stocks No. 1 – No. 45, Hj-6 (Syn 13), Stock 2C,
(ii) Stocks LL SCO-1 – LL SCO-10, AL SCO 1-AL SCO 10,
(iii) Stock Cal-1 – Cal-10,

and using the 67% criterion (see above) can justify saying they represent 3 syngens. Group 1 mates with known syngen 13. There is an apparent low electrophoretic identity between stocks from different geographical sites and this result looks promising for syngen identification in *P. caudatum*. Even allowing for a low degree of intra-syngenic variation it offers hope, which the work with the known syngens did not, for using electrophoretic patterns in breeding groups and species identification. Do these results suggest that there may be geographical differentiation in terms of enzyme polymorphism? Perhaps the Japanese syngens are unique, although it is recorded that syngens 1 and 3 occur in the U.S.A. It seems unlikely that syngens are restricted to one geographical site and indeed syngen 13 has been identified now in England (see below). The surprising results here were the electrophoretic differences in the same syngen (13) in different countries. Geographical differentiation seems to occur. Lack of geographical differentiation was reported in *P. aurelia* between stocks in one breeding group from habitats as far apart as Japan and Scotland (Allen *et al.*, 1971). Clearly only breeding tests can solve this problem and electrophoretic analysis is not promising.

We need to consider these results in terms of natural selection and the importance of the enzymes in evolution and in ecological terms. We have carried out studies using the highly polymorphic antigen system and enzymes in *P. aurelia* isolated from a Norfolk pond.

STUDIES ON *PARAMECIUM AURELIA* IN NATURE

Some of the apparent paradoxes mentioned above, e.g. the apparent variability of enzymes in species 2 of *P. aurelia* and the apparent similarity of some enzymes in *P. caudatum* and *P. aurelia* (1 and 5) (Tait, 1978; Beale and Knowles, 1976) and the low level of inter-syngenic similarity in *Tetrahymena* might be resolved by investigating the behaviour of enzymes in different species of protozoa in nature. Earlier studies of antigenic polymorphism in *P. aurelia* in species 1 and 9 showed that these loci did vary in a single pond, and, indeed, heterozygotes could be detected. The latter was surprising in the face of the occurrence of the one generation process of inbreeding (autogamy) in these ciliates. It indicated mating did occur (Pringle, 1956; Pringle and Beale, 1960).

Four breeding groups were defined in this pond but in only one, species 2, did polymorphism occur. Breeding tests have identified 6 ESA alleles, 2 ESB alleles and 6 c and 6 e antigen alleles over a thirteen-year period. The number of alleles interestingly equates with the numbers detected in the whole world-wide distribution of *P. aurelia*. The frequency of the various alleles varies erratically from month to month, but the same alleles appear each year. Recently, cells have been isolated again after a five year gap and the same erratic changes take place. With one antigen allele, there was an increase in its frequency which correlated with a springtime rise in temperature in the pond. However, the gene frequency fluctuations are unpredictable. Attempts are being made to use these data to correlate the genetic variability with the mating system. For example, the frequency of heterozygotes at loci fluctuates wildly from 0 to 60% of the isolates over a five year period and each locus behaves independently.

In general, the antigen loci are much more polymorphic than the enzyme loci. Only the esterase alleles have shown much variation so far. Initial studies on pond samples of *P. caudatum*, however, indicate that enzymes which are polymorphic may be invariant in *P. aurelia* and vice versa. There appear to be three breeding groups of *P. caudatum* which fluctuate in their frequency from year to year and which have "private" variants of some enzymes. Each group has variants of the three enzymes, butyrl esterase, phosphoglucose isomerase and superoxide dismutase, but these make up a small percentage of all isolates. So, once again, there are differences between enzymes and species differences to explain in this natural habitat.

These studies have not as yet taken seriously the collection method of the cells, since it seems there may be spatial and temporal distribution considerations which vary for different protozoa. Inbreeders may clump together in restricted habitats whilst outbreeders have a wider distribution (Landis, 1982).

CONCLUSIONS

The use of electrophoretic techniques on the free-living protozoa has certainly raised many questions. It has offered, along with the mating type test which is laborious and time consuming and in itself not always conclusive (Nyberg, 1981), another method for detecting genetic variation in natural populations. It has also shown up differences in some breeding groups on a large scale (*Tetrahymena*) and no differences at all in four groups of *P. caudatum*. With the variations within groups, it has been possible to detect other genetic loci in various crosses.

As far as species identification is concerned, the technique has run into some problems both of a technical nature (detection of enzymes) and of a biological nature (similar variations of the same enzymes within as between groups). However, allied with breeding tests and also perhaps with temperature tolerance tests (Nyberg, 1981), these problems may be overcome. On the positive side, the technique has shown how cultures have become contaminated with bacteria, how mislabelling of stocks has occurred and how dangerous it is to extrapolate from one organism to another. It has illustrated a matter of fundamental biological interest to cell biologists, geneticists and others – how, from a base of biochemical diversity, you can construct cells which morphologically, to the human eye anyway, look and are architectured alike.

REFERENCES

Adams, J. and Allen, S. L. (1975). Polymorphism and differentiation in *Paramecium*. *In* "Proc. 3rd Int. Isozyme Conf. Vol. IV Genetics and Evolution" (C. L. Markert, ed.), pp. 867–882. Academic Press, New York.

Agatsuma, T. and Tsukii, Y. (1980). Genetic control of lactate dehydrogenase isozymes in *Paramecium caudatum*. *Biochem. Genet.* **18**, 77–85.

Allen, S. L. and Gibson, I. (1971). Intersyngenic variations in the esterases of axenic stocks of *Paramecium aurelia*. *Biochem. Genet.* **5**, 161–181.

Allen, S. L. and Gibson, I. (1975). Syngenic variations for enzymes of *Paramecium aurelia*. *In* "Proc. 3rd Int. Isozyme Conf. Vol IV Genetics and Evolution" (C. L. Markert, ed.), pp. 883–899. Academic Press, New York.

Allen, S. L. and Golembiewski, P. A. (1972). Inheritance of esterase A and B in syngen 2 of *Paramecium aurelia*. *Genetics* **71**, 455–469.

Allen, S. L. and Nerad, T. A. (1978). Effect of acetate on esterase C activity during the growth cycle of *Paramecium*. *J. Protozool.* **25**, 273–279.

Allen, S. L. and Weremiuk, S. L. (1971). Intersyngenic variations in the esterases and acid phosphatases of *T. pyriformis*. *Biochem. Genet.* **5**, 119–133.

Allen, S. L., Byrne, B. C. and Cronkite, D. L. (1971). Inter-syngenic variations in the esterases of bacterized *Paramecium aurelia*. *Biochem. Genet.* **5**, 135–150.

Allen, S. L., Farrow, S. W. and Golembiewski, P. A. (1973). Esterase variations between the 14 syngens of *Paramecium aurelia* under axenic conditions. *Genetics* **73**, 561–573.

Beale, G. H. and Knowles, J. K. (1976). Interspecies transfer of mitochondria in *Paramecium aurelia*. *Mol. Gen. Genet.* **143**, 197–201.

Borden, D., Miller, E. T., Nanney, D. L. and Whitt, E. S. (1973a). The inheritance of enzyme variants for tyrosine aminotransferase, NADP-dependent malate dehydrogenase, NADP-dependent isocitrate dehydrogenase, and tetrazolium oxidase in *Tetrahymena pyriformis* syngen 1. *Genetics* **74**, 595–603.

Borden, D., Miller, E. T., Whitt, E. S. and Nanney, D. L. (1977). Electrophoretic analysis of evolutionary relationships in *Tetrahymena*. *Evolution, Lancaster, Pa.* **31**, 91–102.

Borden, D., Whitt, E. S. and Nanney, D. L. (1973b). Electrophoretic characterisation of classical *Tetrahymena pyriformis* strains. *J. Protozool.* **20**, 693–700.

Cavill, A. and Gibson, I. (1972). Genetic determination of esterases of syngen 1 and 8 in *Paramecium aurelia*. *Heredity* **28**, 31–37.

Finger, I. (1974). Surface antigens of *Paramecium aurelia*, *In* "Paramecium – A current survey" (W. J. Van Wagtendonk, ed.). Elsevier Press, Amsterdam.

Gibson I. and Cavill, A. (1973). Effect of a bacterial product on a *Paramecium* esterase. *Biochem. Genet.* **8**, 357–364.

Landis, W. (1982). The spatial and temporal distribution of *Paramecium bursaria* in the littoral zone. *J. Protozool.* **29**, 159–161.

Nanney, D. L. and McCoy, J. (1976). Characterisation of the species of the *Tetrahymena pyriformis* complex. *Trans. Am. microsc. Soc.* **95**, 664–682.

Nanney, D. L., Cooper, L., Simon, E. M. and Whitt, E. S. (1980). Isozymic characterisation of three mating groups of *Tetrahymena pyriformis*. *J. Protozool.* **27**, 451–459.

Nyberg, D. (1981). Three new 'biological' species of *Tetrahymena* (*T. hegewischi* n. sp., *T. sonneborni* n. sp., *T. nipissingi* n. sp.) and temperature tolerance of members of the 'pyriformis' complex. *J. Protozool.* **28**, 65–69.

Preer, J. R. Jr., Prees, L. B. and Rudman, B. M. (1981). m-RNAs for the immobilisation antigens of *Paramecium*. *Proc. Nat. Acad. Sci., U.S.A.,* **78**, 6776–6778.

Pringle, C. R. (1956). Antigenic variation in *Paramecium aurelia*, Variety 9. *Z. indukt Abstamn u. Vererb. Lehre* **87** 421–430.

Pringle, C. R. and Beale, E. H. (1960). Antigenic polymorphism in a wild population of *Paramecium aurelia*. *Genet. Res.* **1**, 962–968.

Rowe, E., Gibson, I. and Cavill, A. (1971). The effect of growth conditions on the esterases of *Paramecium aurelia*. *Biochem. Genet.* **5**, 151–159.

Sonneborn, T. M. (1957). Breeding systems, reproductive methods and species problems in protozoa. *In* "The Species Problem" (E. Mayr, ed.)

pp. 155–324. American Association for the Advancement of Science, Washington D.C.

Sonneborn, T. M. (1974). The *Paramecium aurelia* complex of 14 sibling species. *Trans. Am. microsc. Soc.* **94**, 155–178.

Tait, A. (1968). Genetic control of B-hydroxybutyrate dehydrogenase in *P.aurelia*. *Nature, Lond.* **219**, 941.

Tait, A. (1970a). Enzyme variation between syngens in *Paramecium aurelia*. *Biochem. Genet.* **4**, 461–470.

Tait, A. (1970b). Genetics of NADP isocitrate dehydrogenase in *Paramecium aurelia*. *Nature, Lond.* **225**, 151–182.

Tait, A. (1978). Species identification in Protozoa; glucophosphate isomerase variation in the *Paramecium aurelia* group. *Biochem. Genet.* **16**, 945–955.

5 | The Circumboreal Barnacle *Balanus balanoides* (L.) and its Subpopulations

M. W. FLOWERDEW

N.E.R.C. Unit of Marine Invertebrate Biology, University College of North Wales, Marine Science Laboratories, Menai Bridge, Gwynedd LL59 5EH, U.K.

Abstract: An electrophoretic comparison of a population of the circumboreal cirripede *Balanus balanoides* from Alaska was made with two (possibly three) previously determined populations in the North Atlantic Ocean. The Alaskan population exhibited significantly different allele frequencies at two polymorphic loci when compared with the North Atlantic populations, indicating three (possibly four) populations throughout the entire geographical range of the species. Nei's coefficient of identity between the Alaskan and European populations, I = 0·814, indicated a possible taxonomic split at subspecies level between two populations at the extremes of the species range of distribution.

The life history of the barnacle is compared with that of the mussel, *Mytilus edulis*. The question is posed as to why *B. balanoides* should be genetically conservative when compared with *M. edulis*, when both species are marine invertebrates exhibiting a similar life-history.

INTRODUCTION

Balanus balanoides (L.) is an abundant intertidal cirripede found on rocky circumboreal shores. It is likely that the distribution was once completely circumpolar (Feyling-Hanssen, 1953) but has been discontinuous since the Pleistocene. Around the eastern Atlantic seaboard, the distribution extends from Novaya Zemyla, Iceland,

Systematics Association Special Volume No. 24, "Protein Polymorphism: Adaptive and Taxonomic Significance", edited by G. S. Oxford and D. Rollinson, 1983, Academic Press, London and New York.

Greenland and down European coasts to the mouth of the Gironde in France where the Lucitanean fauna of the Gulf of Gascony predominates; there is an isolated population in north-west Spain (Hutchins, 1947; Fischer-Piette and Prenant, 1956; Petersen, 1966). In western Atlantic waters, the distribution extends from the Canadian Arctic, south to Cape Hatteras (Wells *et al.*, 1960). On the Pacific coast, *B. balanoides* is present in the Bering Straits as far south as Sitka and in the Sea of Okhotsk, (Hiro, 1935; Nilsson-Cantell, 1978).

The life-cycle of *B. balanoides* exhibits a series of six planktonic larval stages over a twenty-one to thirty-five day period (Barnes and Barnes, 1958) followed by a cypris stage which metamorphoses to the adult after a period of exploration for a suitable substrate. Lucus *et al.* (1979) have shown that 2·5 to 4 weeks is the maximum time at 10°C that cyprids can be maintained in the plankton if metamorphosis is to be successfully completed. Thus, the larval stages may be pelagic from five to nine weeks. Larvae can be transported along coasts with resultant gene flow but it seems unlikely that there would be genetic exchange between the east and west Atlantic and the Atlantic and Pacific populations and the development of races or subspecies would be expected. A number of subspecies of *B. balanoides* have been described since the publication of Darwin's monograph on the Cirripedia (1854), but there is now little doubt that these morphological variants are due to plasticity of shell growth and not genotypic differences.

However, *Balanus balanoides* does show small but consistent differences in reproductive phenology, egg size and rate of larval development at uniform temperature between populations from the east and west coasts of the Atlantic (Barnes, 1958; Barnes and Barnes, 1959; Crisp, 1959a, b). These differences were retained for at least one year when samples were transplanted from North America to Europe and *vice versa* (Crisp, 1964, 1968). The number of populations examined on both sides of the Atlantic has been increased by Barnes and Barnes (1976). Their results confirm those of Crisp (1964) that the timing of fertilization and rate of larval development suggest that European *B. balanoides* are racially distinct from those in the U.S.A. population. Crosses between North American and European samples of *B. balanoides* have produced offspring which have been sucessfully metamorphosed and settled in the laboratory (Crisp and Flowerdew, unpublished observations). Flowerdew and Crisp (1975) have recorded differences in allele frequencies at an esterase locus between samples from New Brunswick and North Wales, providing further evidence for separate races. More recently, the ranges of genetically separated populations of *B. balanoides* around the North Atlantic seaboard have been established (Flowerdew, 1983) using samples taken from as far south as Indian River Inlet (38° 45′N), U.S.A., around the North American seaboard, Iceland, Norway, Denmark, Holland, France and the British Isles. Two genetically distinct populations were found, one stretching up the east coast of the U.S.A. to Newfoundland and as far as Iceland and the other encompassing Europe as far north as southern Norway and down to Brest in north-west France.

The work presented here is an extension of the investigation of populations of *B. balanoides* to include a sample from the Pacific Ocean.

MATERIALS AND METHODS

A sample of *Balanus balanoides* attached to rocks was air-freighted from Port Valdez, Alaska (61° 06′N, 146° 28′W). Samples were prepared and electrophoresed as described in Flowerdew (1983). The enzymes examined are contained in Table III. Malate dehydrogenase and α-glycerophosphate dehydrogenase were run in 0·15 M citrate, 0·254 M NaH_2PO_4 buffer, pH 5·9 (gel, 1:40 dilution); sorbitol dehydrogenase and malic enzyme were run in 0·3 M boric acid buffer, pH 8·6 (gel, 0·142 M Tris); the other enzymes were examined using 0·1 M Tris, 0·1 M maleic acid, 0·01 M EDTA (Na_2), 0·01 M $MgCl_2.6H_2O$ buffer, pH 7·4 (gel, 1:10 dilution). Recipes for the staining solutions were taken from Harris and Hopkinson (1976) but using 0·8% Meldola's Blue solution instead of PMS except in the preparation for superoxide dismutase.

RESULTS AND DISCUSSION

In a recent investigation of *Balanus balanoides* around the North Atlantic seaboard, Flowerdew (1983) used data from nine loci, two of which, *MPI-1* and *GPI* were polymorphic, to distinguish two, possibly three distinct populations. Contingency comparisons of allele frequencies indicated a North American population extending to and including Iceland and a European population. However, detrended correspondence analysis (Hill and Gauch, 1980) ranked the allele proportion data into three clusters, suggesting that *B. balanoides* from

Newfoundland and Iceland might be a sub-population of the North American population. A comparison between these North Atlantic populations and the Alaskan population at the two polymorphic loci is contained in Table I. Contingency comparisons of the allele frequencies at the *GPI* and *MPI-1* loci showed *B. balanoides* from Alaska to be significantly dissimilar to the U.S.A., Iceland/Newfoundland and European samples (Table II). Thus, it appears that throughout its entire range of distribution, there are only three, possibly four, distinct genetically separate populations of *B. balanoides*, namely the Pacific, U.S.A./Newfoundland/Iceland and European populations.

Table I. Comparison of allele proportions (S.E.) at two polymorphic loci in populations of *B. balanoides*. Data from the Atlantic populations taken from Flowerdew (1983)

	Europe	Iceland Newfoundland	U.S.A.	Alaska
GPI^{182}	0·419 (0·008)	0·238 (0·021)	0·282 (0·021)	0·084 (0·025)
GPI^{100}	0·556 (0·008)	0·760 (0·021)	0·714 (0·021)	0·912 (0·025)
GPI^{others}	0·025 (0·004)	0·002 (0·0002)	0·004 (0·003)	0·004 (0·006)
$MPI\text{-}1^{110}$	0·424 (0·008)	0·782 (0·020)	0·616 (0·025)	0·898 (0·026)
$MPI\text{-}1^{100}$	0·517 (0·008)	0·194 (0·018)	0·367 (0·030)	0·081 (0·024)
$MPI\text{-}1^{others}$	0·059 (0·005)	0·024 (0·007)	0·015 (0·005)	0·020 (0·012)

In a more extensive analysis, eighteen loci were compared between the Alaskan sample and a sample from the Menai Straits, North Wales (Table III). These data produced a Nei's coefficient of identity I = 0·814 and genetic distance D = 0·206 (Nei, 1972). This compares with I = 0·998 and D = 0·024 for a comparison of similar loci between the Menai Straits and Honningsvåg (71°N, 26° 20′E), North Norway (unpublished observations). The Menai Straits and Honningsvåg samples were taken from almost the extremes of the range of distribution of *B. balanoides* within Europe and their indices are typical of conspecific populations. However, the value of I = 0·814 between the Alaskan and the Menai Straits samples indicates a taxonomic identity less than that expected for conspecific populations but suggests taxonomic differentiation which might be of subspecies status (Ayala *et al.*, 1974; Thorpe, 1979).

The life-cycle of *B. balanoides* exhibits a long-lived larval stage, the

Table II. Contingency comparisons of the distribution of allele frequencies at the *MPI-1* and *GPI* loci in *B. balanoides* taken from samples covering the species' geographical range.

		Alaska vs East U.S.A. vs. Newfoundland/ Iceland vs. Europe	Alaska vs. East U.S.A.	Alaska vs. Newfoundland/ Iceland	Alaska vs. Europe
MPI-1	χ^2	474·563	124·973	15·400	208·403
	d.f.	48	2	2	2
	P	<0·001[a]	<0·001[a]	<0·001[a]	<0·001[a]
GPI	χ^2	267·442	45·582	31·292	104·438
	d.f.	24	4	2	1
	P	<0·001[a]	<0·001[a]	<0·001[a]	<0·001[a]

d.f. degrees of freedom
[a] Significant heterogeneity ($P < 0{\cdot}05$).

longevity of which is uncertain in the wild but probably lasts up to 35 days (Barnes and Barnes, 1958) followed by a cypris stage, the energetics of which determines that successful settlement and metamorphosis must be achieved within twenty-eight days (at 10°C) (Lucus *et al.*, 1979). Thus, the larval stages may be pelagic for five to nine weeks during which time they could be transported many kilometres (Crisp, 1977) depending on surface currents and wind direction. Cirripede larvae have been recorded up to 300 km from shore in the Norwegian Sea (Mileikovsky, 1968) and a similar attenuation of the larvae of littoral cirripede species has been reported off S.W. England (Crisp and Southward, 1958). Thus the large population sizes exhibited by *B. balanoides* is as anticipated for a marine invertebrate species with a prolonged larval phase and able to settle in a wide range of intertidal environments providing suitable substrata. Work on other balanomorph species (Flowerdew, unpublished) indicates they too exhibit large population sizes: no significant differences in allele proportion at 15 loci have been found in *Elminius modestus* throughout its range of distribution in Europe (Northern Spain to Scotland) nor between populations of *Balanus crenatus* sampled from the Irish Sea and the North Sea.

A worthwhile comparison can perhaps be made between *B. balanoides*

Table III. Comparison of allele proportions (S.E.) at 18 loci in *Balanus balanoides* from the Menai Straits, U.K. and Port Valdez, Alaska. n = number of animals examined, H_o = observed heterozygosity, H_ε = expected heterozygosity.

Locus	Allele	Port Valdez					Menai Straits				
				n	H_o	H_ε			n	H_o	H_ε
SORDH	112	0·867	(0·044)	41	0·067	0·231	–		50	–	–
	100	0·133	(0·044)				1·000				
GPD		1·000		48	–	–	1·000		50	–	–
MDH	145	0·010	(0·014)	48	0·021	0·022	0·023	(0·018)	67	0·050	0·057
	100	0·990	(0·014)				0·971	(0·021)			
	77	–					0·007	(0·010)			
ME		1·000		92	–	–	1·000		98	–	–
ICD-1	112	–		129	–	–	0·008	(0·005)	133	0·113	0·107
	100	–					0·944	(0·014)			
	86	–					0·015	(0·008			
	77	1·000					0·034	(0·011)			
ICD-2		1·000		129	–	–	1·000		133	–	–
PGD	100	1·000		129	–	–	0·991	(0·012)	58	0·017	0·018
	75	–					0·009	(0·012)			
SOD		1·000		130	–	–	1·000		133	–	–
PGM-1		1·000		123	–	–	1·000		151	–	–

PGM-2	114	0·018	(0·009)	123	0·033	0·035	0·010	(0·008)	151	0·053	0·064
	100	0·982	(0·009)				0·967	(0·015)			
	86	–					0·023	(0·012)			
PGM-3		1·000		123	–	–	1·000		151	–	–
HEX-1		–		63	–	–	1·000		89	–	–
HEX-2	130	0·032	(0·021)	63	0·000	0·061	0·077	(0·029)	89	0·060	0·142
	100	0·968	(0·021)				0·923	(0·029)			
β-GUS	100	1·000		69	–	–	0·970	(0·017)	101	0·059	0·058
	80	–					0·030	(0·017)			
ALD-1		1·000		69	–	–	1·000		133	–	–
ALD-2		1·000		69	–	–	1·000		133	–	–
MPI-1	115	0·016	(0·011)	124	0·186	0·185	0·020	(0·011)	150	0·567	0·652
	110	0·898	(0·025)				0·427	(0·040)			
	106	0·004	(0·006)				0·003	(0·005)			
	100	0·081	(0·024)				0·503	(0·041)			
	93	–					0·043	(0·017)			
	83	–					0·003	(0·005)			
GPI	300	–		125	0·128	0·161	0·004	(0·003)	382	0·482	0·523
	261	–					0·008	(0·005)			
	222	–					0·010	(0·005)			
	182	0·084	(0·025)				0·434	(0·023)			
	140	–					0·004	(0·003)			
	100	0·912	(0·025)				0·537	(0·026)			
	30	0·004	(0·006)				0·014	(0·010)			

and another marine invertebrate species, *Mytilus edulis* which has received a great deal of attention in the literature. Like *B. balanoides*, adult *M. edulis* are sessile (or generally so) and almost exclusively intertidal. Unlike *B. balanoides*, fertilization is external, but like the barnacle, the mussel exhibits a prolonged larval stage, lasting up to seven weeks (Seed, 1976) or longer since *Mytilus* larval development can be arrested for weeks at low temperatures to be completed with increasing temperatures (Beaumont and Budd, 1982). Dispersal of *Mytilus edulis* post-larvae can be greatly enhanced by bysso-pelagic drifting (Sigurdsson *et al.*, 1976). Thus, it seems that the pelagic phase in *M. edulis* is probably longer than that of *B. balanoides* and hence, its dispersal potential is greater. Indeed, Mileikovsky (1968) has found bivalve larvae (and post-larvae), including *Mytilus edulis*, in abundance up to 10^3 km offshore in the Norwegian Sea with barnacle larvae attenuating at 300 km. Thus, if population size is a function of dispersal potential in marine invertebrates, the population sizes of *M. edulis* should be at least as large as those for *B. balanoides*. *M. edulis* exhibits little genetic variation around the British Isles (Ahmad *et al.*, 1977; Burfitt, pers. comm.) but unlike *B. balanoides*, *M. edulis* exhibits several genetically distinct populations within European waters, including Iceland (Burfitt, pers. comm.), and as we have heard at this Symposium, (Skibinski, 1983), there is dispute over the taxonomic relationship between some variants of the *M. edulis*/*M. galloprovincialis* complex. Thus not only are there several populations of *M. edulis* over the geographical range where only one *B. balanoides* population is found, differences in morphology, mantle colour and population genetics are sufficiently great for there to be disagreement as to the taxonomic relationship of the "galloprovincialis" form and "edulis" form of *Mytilus*. In *B. balanoides*, however, phenotypic differences between populations appear to be sufficiently great to indicate a possible taxonomic split (into sub-species) only at the extremes of the range of the species' distribution, namely U.K. and Alaska.

Thus, it seems for marine invertebrate species with a prolonged larval stage, able to settle in a variety of environments over a wide geographical range, genetic differentiation between populations is not a simple function of dispersal potential. But the answer to the question as to why *Balanus balanoides* is genetically conservative when compared to *Mytilus edulis* would go along in helping to understand the functional significance of isozymes.

ACKNOWLEDGEMENTS

Thanks are due to Pami Rucker of the Marine Science Institute, Fairbanks, Alaska for kindly collecting and sending the sample from Port Valdez, and to Ms J. Forrest for secretarial assistance.

REFERENCES

Ahmed, M., Skibinski, D. O. F. and Beardmore, J. A. (1977). An estimation of the amount of genetic variation in the common mussel, *Mytilus edulis, Biochem. Genet.* **15**, 833–846.

Ayala, F. J., Tracey, M. L., Hedgecock, D. and Richmond, R. C. (1974). Genetic differentiation during the speciation process in *Drosophila. Evolution, Lancaster, Pa.* **28**, 576–592.

Barne, H. (1958). Regarding the southern limits of *Balanus balanoides* L. *Oikos* **9**, 139–157.

Barnes, H. and Barnes, M. (1958). The rate of development of *Balanus balanoides* (L.) larvae. *Limnol. Oceanogr.* **3**, 29–32.

Barnes, H. and Barnes, M. (1959). The effectof temperature on the oxygen uptake and rate of development of the egg masses of two common cirripedes *Balanus balanoides* (L.) and *Pollicipes polymerus*. J. B. Sowerby. *Kieler Meeresforsch.* **15**, 242–251.

Barnes, H. and Barnes, M. (1959). The effect of temperature on the oxygen of *Balanus balanoides* (L.) from a number of European and American populations and the designation of local races. *J. exp. mar. Biol. Ecol.* **24**, 251–269.

Beaumont, A. R. and Budd, M. D. (1982). Delayed growth of mussels (*Mytilus edulis*) veliger at low temperatures. *Mar. Biol.* **71**, 97–100.

Crisp, D. J. (1959a). Factors influencing the time of breeding of *Balanus balanoides. Oikos* **10**, 275–289.

Crisp, D. J. (1959b). The rate of development of *Balanus balanoides* (L.) embryos *in vitro. J. anim. Ecol.* **28**, 119–132.

Crisp, D. J. (1964). Racial differences between North American and European forms of *Balanus balanoides. J. mar. biol. Assoc. U.K.* **44**, 33–45.

Crisp, D. J. (1968). Differences between North American and European populations of *Balanus balanoides* revealed by transplantation. *J. Fish. Res. Bd. Can.* **25**, 2633–2641.

Crisp, D. J. (1977). Genetic consequences of different reproductive strategies in marine invertebrates. *In* "Marine Organisms, Genetics, Ecology and Evolution" (B. Battaglia and J. A. Beardmore, eds.) pp. 257–274. [N.A.T.O. Conference Series IV. Marine Science Vol. 2.] Plenum Press, New York.

Crisp, D. J. and Southward, A. (1958). The distribution of intertidal organisms along the coast of the English Channel. *J. mar. biol. Assoc. U.K.* **37**, 157–208.

Darwin, C. (1854). "A monograph on the subclass Cirripedia. Vol. 2. The Balanidae, the Verrucidae". Ray Society, London.

Feyling-Hanssen, R. W. (1953). The barnacle *Balanus balanoides* (Linné, 1776) in Spitsbergen. *Norsk Polarinstitutt. Skr.* **No. 98**, 1–64.

Fischer-Piette, E. and Prenant, M. (1956). Distribution des cirripèdes intercotidaux d'Espagne septentrionale. *Bull. Cent. Étud. Rech. Sci. Biarritz* **1**, 7–19.

Flowerdew, M. W. (1983). Electrophoretic investigation of populations of the cirripede *Balanus balanoides* (L.) around the North Atlantic seaboard. *Crustaceana*, in press.

Flowerdew, M. W. and Crisp, D. J. (1975). Esterase heterogeneity and an investigation into racial differences in the cirripede *Balanus balanoides* using acrylamide gel electrophoresis. *Mar. Biol.* **33**, 33–39.

Harris, H. and Hopkinson, D. A. (1976). "Handbook of Enzyme Electrophoresis in Human Genetics". North-Holland Publishing, Oxford.

Hill, M. O. and Gauch, H. G. (1980). Detrended correspondence analysis: an improved technique. *Vegetatio* **42**, 47–58.

Hiro, F. (1935). The fauna of Akkeshi Bay II. Cirripedia. *J. Fac. Sci. Hokkaito Univ., ser. 6, Zoology* **4**, 213–229.

Hutchins, L. W. (1947). The basis for temperature zonation in geographical distribution. *Ecol. Monogr.* **17**, 325–335.

Lucus, M. I., Walker, G., Holland, D. L. and Crisp, D. J. (1979). An energy budget for the free-swimming and metamorphosing larvae of *Balanus balanoides* (Crustacea:Cirripedia). *Mar. Biol.* **55**, 221–229.

Mileikovsky, S. A. (1968). Distribution of pelagic larvae of bottom invertebrates of the Norwegian and Barents Sea. *Mar. Biol.* **1**, 161–167.

Nei, M. (1972). Genetic distance between populations. *Am. Nat.* **106**, 283–292.

Nilsson-Cantell, C. (1978). "Marine Invertebrates of Scandinavia, No. 5 Cirripedia, Thoracica and Acrothoracica." Universitetsfovlaget, Oslo.

Petersen, G. H. (1966). *Balanus balanoides* (L.) (Cirripedia). Life-cycle and growth in Greenland. *Meddr. Grønland.* **159**, 1–116.

Seed, R. (1976). Ecology. *In* "Marine Mussels: Their Ecology and Physiology" (B. L. Bayne, ed.), pp. 16–65. Cambridge University Press, Cambridge.

Sigurdsson, J. B., Titman, C. W. and Davies, P. A. (1976). The dispersal of young post-larval bivalve molluscs by byssus threads. *Nature, Lond.* **262**, 386–387.

Skibinski, D. O. F. (1983). Natural selection in hybrid mussel populations. *In* "Protein Polymorphism: adaptive and taxonomic significance" (G. S. Oxford and D. Rollinson, eds), pp. 283–298. Academic Press, London.

Thorpe, J. P. (1979). Enzyme variation and taxonomy: the estimates of sampling errors in measurements of interspecific genetic similarity. *Biol. J. Linn. Soc.* **11**, 369–386.

Wells, H. W., Wells, M. J. and Grey, I. E. (1960). On the southern limits of *Balanus balanoides* (L.) in the Western Atlantic. *Ecology* **11**, 587–580.

6 | Evolutionary and Taxonomic Significance of Protein Variation in the Brown Trout (*Salmo trutta* L.) and other Salmonid Fishes

A. FERGUSON and C. C. FLEMING★

Department of Zoology, The Queen's University, Belfast BT7 1NN, Northern Ireland, U.K.

Abstract: Genetic variation within and among 116 British and Irish brown trout populations was examined by electrophoresis of 20 enzymes representing 60 or more loci. The proportion of loci polymorphic ranged from 0 to 21% in individual populations with an overall value of 23%. The mean expected heterozygosity was 0·038 (range 0–0·062). Of the 21 variant alleles at the 14 polymorphic loci, most were present in less than half of the populations and five were restricted to individual populations. Genetic evidence was obtained for the existence of reproductively isolated sympatric populations in Lough Neagh and Lough Melvin. Anadromous populations separated by as little as 2 km were also found to be genetically distinct. The *LDH-5(105)* allele was found in substantial frequency in only a small number of populations. This allele would appear to be the ancestral type and has predominated only in populations which are isolated from present-day migratory trout, the latter being characterized by the presence of the *LDH-5(100)* allele. The congeneric placing of *Salmo gairdneri* with *S. trutta* and *S. salar* is questioned. Enzyme variants serve to delimit adaptively distinct brown trout populations and are particularly valuable in the identification of sympatric populations.

★ Present address: Department of Agriculture of Northern Ireland, Agricultural Entomology Research Division, Nematology Laboratory, Felden, Mill Road, Newtownabbey, BT36 7ED, Northern Ireland, U.K.

Systematics Association Special Volume No. 24, "Protein Polymorphism: Adaptive and Taxonomic Significance", edited by G. S. Oxford and D. Rollinson, 1983, Academic Press, London and New York.

INTRODUCTION

Electrophoretic analysis of protein, and especially enzyme, variants has been used widely in the study of many aspects of intra-specific and species-level taxonomy. The simple genetic control of such variants enables a direct approach to the investigation of genetic variation within and among populations. There are, however, a number of problems inherent in this approach (Ferguson, 1980). An in-depth study of one taxon, on which a wide range of biological information is available, may help clarify some of these problems.

The brown trout occurs as a native fish in Europe, North-West Asia and North Africa and has been artificially introduced throughout the world. In his catalogue of fishes in the British Museum, Günther (1866) described ten species of trout in the British Isles. The adoption of the polytypic species concept led twentieth-century ichthyologists, e.g. Regan (1911), to group all these forms as a single species, *Salmo trutta* L.

The importance of an adequate taxonomic basis for ecological research and fisheries management is becoming increasingly recognized. In many cases, the species is much too gross a classification and even sub-species and races may not be sufficiently discriminatory for proper fisheries management (Everhart and Youngs, 1981). There is a tendency for many fisheries biologists to assume that a given species is composed of genetically similar individuals which will respond identically under all conditions. However, in recent years, the use of natural strains for creative management has received more attention (Behnke, 1981) as has the need for conservation of these unique gene pools (Ryman, 1981).

The present brown trout species encompasses a diversity of populations which are distinct in their life history, feeding behaviour, morphology, growth rate and potential, age of maturation, time and place of spawning, age of migration etc. A major question concerns the relative contribution of genetic and environmental influences to this variability. In brown trout, as with other salmonid fishes, there is a strong innate homing to their natal area for reproduction. Thus, reproductive isolation can be maintained even between sympatric populations. An adequate taxonomy is therefore necessary to make clear to fisheries biologists the genetic heterogeneity of brown trout populations. Also, for the conservation of unique populations for management and other purposes, it may be necessary to draw attention to them through their recognition as distinct taxa.

MATERIALS AND METHODS

Since 1975, samples of brown trout have been collected from 116 geographically discrete sites in the British Isles (Ireland, 92; England, 12; Scotland, 9; Wales, 3) and from several fish farm stocks, representing in excess of 7000 individuals. Rainbow trout, *Salmo gairdneri* Richardson, were available from a local commercial source; Atlantic salmon, *Salmo salar* L., from various Irish rivers; Arctic charr, *Salvelinus alpinus* (L.), from Lough Melvin; and pollan, *Coregonus autumnalis pollan* Thompson, from Lough Neagh. Samples of skeletal muscle, liver, heart, eye and brain were routinely taken and analysed by starch gel electrophoresis and, to a limited extent, agarose and polyacrylamide isoelectric focusing (Taggart *et al.*, 1981). The following 20 enyzmes gave good electrophoretic resolution and have been studied in detail: adenylate kinase (AK); alcohol dehydrogenase (ADH); aldolase (ALD); aspartate aminotransferase (AAT); creatine kinase (CK); diaphorase (DIA); fumarase (FUM); glucose-6-phosphate dehydrogenase (G-6-PDH); glycerol-3-phosphate dehydrogenase (G-3-PDH); isocitrate dehydrogenase (IDH); lactate dehydrogenase (LDH); malate dehydrogenase (MDH); malic enzyme (ME); peptidase (PEP); phosphoglucomutase (PGM); 6-phosphogluconate dehydrogenase (6-PGDH); phosphoglucose isomerase (PGI); phosphomannose isomerase (PMI); sorbitol dehydrogenase (SDH); superoxide dismutase (SOD). Due to the tetraploid origin of salmonid fishes, from two to six loci code for most enzymes in brown trout. The nomenclature used is based on that of Allendorf and Utter (1979). Hyphenated numerals designate multiple loci which are numbered in order of increasing electrophoretic mobility from the cathodal end of the gel. Allelic variants at a particular locus are denoted in parentheses by the mobility of the homomeric band relative to a standard allele (usually the most frequent one) which is designated 100. (For further details, see Taggart *et al.*, 1981).

GENETIC VARIATION IN BROWN TROUT

The enzymes examined represent at least 60 loci of which 14 (23%) have been definitively identified as polymorphic with Mendelian disomic inheritance being confirmed by breeding studies (Taggart, 1981; Taggart *et al.*, 1981). The proportion of loci polymorphic in individual natural populations ranged from 0 to 21%. Table I shows

Table I. Distribution and frequency (mean, standard deviation, and maximum) of enzyme variants in brown trout populations.

Locus	Variant allele	No. of populations examined[a]	No. of populations with variant	Frequency Mean	S.D.	Max.	Inheritance studied[b]
$_s$*AAT-1, 2*[c]	140	87	68	0·19	0·16	0·61	+
$_s$*AAT-1, 2*	45	87	5	0·11	0·08	0·20	+
$_s$*AAT-4*	74	99	88	0·24	0·16	0·65	+
CK-2	115	116	31	0·24	0·10	0·50	+
CK-2	75	116	3	0·03	0·03	0·07	
DIA	120	93	3	0·04	0·01	0·05	+
DIA	90	93	40	0·07	0·06	0·25	+
G-3-PDH-2	120	116	1			0·09	
G-3-PDH-2	50	116	111	0·32	0·21	0·83	+
G-3-PDH-2	30	116	1			0·02	
$_s$*IDH-1*	160	114	90	0·16	0·15	0·98	+
$_s$*IDH-2*	130	114	16	0·07	0·10	0·43	+
LDH-1	240	116	1			0·27	+
LDH-5	105	116	60	0·29	0·28	1·00	+
$_s$*MDH-2*	152	116	115	0·32	0·17	0·93	+
$_s$*MDH-3, 4*	134	116	1			0·14	
$_s$*MDH-3, 4*	125	116	43	0·09	0·07	0·14	
$_s$*MDH-3, 4*	75	116	[d]				+
PGI-2	135	116	54	0·10	0·12	0·46	+
PGI-2	65	116	1			0·40	
PGI-3	110	116	20	0·05	0·03	0·61	+
PGI-3	85	116	1			0·02	

[a] Refers to samples > 20 from a discrete geographical area (excluding hatchery stocks).
[b] Indicates verification by inheritance studies.
[c] Inheritance study shows both loci polymorphic for this allele.
[d] The presence of additional non-genetic variability precludes accurate assessment of allelic frequencies.

the loci found to be polymorphic and the relative mobility designations of the variant alleles (i.e. other than the 100 allele). The mean expected heterozygosity for natural populations was 0·038 (range 0–0·062). This compares with $\bar{H}_e$ of 0·059 for the rainbow trout (*Salmo gairdneri*) (Allendorf and Phelphs, 1981) and of 0·023 for Atlantic salmon (*Salmo salar*) (Ståhl, 1981).

Of the 21 variant alleles (at frequencies of 0·02 or greater), the *MDH-2(152)* and the *G-3-PDH-2(50)* alleles were found in almost all populations (Table I). It is of interest to note that both of these enzymes are polymorphic in most of the salmonid species which have been examined in sufficient detail. Also, the *MDH-2(100/100)* allozyme of brown trout has the same electrophoretic mobility as the common allozymes of Atlantic salmon and rainbow trout. The *MDH-2(152/152)* allozyme of brown trout is of equivalent mobility to a variant of Atlantic salmon (Fleming, 1982). In the case of *G-3-PDH-2*, both the *G-3-PDH-2(100/100)* and *G-3-PDH-2(50/50)* allozymes have the same mobility in brown trout and Atlantic salmon (Crozier, 1982). The existence of presumably the same ancestral alleles in various species at these highly polymorphic loci may suggest that they are maintained by selection. Most variant alleles were found in less than half of the populations examined and five alleles were restricted to individual populations even though occurring at substantial frequencies in the case of *LDH-1(240)* and *PGI-2(65)*.

Much of the genetic variation in brown trout is thus due to variation among populations. As this may be a common feature for other species, it is important that conclusions on genetic variability are not based on a small number of populations. However, by contrast, in the rainbow trout, Allendorf and Phelps (1981) note that only 8% of the genetic differentiation is due to differences among populations and 92% is the result of within population variability.

GENETIC STRUCTURE OF BROWN TROUT POPULATIONS

1. *Accuracy and stability of allelic frequencies*

The usefulness of electrophoretic data for defining populations relies on first, that allelic frequencies can be estimated accurately and, secondly, that they are stable over time. In the examination of different natural populations, sample sizes were between 30 and 200 in almost all cases. Based on multiple samples from different parts of a river in which all available evidence would suggest that only a single panmictic population is present (Table II) and on similar information from other lake and river systems (Ferguson and Mason, 1981; Crozier, 1983), it would appear that a sample of 30 or more gives a reasonable estimate of allelic frequencies when dealing with a di-allelic polymorphism.

Table II. Allelic frequencies in brown trout samples taken at different times and places from the Glynn River. (0+ refers to fish in their first year, 1+ to fish in their second year, etc.)

		Frequency of allele		
Sample	*n*	*PGI-2(135)*	*G-3-PDH-2(50)*	*MDH-2(152)*
1978 2+	45	0·23	0·23	
1979 1+	60	0·25	0·19	
1980 0+	31	0·21	0·17	0·22
1980 1+	38	0·27	0·22	0·23
1980 2+	30	0·28	0·15	0·20
1980 river mouth	37	0·29	0·17	0·22
1980 3 km upstream	44	0·26	0·17	0·28
1980 5 km upstream	48	0·34	0·20	0·18
1982 1+	37	0·22	0·22	0·27
1982 2+	37	0·29	0·15	0·33
mean (± S.D.)		0·26 (± 0·04)	0·19 (± 0·03)	0·24 (± 0·05)

2. *Sympatric populations*

Genetic heterogeneity, in conjunction with other information, has demonstrated the existence of sympatric reproductively isolated populations in two Irish lakes, Lough Neagh and Lough Melvin. In the Lough Neagh river systems, two types of migratory brown trout, in addition to non-migratory populations, appear to be present (Crozier, 1983).

In Lough Melvin of north west Ireland, three morphologically distinct types of trout, known locally as gillaroo, sonaghen and ferox, have been recognized by anglers and naturalists for over 200 years. The deficit of heterozygotes in pooled samples and the significant heterogeneity in allelic frequencies at a number of loci (Table III) would suggest that these are not merely ecophenotypes as previously thought but represent genetically distinct populations with a very high degree of reproductive isolation (Ferguson and Mason, 1981).

Table III. Frequencies for variant alleles in sonaghen, gillaroo and ferox of Lough Melvin at loci showing significant heterogeneity (information is not available for some alleles in ferox as indicated by –).

Locus	Variant allele	Frequency in		
		sonaghen (n = 249)	gillaroo (n = 153)	ferox (n = 42)
sAAT-1, 2	140	0·36	0·17	–
CK-2	115	0·33	0·00	–
G-3-PDH-2	50	0·23	0·08	0·06
LDH-1	240	0·00	0·27	0·00
LDH-5	105	0·01	0·01	0·61
PGI-2	135	0·35	0·05	0·04

Recent work (A. Ferguson and J. Taggart, unpublished), has shown that the sonaghen breeds in the inflowing rivers and the gillaroo in the outflowing river. The frequencies of characteristic alleles in samples of 0+ trout from the inflowing rivers were what would be expected if only sonaghen were present and, similarly, for the outflowing river and gillaroo. Also, of 500+ mature trout which were examined in the rivers in November and December, 1981, only gillaroo were found in the outflowing river and only sonaghen in the inflowing rivers. The spawning behaviour of the ferox is not known for certain but there is some indication that it may breed earlier in the autumn than the other types and possibly in deeper water.

Initial evidence suggests that the three types are distinct in their food preferences, with the gillaroo feeding predominantly on molluscs and Trichoptera whereas planktonic organisms form a significant proportion of the food of sonaghen. After the age of about three years, the ferox is mainly piscivorous with Arctic charr (*Salvelinus alpinus*) being the main prey. There are also differences in growth patterns with the ferox being a later-maturing, high growth potential type of trout.

3. Inter-population variation

Considerable allelic frequency heterogeneity was found among sample sites at many loci with extremes of allelic frequencies often being found in adjacent geographical areas. The heterogeneity seen among anadromous ("sea trout") populations points to the accurate homing of brown trout to their natal areas for breeding. For example, the *PGI-2(65)* allele was found in only one anadromous population, even though occurring at a frequency of 0·4, and being absent from stocks in rivers as little as 2 km away.

The distribution of the *LDH-5(105)* allele is of particular interest (Fig. 1). This variant allele was found in about half (60) of the popula-

Fig. 1. Geographical variation in frequency of *LDH-5* alleles (plain = 100; shaded = 105) in selected British and Irish brown trout populations. The frequency of each allele is proportional to the area of the circle occupied by its symbol.

tions screened but only in eight of these did its frequency exceed 0·20. The *LDH-5(105/105)* allozyme has the same electrophoretic mobility as the common *LDH-5* allozyme of rainbow trout, Atlantic salmon, Arctic charr and the Pacific salmons (*Oncorhynchus* sp.). It would appear, then, that *LDH-5(105)* is the ancestral allele and the *LDH-5(100)* of brown trout is the result of a mutation occurring in the brown trout lineage. A common feature of all populations which show a high frequency of the *LDH-5(105)* allele is that they are isolated from migratory brown trout, usually by impassable water falls. It is proposed that in immediate post-glacial times, rivers and lakes in Britain and Ireland were colonized by migratory brown trout which were fixed for the *LDH-5(105)* allele. In more recent times, migratory brown trout, which were characterized by the *LDH-5(100)* allele, and which were possibly of more southern origin, colonized those areas of freshwater to which they had access and replaced the "ancestral" type. Artificial stocking in the last century has also added to this process. More evidence from other remote brown trout populations is required to complete the picture and to give an indication of times of colonization. On the basis of transferrin allelic frequencies, Payne *et al.* (1971) have proposed the existence of two races of Atlantic salmon in Britain and Ireland and attribute the formation of these to isolation by the final or Würm III phase of the last glaciation. They suggest that the "boreal" race was formed in a North Sea refuge and, at the end of the glaciation, colonized the rivers of Scotland and north and west Ireland, while the "celtic" race remained in the non-glaciated area to the south. It is possible that a similar hypothesis may explain the origin of the two brown trout "races". At any rate, the *LDH-5* alleles would appear to provide useful markers for following the post-glacial colonization of brown trout. There is also some evidence that brown trout populations with a high *LDH-5(105)* frequency have a higher growth potential under suitable conditions than do those with high *LDH-5(100)* frequencies.

The overall variation can be summarized by calculating, on the basis of all loci examined, a coefficient of genetic identity (Nei, 1975) between all possible pairs of populations. Based on pairwise comparisons of mean genetic identity values over 40 loci, a UPGMA cluster analysis (Sneath and Sokal, 1973) was carried out and the resulting dendrogram for some populations is shown in Fig. 2. The most obvious feature of this is the major dichotomy between the

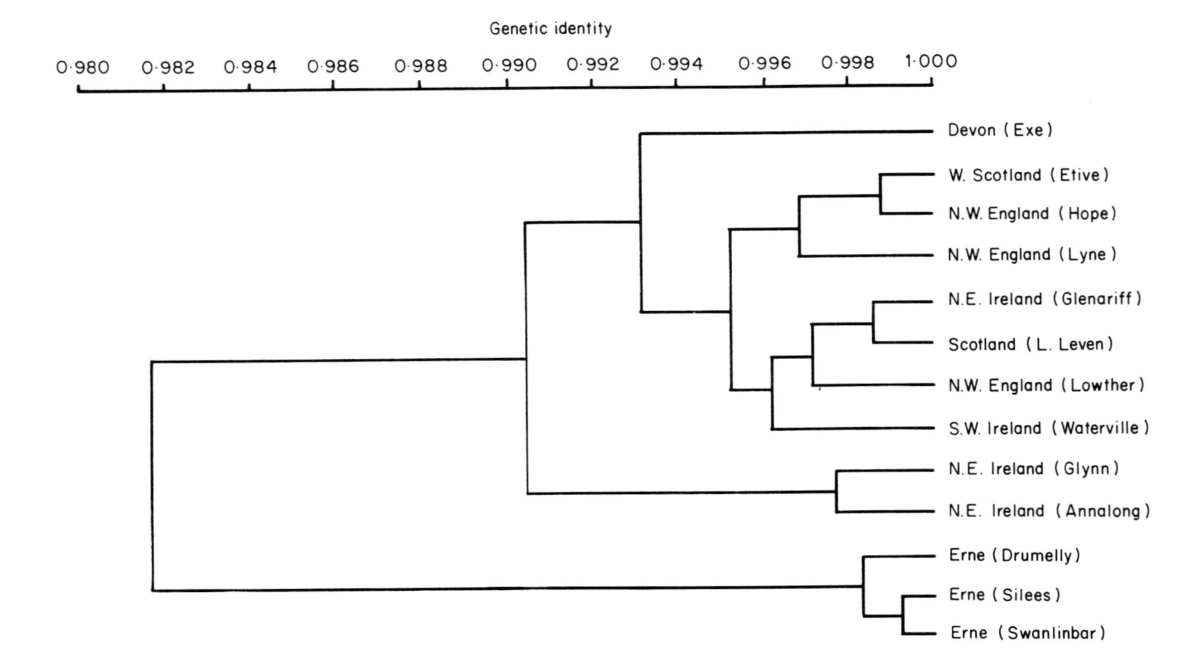

Fig. 2. Dendrogram showing the relationships of some British and Irish brown trout populations, generated according to UPGMA cluster analysis and based on mean genetic identity values over 40 loci.

populations of the Erne-Macnean drainage of County Fermanagh and all other populations. This is a reflection of the distinctive allelic frequencies of these populations at a number of loci.

ORIGINS AND SYSTEMATIC IMPLICATIONS

While there are distinct differences among various populations and groups of populations of brown trout, the overall degree of genetic differentiation is relatively small. Nei's genetic identity values between populations range from 0·96 to 0·99 which are comparatively high compared with values of 0·87 to 0·99 reported between geographically separated populations of other vertebrate species (Ferguson, 1980). Ryman *et al.* (1979) reported a genetic identity of 0·975 (based on 54 loci) between two sympatric populations of Lake Bunnersjöarna in Sweden. These two populations are fixed for different LDH-1 alleles, suggesting that reproductive isolation is complete.

Much of the genetic differentiation among brown trout populations has probably occurred within the past 100 000 years and may be attributable to advances and retreats of the ice cover during the last glaciation, and to post-glacial isolation. Isolation in glacial refuges and in post-glacial lakes with different environmental and physical conditions may thus have resulted in the evolution of a diverse range of ecological and behavioural adaptations. In this way, reproductive isolation has been achieved with little overall genetic divergence, at least as detected by current techniques. Once achieved, reproductive isolation is maintained by the innate tendency of brown trout to spawn in their natal areas.

Allelic frequency data from 30 loci were used to calculate mean genetic identity values (Nei, 1975) among brown trout (*Salmo trutta*), rainbow trout (*Salmo gairdneri*), Atlantic Salmon (*Salmo salar*), Arctic charr (*Salvelinus alpinus*) and pollan (*Coregonus autumnalis pollan*). A UPGMA cluster analysis of these values gives the dendrogram shown in Fig. 3. Very similar results were obtained when Rogers' (1972) coefficients of similarity were computed. An interesting feature of the dendrogram is that *S. gairdneri* shows less similarity to *S. trutta* and *S. salar*, its supposed congeners, than to *S. alpinus* and *C. a. pollan* which belong to different genera. Also, Nei's genetic identity value of 0·236 between *S. trutta* and *S. gairdneri* is of a level expected between

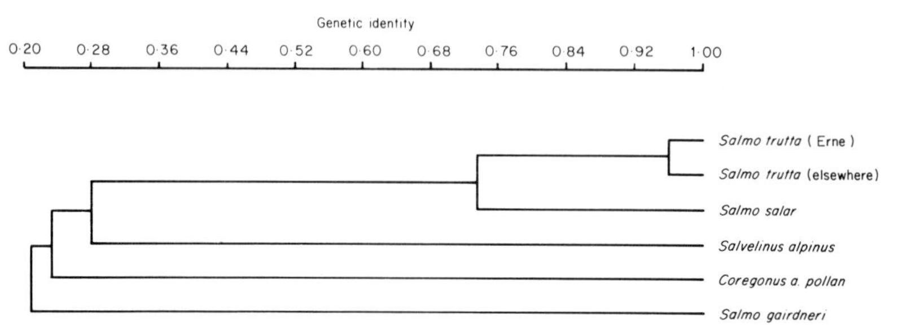

Fig. 3. Dendrogram showing the relationships of five salmonid species, generated according to UPGMA cluster analysis and based on mean genetic identity values over 30 loci.

members of different genera (Ferguson, 1980). When the comparison of salmonid species was extended to include *S. clarki* and the *Oncorhynchus* species, using available published information (for references see Fleming, 1983) and similarities computed in like fashion to those of Utter *et al.* (1973), the dendrogram shown in Fig. 4 was produced. It can be seen that *S. gairdneri* shows greater similarity to the other salmonid species indigenous to the Pacific ocean drainages than it does to its supposed Atlantic congeners *S. trutta* and *S. salar*. Behnke (1968, 1972) recognized the distinctness of *S. gairdneri* by placing it

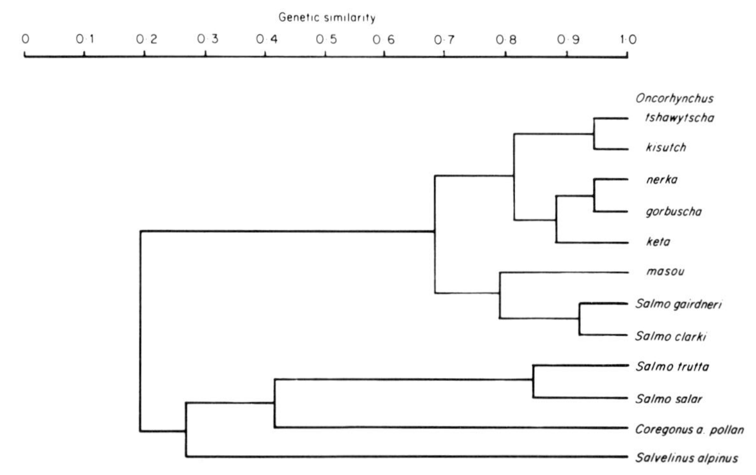

Fig. 4. Dendrogram showing the relationships of 12 salmonid species, generated according to UPGMA cluster analysis and based on similarities over 18 loci, computed from band mobilities in similar fashion to that described by Utter *et al.* (1973).

together with the other N. American *Salmo* species in the sub-genus *Parasalmo*. The data presented here would suggest that *S. trutta* and *S. gairdnei* would be more appropriately classified in separate genera. Alternatively, it is possible that the present classification is correct and the similarities within the Atlantic and Pacific groups are the result of convergence rather than phylogeny.

CONCLUSION

Electrophoretically determined protein variants form useful markers for the delimitation of brown trout populations which possess distinctive adaptive traits and are particularly valuable in the identification of genetically distinct sympatric populations and sibling species. However, the extent to which the observed enzyme variation is directly related to this phenotypic variation and is not simply an indication of widespread changes in the genome as a whole is still an open question. Electrophoretic studies thus form a suitable basis for the intra-specific taxonomy of the brown trout where the aim is the delimitation of separate adaptive complexes. The identification of unique stocks which should be conserved for their potential management value as well as for their inherent scientific interest is a matter of urgency as natural populations of brown trout are continually being lost due to eutrophication, pollution, exploitation, introductions, hydroelectric power station construction, etc. The rapidity with which electrophoretic surveys of populations can be carried out is important in this respect. However, it must be stressed that once genetically unique populations have been identified in this way, detailed ecological and behavioural studies are then necessary to document fully the attributes of these populations.

ACKNOWLEDGEMENTS

We thank our colleagues W. Crozier, T. Henry and J. Taggart for use of their unpublished data. Support from the Natural Environment Research Council for part of this study and the award of a Postgraduate Studentship to C.F. by the Department of Education N.I. is gratefully acknowledged. We are especially grateful to the following for their assistance in obtaining samples: D. Anderson, T. B. Bagenal, W. Baird, N. Campbell, J. F. Craig, D. T. Crisp, A. Cummins, P. Dando, D. Duncan, E. Fahy, J. Fletcher, T. Gallagher, R. Gibson, R. Hamilton, J. Kernaghan, A. Kilgore, A. McGurdy, S. Maclennon, C. Maguire, F. Magee, M. O'Grady, G. O'Neill, J. B. Rogers, J. Thorpe, A. Walker, R. G. Weaver.

REFERENCES

Allendorf, F. W. and Phelps, S. R. (1981). Isozymes and the preservation of genetic variation in salmonid fishes. *In* "Fish Gene Pools" (N. Ryman, ed.) *Ecol. Bull.* (Stockholm) **34**, 37–52.

Allendorf, F. W. and Utter, F. M. (1979). Population Genetics. *In* "Fish Physiology" (D. J. Randall and J. R. Brett, eds), Vol. VIII, pp. 407–454. Academic Press, London.

Behnke, R. J. (1968). A new subgenus and species of trout, *Salmo* (*Platysalmo*) *platycephalus*, from southcentral Turkey, with comments on the classification of the subfamily Salmoninae. *Mitt. Hamburg Zool. Mus. Inst.* **66**, 1–15.

Behnke, R. J. (1972). The systematics of salmonid fishes of recently glaciated lakes. *J. Fish. Res. Bd. Can.* **29**, 639–671.

Behnke, R. J. (1981). Systematic and zoogeographical interpretation of Great Basin trouts. *In* "Fishes in North American Deserts" (R. J. Naiman and D. L. Soltz, eds), pp. 95–124. Wiley, New York.

Crozier, W. W. (1983). "Population Biology of Lough Neagh Brown Trout (*Salmo trutta* L.)". Ph.D. Thesis, The Queen's University of Belfast.

Everhart, W. H. and Youngs, W. D. (1981). "Principles of Fishery Science" (2nd edn.). Comstock, London.

Ferguson, A. (1980). "Biochemical Systematics and Evolution". Blackie, Glasgow.

Ferguson, A. and Mason, F. M. (1981). Allozyme evidence for reproductively isolated sympatric populations of brown trout *Salmo trutta* L. in Lough Melvin, Ireland. *J. Fish Biol.* **18**, 629–642.

Fleming, C. C. (1983). "Population Biology of Anadromous Brown Trout (*Salmo trutta* L.) in Ireland and Britain". Ph.D. Thesis, The Queen's University of Belfast.

Günther, A. (1866). "Catalogue of the Physostomi in the Collection of the British Museum". British Museum, London.

Nei, M. (1975). "Molecular Population Genetics and Evolution". North-Holland, Amsterdam.

Payne, R. H., Child, A. R. and Forrest, A. (1971). Geographic variation in the Atlantic salmon. *Nature, Lond.* **231**, 250–252.

Regan, C. T. (1911). "The Freshwater Fishes of the British Isles". Methuen, London.

Rogers, J. S. (1972). Measures of genetic similarity and genetic distance. *Univ. Texas Stud. Genet.* **7**, 145–153.

Ryman, N. (ed.) (1981). Fish gene pools. *Ecol. Bull.* (Stockholm) 34.

Ryman, N., Allendorf, F. W. and Stahl, G. (1979). Reproductive isolation with little genetic divergence in sympatric populations of brown trout (*Salmo trutta*). *Genetics* **92**, 247–262.

Sneath, P. H. A. and Sokal, R. R. (1973). "Numerical Taxonomy". Freeman, San Francisco.

Ståhl, G. (1981). Genetic differentiation among natural populations of Atlantic salmon (*Salmo salar*) in Northern Sweden. *In* "Fish Gene Pools" (N. Ryman, ed.) *Ecol. Bull.* (Stockholm) **34**, 95–105.

Taggart, J. B. (1981). "An Electrophoretic Study of Genetic Variation in Irish Brown Trout (*Salmo trutta* L.)". Ph.D. Thesis, The Queen's University of Belfast.

Taggart, J. B., Ferguson, A. and Mason, F. M. (1981). Genetic variation in Irish populations of brown trout (*Salmo trutta* L.): Electrophoretic analysis of allozymes. *Comp. Biochem. Physiol.* **69B**, 393–412.

Utter, F. M., Allendorf, F. W. and Hodgins, H. O. (1973). Genetic variability and relationships in Pacific salmon and related trout based on protein variations. *Syst. Zool.* **22**, 257–270.

Part II | Protein Variation and Phylogenetic Reconstruction

7 | Protein Variation and Phylogenetic Reconstruction

J. C. AVISE

Department of Molecular and Population Genetics,
University of Georgia, Athens,
Georgia 30602, U.S.A.

Abstract: In recent years, evolutionists have argued at length about principles and methods of phylogenetic inference. Although much of the debate has concerned conventional systematic data from morphology, most of the philosophical and methodological issues apply with equal force to analyses of molecular characters as well. Here, I discuss four fundamental considerations that apply to acquisition and analysis of protein and other molecular data: (1) the general dichotomy between qualitative versus quantitative methods of data analysis; (2) the distinction between character-state phylogenies and taxa phylogenies; (3) the observation that not all molecular characters are equally informative to phylogeny reconstruction; and (4) for studies in comparative evolution, the frequently overlooked desirability of greater standardization in data analysis.

INTRODUCTION

The last two decades have witnessed a burgeoning interest in principles underlying methodologies of phylogenetic reconstruction and systematics. In the early 1960s, the rise of numerical phenetics promised to convert a traditional intuitive practice of systematics to a hard science based on quantitative assessments of overall (phenetic) similarity among organisms (Sneath and Sokal, 1973). In the 1970s, before

Systematics Association Special Volume No. 24, "Protein Polymorphism: Adaptive and Taxonomic Significance", edited by G. S. Oxford and D. Rollinson, 1983, Academic Press, London and New York.

numerical phenetics took full hold, it was challenged by the equally revolutionary cladistic school of thought, which grew in popularity following the 1966 English translation of Hennig's "Phylogenetic Systematics". Cladists generally focus on branching sequences in evolutionary geneologies, and emphasize the fundamental distinction between ancestral (plesiomorphic) and shared-derived (synapomorphic) traits in determining those branching patterns. In recent years, the cladistic school has further splintered into various subgroups, including a discipline (numerical cladistics or quantitative phyletics) whose concerns of manipulating data from large numbers of characters partially overlap those of the early pheneticists. In sum, these developments have too frequently led to a polarization of views, and to unusually rancorous debates (see, for example, the "Points-of-View" section of almost any recent issue of Systematic Zoology). Future historians of science will decide whether the turmoils represent birth pains of an exciting new field, or death throes of the very old field of systematics struggling once again for respectability as a biological discipline. Most likely they reflect neither extreme.

The field of molecular evolution developed and expanded during the same two decades, but surprisingly remained relatively untouched by the systematic debate (but see Fitch, 1975 and references therein), probably for at least two reasons. First, molecular evolutionists were justifiably preoccupied with early findings that rates of evolutionary divergence in macromolecules appeared rather constant (Wilson *et al.*, 1977). The supposed validity of the "molecular clock" provided justification for the common practice of using phenetic clustering procedures (which assume constant rates of change) to summarize protein distances into phenograms which were then loosely interpreted as phylogenies. Secondly, as noted by Selander (1982), molecular evolution never did have classification as its major objective, but was instead concerned primarily with processes and mechanisms of evolutionary change. Nonetheless, the data of molecular evolution are relevant to systematics, and it is now widely appreciated that most of the difficulties arising in analysis of traditional systematic data apply with equal force to molecular characters. In addition, it can be debated whether *any* study of "processes and mechanisms" of evolutionary change can proceed without reference to phylogeny.

A distinction should be made between phylogeny reconstruction and classification, although both issues are interwoven in the current systematic debate. Biological classification is, in my opinion, a subjective and arbitrary (but nonetheless extremely important) enterprise which we may choose to base wholly or in part on branching topologies of phylogenetic trees (Mayr, 1981), depending on our goals. On the other hand, any assemblage of real organisms has a single phylogenetic history. Since different data analyses frequently yield different phylogenetic reconstructions, which method we choose is in principle not an arbitrary matter.

Here I will restrict consideration to phylogeny reconstruction rather than classification. In this brief overview, I cannot hope to summarize the voluminous literature of phylogenetic inference, nor will I provide detailed analyses of particular protein data sets, of which there are many examples in current journals. Rather, I would like to emphasize a few fundamental distinctions and considerations that apply to acquisition and analysis of protein data. Some of these fundamental considerations have been underemphasized, if not entirely neglected, in discussions of molecular phylogeny.

QUANTITATIVE VERSUS QUALITATIVE ANALYSIS

Evolutionary changes in the structure of biological macromolecules map the stream of heredity that is phylogeny (Simpson, 1945). However, these long-term changes cannot be observed directly. For molecular evolutionists, the problem of phylogenetic reconstruction is the problem of inferring past sequences of change from the observable structural differences among proteins and DNA's of extant organisms. A number of algorithms can be employed to convert genetic data into estimates of phylogenetic trees.* The particular algorithm(s) chosen depend upon the goals and assumptions of the analysis, and upon the nature of the genetic information available.

Some molecular assays such as those involving immunological comparisons of proteins, or thermal stabilities of hybrid DNA's, provide raw data only in the form of numerical or quantitative distances. If comparisons are made among many organisms (operational taxonomic units or OTU's), a distance matrix is generated. Other molecular assays such as protein electrophoresis, or amino acid and

* I will use "phylogenetic tree" or "tree" in the broadest sense of any graphical estimate of the phylogeny of taxa, including "cladograms" and "phenograms" of some authors.

DNA sequencing, provide raw data in the form of discrete or qualitative character states. Any qualitative data base can be converted to a quantitative distance matrix, but the converse is not true. Such a conversion to a distance matrix is done, however, at the risk of significant information loss. This is because distance analyses are typically based on all available data, irrespective of whether or not each particular datum is geneologically informative. For these reasons, it seems desirable to clearly distinguish between phylogenetic analyses that manipulate distance matrices (for convenience, I refer to these as quantitative or distance analyses) from those that manipulate the raw or coded character states themselves (qualitative analyses).

1. *Quantitative analyses*

Three distinct algorithms have been most commonly employed to construct trees from distance matrices derived from protein data: the unweighted pair-group clustering method (UPGMA; Nei, 1975); the F-M procedure (Fitch and Margoliash, 1967); and a distance-Wagner approach (Farris, 1972; Swofford, 1981).

The UPGMA method is by far the simplest, conceptually and computationally. OTU's exhibiting the smallest genetic distance in a matrix are joined first to form a "cluster". Successive scans of the matrix identify next-smallest mean distances, and these form the basis for adding new OTU's to the growing tree, and for joining together clusters previously generated. The most significant criticism of UPGMA is the underlying assumption of homogeneity in rate of evolution in all branches. Li (1981) has recently proposed a modification of UPGMA which partially relaxes this assumption. The F-M and distance-Wagner algorithms are computationally much more difficult, but both relax strong assumptions of uniformity of evolutionary rates.

Much debate concerns which of these (or other) algorithms produces the "best" tree. One criterion for choice is "goodness-of-fit", a measure of the extent to which a tree distorts values in the original matrix. Distances between OTU's read from a tree (output distances) can be compared to original input distances from the matrix by any of several suggested statistics (Sneath and Sokal, 1973; Fitch and Margoliash, 1967; Farris, 1972; Prager and Wilson, 1978). For example, Prager and Wilson's (1978) "F" statistic equals

$$100 \sum_{i=1}^{n} (I_i - O_i) / \sum_{i=1}^{n} I_i,$$

where for n pairwise comparisons of species I and O are input and output distances, respectively. Smaller values of F indicate better fit, i.e. less homoplasy in the data.

Empirical comparisons of methods by goodness-of-fit criteria have produced equivocal results. For example, from an analysis of eleven protein data sets by each of the above tree-generating methods, Prager and Wilson (1978) found the F-M procedure superior in eight cases, and the distance-Wagner approach superior in three cases. Although UPGMA won no single contest, it did average second best (to F-M) in overall performance as judged by F. Similarly, Avise *et al.* (1980b) found a slightly better fit for F-M trees than for UPGMA or distance-Wagner trees in an analysis of protein-electrophoretic distances among species of sparrows. On the other hand, in an electrophoretic survey of *Rhagoletis* flies, Berlocher (1981) found better goodness-of-fit in distance-Wagner trees than in UPGMA trees.

The distance-Wagner algorithms were expressly designed to maximize goodness-of-fit, so it is surprising that they do not always outperform other methods by this criterion. One possible explanation relates to choice of distance measures. It is now recognized that distance-Wagner algorithms should be applied only to "metric" distances (Berlocher, 1981) satisfying the "triangle inequality" (which states that no side of a triangle of distances can surpass the sum of the other two sides). Nei's (1972) genetic distance statistic (D), widely used in summaries of protein electrophoretic data, is not a metric. However, Nei's D has other highly desirable properties: it is unusual among electrophoretic distance statistics in estimating a parameter (accumulated number of codon substitutions per locus) of biological significance; and at low and intermediate values, it may be linearly related to time (Nei, 1981).

A second possible criterion for choice of "best" distance method involves examination of congruence among trees derived from different sets of data (i.e. morphology versus allozymes). Since a given assemblage of OTU's has a single phylogenetic topology along which all characters have evolved, methods of data analysis producing more highly congruent trees might be judged superior (Farris, 1971). Several statistics for estimating various aspects of congruence have been suggested (Farris, 1973; Mickevich, 1978; Colless, 1980; Nei, 1981), but few applications to protein distance summaries have as yet been conducted. Berlocher (1981) found greater congruence among distance-Wagner trees than among UPGMA trees for morphological and allozyme data in *Rhagoletis*. Prager and Wilson (1976) found strong congruence among F-M trees derived from amino acid sequences and immunological comparisons of different proteins in cracid birds.

A third criterion for "best" distance method is potentially powerful, but has seldom been employed. It involves computer simulation of evolutionary change along specified branching topologies (model trees). The output of the simulations, distributions of characters among taxa, is then used to estimate phylogeny by any of the above algorithms. Results can be evaluated by the extent to which each reconstruction distorts topology and/or branch lengths of the model tree. In an example of this approach, Tateno and Nei (see Nei, 1981) simulated evolutionary changes in allele frequencies along a tree, assuming random genetic drift and an infinite-allele model of mutation. In general, the F-M algorithms proved more reliable than UPGMA or distance-Wagner analyses for obtaining tree topologies, while UPGMA most accurately estimated branch lengths. Further applications of this approach with differing assumptions for evolutionary change may be desirable.

2. *Qualitative analyses*

Some molecular techniques, such as protein electrophoresis, provide raw data in the form of distributions of qualitative character states (electromorphs in this case) among taxa. Since the evolutionary process consists of changes in the genetic character states themselves, there is potential phylogenetic information to be gained by considering these states directly, rather than first submerging data into a summary distance statistic. Character state evolution can include convergences, reversals, and parallel retentions of ancestral states, phenomena which cannot be elucidated solely by reference to a distance matrix. Thus qualitative analyses at least offer hope of distinguishing geneological relationship from non-geneological noise.

Among the numerous specific approaches to qualitative analyses, most involve in one way or another the principle of parsimony. This concept has a long-standing justification, which lies in attempts to most efficiently utilize available information (Dobzhansky and Eppling, 1944; Cavalli-Sforza and Edwards, 1967; Fitch, 1971). A maximum parsimony tree is one which requires the smallest possible number of character-state transitions or steps, summed across all branches. The number of conceivable trees for moderate or even small numbers of OTU's is astronomical (i.e. for 10 OTU's at least 34 000 000 trees are possible – Felsenstein, 1978a). It is methodologically impossible to examine all of these, but many algorithms have been introduced for obtaining most-parsimonious network approximations (reviews in Felsenstein, 1978b, 1979). Different algorithms depend on mode of evolution assumed (Felsenstein, 1979). For example, the original Camin and Sokal (1965) parsimony procedure assumes no evolutionary reversals, while Farris' (1970) Wagner approach (not to be confused with distance-Wagner analysis) does allow reversals. Needless to say, extensive discussions have concerned which of the many methods is most appropriate. However, Felsenstein (1978b) points out that any parsimony method may still fail to identify the true phylogeny, particularly when parallel changes outnumber nonparallel changes; and Colless (1980) reminds us that trees of virtually identical overall length (number of character state transitions) can often differ substantially in topology.

Qualitative parsimony procedures have a strong tradition of use in molecular evolution, for example in analyses of data on overlapping chromosomal inversions (Dobzhansky, 1970; Carson and Kaneshiro, 1976; White, 1978), and for recent data on restriction site polymorphisms in mitochondrial DNA (Avise *et al.*, 1979; Lansman *et al.*, 1981). However, qualitative analyses have only infrequently been applied to protein data, probably for two reasons: (1) as already emphasized, some protein assays (i.e. microcomplement fixation) provide raw data only as numerical distances, and (2) even for those assays providing qualitative information, the phylogenetic order (transformation series) of character states is often unavailable from inspection of the states themselves. Thus, for example, an electrophoretic survey might reveal five electromorphs (states) of the protein transferrin (character). Transferrin represents a qualitative multistate character (*sensu* Sneath and Sokal, 1973, p. 149) whose states cannot be evolutionarily ordered by reference solely to electrophoretic mobility. The success of qualitative analyses of chromosome inversion data and restriction maps of mitochondrial DNA rests on the ability to provide highly parsimonious summaries of the evolutionary transformations among their respective character states.

Transformation series among character states are valuable for reconstructing evolutionary networks, but such networks usually remain unrooted unless ancestral conditions are known or hypothesized. The distinction of ancestral from derived states of characters is fundamental to modern phylogenetic reasoning (Hennig, 1966). Since primitive characters may be evolutionarily retained by distantly related extant species, common ancestries (clades) within an assemblage of taxa under study must be defined solely by possession of shared-derived (synapomorphic) states. For determination of branching topologies (but not necessarily branch lengths) of phylogenetic trees, non-synapomorphic states are misleading or irrelevant.

Problems of phylogenetic ordination apply to amino acid (or DNA) sequence data as well. Fitch (1971) provides a qualitative parsimony procedure for determining ancestral character states from amino acid sequences of extant organisms, but the procedure necessitates *a priori* specification of the phylogenetic relationships among the taxa represented. Thus Fitch's approach (and others built upon it – see Klotz *et al.*, 1979) are really more concerned with finding most-parsimonious assignments of genetic information to a given tree, a problem which is subordinate to finding an unknown most-parsimonious tree from the genetic data when ancestral states are unknown (Fitch, 1971).

Several general criteria have been suggested to decide whether a particular character state is ancestral or derived (reviews in deJong, 1980; Stevens, 1980). For example, in "in-group analysis" the primitive state is taken to be that state which occurs most frequently in the group being examined. Other proposed criteria for recognizing primitiveness include presence in fossils, early appearance in ontogeny, correlation with other primitive states, and so on. Each of these criteria has been seriously and justifiably criticized (Stevens, 1980). Only "out-group analysis" provides a particularly compelling rationale for estimation of character state polarity. In "out-group analysis", a character state is considered plesiomorphic if it is shared with members of a group closely related to, but phylogenetically distinct from, the assemblage under study. In other words, presence of a given character state both in the out-group and in one or more OTU's under study is taken as evidence that the common ancestor of all possessed that trait.

Naturally, this inference will be incorrect if the character state has arisen more than once by convergence.

In exploratory studies, Patton and Avise (1983; Avise *et al.*, 1980a) attempted qualitative Hennigian analyses of several sets of protein-electrophoretic data, using out-group criteria to indirectly infer ancestral and derived electromorphs (Fig. 1; Table I). In each case, the qualitative cladistic trees provided fits to "model phylogenies" which were at least as strong as those resulting from quantitative phenetic-clustering or distance-Wagner approaches. These attempts also revealed some of the major strengths and weaknesses of the qualitative Hennigian analysis that are likely to apply to most protein-electrophoretic data. Major strenths are: (1) the analyses are simple, and for moderate size data sets can be performed by hand; (2) newly acquired data can readily be added to a tree, without need to recalculate distance matrices; and (3) outputs of the qualitative analyses explicitly define character states along all branches of the tree, and hence provide phylogenetic trees which are extremely rich in empirical content (*sensu* Popper, 1968). These advantages are, however, counterbalanced by several serious reservations about the approach: (1) from a given data base, numerous equally-parsimonious Hennigian trees could sometimes be generated; (2) only a small proportion (less than 20%) of all electromorphs in a study appeared to be synapomorphs actively contributing to clade identification (by cladistic reasoning, however, these are the only phylogenetically informative character states); (3) rare

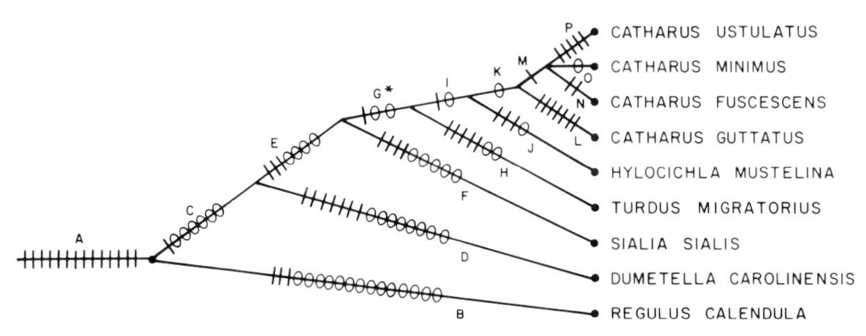

Fig. 1. Example of a phylogenetic tree based on qualitative Hennigian analysis of electromorph distributions among species of thrushes and allies (from Avise *et al.* 1980a). Circles refer to electromorphs whose ancestral states remain undetermined, and lines crossing branches refer to derived electromorphs, some of which provisionally define clades (see Table I).

Table I. Electromorphs inferred to exist along branches of Fig. 1.

Branch	Electromorphs
(A)	PGD(60); PEP-2(110); IDH-1(100); IDH-2(-120); ME-2(-100); EST-3(-100); PT-1(100); GOT-1(100); LDH-2(-100); MDH-1(100); MDH-2(-100); GOT-2(-100); PEP-3(blur); GPD-2(-200).
(B)	IDH-1(50); PGD(98,120); [PT-2(150); PEP-1(blank); ME-1(85); CK-1(80); PGM-2(95, 135); EST-1(80,85); PGI(98); EST-2(D); CK-2(80); LDH-1(-75); Hb(-250); CK-3(200)].
(C)	IDH-2(-100); [PGM(140); EST-1(100); PGI(200,100); Hb(-100); CK-3(100)].
(D)	PGD(70,93,95); IDH-1(120); GPD-2(-20); EST-3(-125); GOT-1(110); [PT-2(blank); ME-1(89); CK-1(97); EST-2(E); LDH-1(50); CK-2(105); PEP-1(92,101)].
(E)	GPD(-100); PEP-3(100); PGM(90); [ME-1(100); PT-2(100); PGD(100); CK-2(100)]
(F)	EST-3(-75); GOT-1(50); PEP-2(2 band); ME-1(90); [PEP-1(90); CK-1(95); EST-2(C,C′); LDH-1(40)].
(G)	PGM(100); [PEP-1(100); CK-1(100)].
(H)	PEP-2(120); EST-3(-50); EST-1(98); ME-1(95); PT-2(50); [LDH-1(65); EST-2(G)].
(I)	PEP-2(100); [LDH-1(100)].
(J)	PT-2(110); PGD(80); ME-1(80); [EST-2(F)].
(K)	[EST-2(A)].
(L)	ME-1(175); GPD-2(100); IDH-1(150); EST-1(90); PGD(150); PEP-1(88,110).
(M)	PGM(150).
(N)	IDH-1(200); PGI(250).
(O)	[EST-2(B)].
(P)	PEP-3(95); PT-1(110); ME-2(-105); IDH-2(-130); ME-1(150).

Letters refer to the stem and branches of the Fig. Line A lists electromorphs ancestral to the taxa examined. Successive lines list derived electromorphs, some of which aid in defining clades. Electromorphs in brackets are those considered unique to a single observed taxon or clade, but whose presumed ancestral electromorphs remain undetermined. (From Avise *et al.* 1980a.)

electromorphs may carry as much weight as abundant electromorphs; and (4) the analyses are critically dependent upon correct determinations of character state polarities, and these in turn are very sensitive to particular choice of taxa as out-group, and to the assumption that shared electromorphs have not arisen independently by convergence.

Other qualitative approaches to data analysis have been suggested. "Compatibility" methods operate under the assumption that phylogenetic relationships are best reflected by distributions of largest collections of mutually compatible character states across taxa (LeQuesne, 1972; Estabrook, 1979). If compatibility methods are applied strictly to presumed synapomorphic states, resulting tree topologies should be similar or identical to Hennigian cladistic trees. Throckmorton (1978) and Straney (1981) have applied modifications of this general pattern of reasoning to qualitative protein electrophoretic data. Both authors provide excellent summaries of phylogenetic reasoning in genetics.

3. Comparison of phylogenetic methodologies

Admittedly, my distinction between qualitative and quantitative analyses becomes strained in some specific applications. For example, Mickevich and Johnson (1976) used Wagner algorithms (Farris, 1970) to generate phylogenetic trees from qualitative (but coded) allozyme and morphological data, and concluded that the Wagner approaches outperformed phenetic methods (this conclusion was seriously challenged by Colless, 1980). However, their output cladograms include no description of particular characters along various branches of the tree. The dichotomy which I wish to emphasize distinguishes between analyses which yield trees with specific character states defined along branches, from those which do not.

Interestingly, the question of which method of data manipulation yields a reconstruction of true phylogeny is probably unanswerable. The definitive answer would require absolute knowledge of a phylogeny against which to compare data summaries, but such a phylogeny is unknown for any assemblage of real taxa. I view the value of qualitative approaches to reside not so much in necessarily providing "correct" tree topologies, but rather in providing readily testable trees. Because discrete character states are defined for all branches, points of ambiguity or dilemma in the tree may be specifically identified as points for further study. For example, suppose that two cladistically distant species (as provisionally determined by other information) nonetheless appear to share a derived protein electromorph as determined by comparison with the out-group. Correct identity of the electromorph could then be questioned. Perhaps an error arose in gel scoring or in data transfer. Or perhaps distinct allelic products were masquerading as a shared electrophoretic band (a large literature on "hidden" protein variation documents the widespread occurrence of this phenomenon – e.g. Coyne, 1976; Johnson, 1976; Singh *et al.*, 1976; McDowell and Prakash, 1976). These possibilities are directly testable. Perhaps the electromorph does indeed represent a single protein, in which case it could (1) have evolved independently more than once (an interesting molecular phenomenon); (2) have been retained in common from a polymorphic ancestor; or (3) actually be a valid synapomorph defining a tree different from that originally proposed (Patton and Avise, 1983). Thus, a wealth of testable hypotheses arises directly from consideration of qualitative character-state distributions, and tests of these hypotheses can feed back in a positive way both on the evaluation of the original method of data summary, and on inferences about the molecular nature of character-state changes themselves. Nice examples of this synergism between qualitative data analysis and molecular understanding are provided by Lansman *et al.* (1983), Templeton (1983), and Aquadro and Greenberg (1983) for mitochondrial DNA, and by Aquadro and Avise (1983) for protein electromorphs.

Nonetheless, quantitative analyses will also continue to enjoy wide use in phylogenetics. Although distance algorithms cannot explicitly distinguish geneological from non-geneological relationships in data, phenetic trees based on quantitative approaches which survey large portions of the genome (i.e. single-copy DNA hybridizations) may nonetheless better summarize organismal phylogeny than can any qualitative approaches based on limited or unreliable character state information.

CHARACTER STATE PHYLOGENIES VERSUS TAXA PHYLOGENIES

Any reconstruction of a phylogeny of taxa is necessarily based upon the data at hand. Technically then, all phylogeny reconstructions are trees of data interrelationships rather than trees of species interrelationships. In quantitative analyses, it is usually assumed that elements of the distance matrix are reliable and unbiased estimates of overall divergence among the taxa exhibiting those distances, so that the inferred species phylogeny mirrors that data phylogeny. In many qualitative parsimony networks, branches interconnect cohorts of character states, and the taxa exhibiting these cohorts are then superimposed over the networks. To the extent that any quantitative or qualitative data bases are biased or unreliable summaries of overall character-state relationships, the reconstructed taxa phylogenies will also be insecure.

To help clarify these distinctions, consider a class of molecular markers commonly suggested as providing the epitome of reliability in phylogeny reconstruction – overlapping chromosomal inversions. Figure 2 shows a most-parsimonious phylogenetic network for about 25 of the gene arrangements along the third chromosome of the sibling species *Drosophila persimilis* and *D. pseudoobscura*. Populations of *D. pseudoobscura* carrying the "Texas" versus "Santa Barbara" gene arrangements (for example) are separated from one another by six sequential evolutionary inversions, while populations of *D. pseudoobscura* carrying "Texas" are only three steps removed from *persimilis* "Klamath". Furthermore, some representatives of *pseudoobscura* and *persimilis* share the identical (and presumably ancestral) gene arrangement "standard". If species boundaries were in fact unknown here, one could not reconstruct them from these third chromosome data alone. (Nonetheless, once the "standard" gene arrangement is known to be ancestral, monophyletic assemblages of taxa (clades) within the network can be defined by shared possession of derived inversions (such as "Santa Cruz") and their descendants.) It is probable that the species boundaries do indeed reflect a real phylogenetic split. Assume (as is

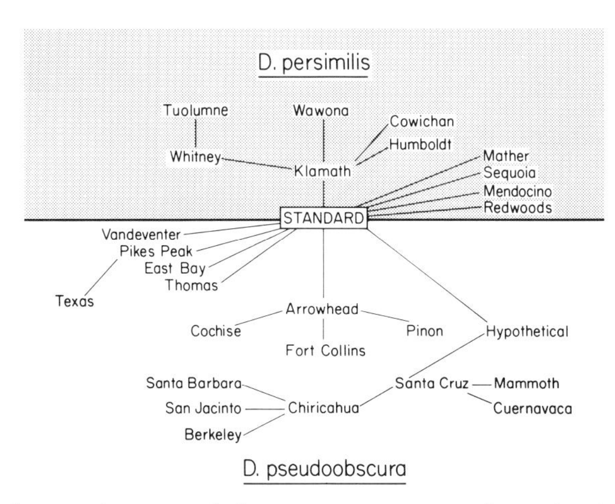

Fig. 2. Phylogenetic network for some inversions along the third chromosome in *Drosophila pseudoobscura* and *D. persimilis* (modified from Dobzhansky, 1970).

likely the case) that all populations within each respective species are monophyletic, i.e. that they have shared a common ancestor following the speciation event. Then it must be the case that there has been heterogeneity across lineages with respect to rate of accumulation of chromosomal changes along the third chromosome. Figure 2 is thus a network of character state relationships, and not ineluctably one of taxa relationships. Farris (1978) further addresses the issue of phylogenetic inference from inversion data.

Pheneticists might argue that in overall genetic composition, populations within either species of *Drosophila* should be closer to one another than to any non-conspecific and, hence, that phenetic clustering algorithms based upon such complete data should reflect the true phylogeny of the taxa. From this perspective, the apparent discrepancy between the inversion phylogeny and the taxa phylogeny arises because of the problem of sampling error in characters examined.

Problems of sampling error apply to quantitative data bases as well, at several levels. Consider for example a species distance matrix generated from protein electrophoretic data. First is the question of the "intra-locus" sampling variance due to the restricted sample sizes used in estimating electromorph frequencies. Secondly is the question of the "inter-locus" variance due to the restricted number of structural genes whose products were sampled. Thirdly, how representative of other classes of genetic characters (i.e. regulatory genes) are the assayed structural genes? And fourthly, how sensitive were the electrophoretic procedures in ability to detect true protein differences? (This subsumes the problem of gel-scoring standards which are likely to vary from laboratory to laboratory.)

The first two questions have been addressed by Nei and Roychoudhury (1974; Nei, 1978). For estimating distances or phylogenetic relationships among taxa, the total variance (the sum of the intra- and inter-locus variances) is relevant. At a given locus, most species pairs empirically appear essentially identical in electromorph frequencies, or else almost completely distinct (Ayala, 1975; Fig. 3). Thus, the variance of genetic distance has an especially large inter-locus component, and in many electrophoretic surveys, the standard error of genetic distance is at least one-fifth to one-half as large as the mean (cf. Parker *et al.*, 1981). It is pointless to argue about alternative phylogenetic reconstructions when they are based upon distance values not significantly different from one another.

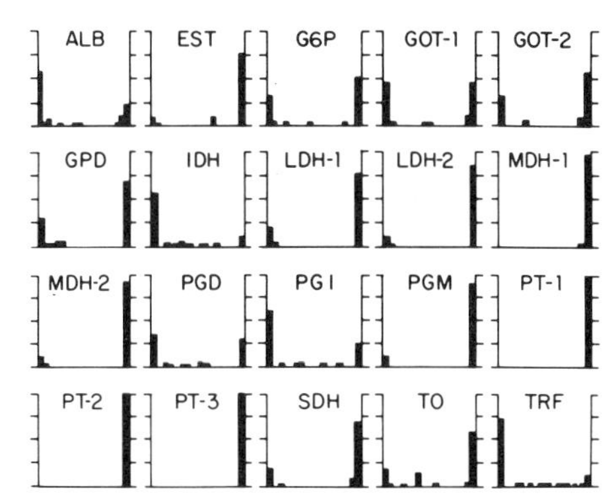

Fig. 3. Frequency distributions of genetic similarities for each of 20 electrophoretically-assayed loci in 120 pairwise comparisons among 16 species of *Peromyscus* (from data in Avise *et al*. 1974). The ordinate is percentage of comparisons, and the abscissa represents level of genetic similarity in increments of 0·05, from 0·00 (left) to 1·00 (right).

The third and fourth questions posed above relate to the representativeness of protein divergence to total genome divergence, and to the sensitivity of a particular protein assay. Issues raised by all four questions must qualify interpretations of phylogenetic reconstructions from protein data. For example, much debate concerns the phylogenetic branching order of man, chimpanzee and gorilla. Available protein electrophoretic data do not resolve the issue. Based on 23 proteins, mean genetic distances between pairs of species of *Homo, Pan* and *Gorilla* range from 0·31 to 0·39, and these distances are not significantly different from one another (Bruce and Ayala, 1979). Bruce and Ayala's estimate for the *Homo-Pan* distance is $D = 0{\cdot}39$, while for the same species, King and Wilson's (1975) estimate based on 44 loci is $D = 0{\cdot}62$. Presumably, the larger distance obtained by King and Wilson is attributable to the more sensitive electrophoretic techniques they employed. Furthermore, King and Wilson suggest that the structural gene differences distinguishing *Homo* and *Pan* are not representative of the biological differences between the species, which may instead be attributed to evolutionary changes in regulatory genes.

A distinction between character-state phylogeny and taxa phylogeny can sometimes assume pragmatic significance. If one explicitly limits interpretation of given data to an inferred character-state phylogeny, then only the fourth source of error listed above retains relevance. When this is done, significant insights into the molecular nature of character state transitions may be revealed. For example, from a qualitative character-state phylogeny Lansman *et al.* (1983) provided quite unambiguous evidence for an unanticipated molecular occurrence – convergent evolution of particular restriction sites in mitochondrial DNA. For the particular characters assayed, this conclusion need not be qualified by concerns about the representativeness of those sites to other genetic characters, nor to concerns about the organisms in which those particular restriction sites appear. It may be hoped that qualitative data phylogenies will eventually permit discoveries about protein character-state evolution as well, although I know of no specific examples as yet.

BETTER DATA ANALYSES VERSUS BETTER DATA

By phylogenetic reasoning, problems associated with sampling of characters can in principle be circumvented by choice of more "informative" characters for analysis. As noted by Wiley (1981, p. 82), "One *true* synapomorphy is enough to define a unique geneological relationship". It is sometimes claimed that amino acid or nucleotide sequences will provide the ultimate phylogenetic data. I believe this is an oversimplification. An ideal character state for defining geneological affinity would (1) have evolved only once so that its presence in extant taxa is not due to convergence, and (2) once evolved, be retained by all descendants so that no extant taxa are inadvertently omitted from geneological reconstruction. These properties are unlikely to hold for particular amino acid or nucleotide sites. Indeed, Throckmorton (1978) has argued that "the closer the material is to the gene itself, the more sensitive it is to . . . perturbing [evolutionary] influences". While it is true that particular amino acid sites are individually unreliable as phylogenetic markers, their composite series (such as the entire sequence produced by a structural gene) can also be considered the basic character state in an analysis. Surely it is more valuable to know that protein states A and B differ by, say, amino acid substitution at sites a, b and c than to simply know that they are different from one another (as might be determined by protein electrophoresis). In other words, higher-order character states for proteins can be retrieved from amino acid sequences, but the converse is not true.

In any event, it is clear that not all genetic data are equally valuable for phylogenetic reconstruction. The most commonly employed protein approaches (amino acid sequencing, immunological cross-reactivity and standard one-dimensional electrophoresis of samples from extant organisms) have been thoroughly reviewed (Wilson *et al.*, 1977). In this section, I want to briefly consider some alternative possibilities for protein assay. These few examples are included merely to indicate that other classes of protein information do exist, and that some alternatives may have special phylogenetic significance in particular applications.

1. Gene duplications

Duplicate structural genes are usually detected by the multiplicity of their protein products in, for example, electrophoretic assays. Ohno (1970) outlines the probable evolutionary course of many duplicate genes. Following the duplication, one locus is essentially freed from evolutionary constraints, while the other retains former function. Eventually, the new gene may accumulate mutations and perhaps assume a new metabolic role, most likely related to that of its sister gene. The potential significance of duplicate genes in phylogeny reconstruction is that duplication of a particular gene is probably a *relatively* rare event (notwithstanding recent molecular evidence for "pseudogenes" – e.g. Schopf, 1981); and that if and when an important new metabolic function is acquired, duplicate gene expression will likely be retained in all descendant species.

Gottlieb and co-workers have used gene duplications, each presumably of unique evolutionary origin, to identify monophyletic assemblages in plants (Gottlieb, 1977, 1982). For example, because all assayed species within the sections *Peripetasma, Phaeostoma, Fibula* and *Eucharadium* of *Clarkia* exhibit two phosphoglucose isomerase (PGI) loci, they appear to constitute a monophyletic grouping, distinct from species in three other sections of the genus which express a single PGI gene (Gottlieb and Weeden, 1979). Similarly, several workers have utilized gene duplications to infer phylogenetic affinities in chordates (Avise and Kitto, 1973; Whitt *et al.*, 1973; Fisher and Whitt, 1978; Fisher *et al.*, 1980). In addition to number of gene copies, different patterns of gene expression during ontogeny and across tissues appear to provide ancillary information about phylogenetic affinity (Ferris and Whitt, 1979; Shaklee and Whitt, 1981; Whitt, 1981).

Since functional redundancy is a characteristic of genes immediately following duplication, loss of expression of one copy of the pair is presumably of no detriment to the organism. Through evolutionary time, silencing of duplicate gene expression appears to be a common occurrence (Ferris and Whitt, 1977). Fishes of the family Catastomidae are thought to have descended from a single tetraploid ancestor alive 50 million years ago. Extant catastomids express an average of 47% of their genes in duplicate, meaning that functional diploidization has occurred at roughly 50% of loci (Ferris and Whitt, 1977). In an ingenious application of phylogenetic methodology, Ferris and Whitt (1978) used the pattern of extinction of duplicate gene expression in catastomids to reconstruct phylogenetic relationships. The character states "presence" versus "absence" of duplicate gene expression in diploidizing phylads are of special phylogenetic significance for two reasons: (1) presence of expression of both loci can reliably be considered the primitive state, and (2) once the derived state (single locus expression) is attained, the process is almost certainly irreversible.

2. *Two-dimensional gels*

Lewontin (1974) and Selander (1976) estimate that at least 100 electrophoretically assayed loci will be required for adequate estimations of average heterozygosity within species, and similar numbers of loci are probably necessary for precise estimates of genetic distance between species as well. Yet most electrophoretic studies survey on the order of 20–30 gene products. It is for this reason that the development of a two-dimensional electrophoretic method for separating many hundreds of proteins at a time (O'Farrell, 1975) was greeted with initial optimism. O'Farrell's technique separates denatured proteins on the basis of charge by isoelectric focusing in the first dimension of the gel, and then on the basis of molecular weight in the second dimension. Two-dimensional (2-DGE) gels also offer the opportunity to assay several classes of proteins (ribosomal, regulatory, membrane-associated and structural) that are not included in conventional one-dimensional separations which primarily have been used to assay water soluble glucose-metabolizing and other enzymes.

Aquadro and Avise (1981) used both conventional (SGE) and 2-DGE electrophoretic approaches to estimate genetic distances between several species of rodents. The correlation between SGE and 2-DGE distances across species was strong, although in absolute values, 2-DGE distances were only about one-half as large as those of SGE. It is uncertain whether differences in magnitudes of genetic distance are attributable to differing sensitivities of 2-DGE and SGE approaches, or to the fact that the techniques sample partially non-overlapping cohorts of proteins. In any event, the study revealed some of the weaknesses of 2-DGE as a phylogenetic assay. Apart from the technical difficulty of the approach (at least relative to SGE), the major drawback is the difficulty of determining homologies of protein spots in the 2-dimensional "constellations" which appear on gels. This problem is especially serious in comparisons of more distantly related species. Because protein homologies are obscure (a problem not generally shared by SGE), qualitative data analyses are effectively precluded. And since quantitative distances between SGE and 2-DGE appear correlated, at the present time there seems to be little justification for advocacy of the more complicated 2-DGE approach in systematics.

3. *Proteins in fossils*

A dramatic announcement was the recent recovery of serum albumin from muscle tissue of a mammoth that had been frozen in soil for 40 000 years (Prager *et al.*, 1980)! Crude extracts from mammoth tissue were injected into rabbits, and the resulting antisera were tested for immunological crossreactivity with albumins of living elephants and other mammals. Although much of the mammoth albumin had undergone post-mortem degradation, sufficient preservation had occurred to reveal extensive and equal crossreaction with albumins from both Indian and African elephants. A small amount of cross-reaction was also obtained with albumin from sea cow (representing the order Sirenia thought to include nearest living relatives of elephants and mammoths), but not in immunological tests with albumins from other mammalian orders.

Using slightly different approaches, Lowenstein (1980) showed that small amounts of collagen and albumin recovered from fossils of *Australopithecus* (dating to 1·9 million years ago), and of other more recent remains of fossil *Homo*, react with antibodies to the respective proteins purified from living humans. And in another application of the approach, Lowenstein *et al.* (1981) quite clearly demonstrate that albumin from the Tasmanian wolf (*Thylacinus*) is most similar to albumins from two other living genera of marsupials, *Dasyurus* and *Dasyuroides*. The wolf albumin was recovered from untanned skins which had been collected early in the century, just before *Thylacinus* went extinct! The excitement raised by these remarkable recoveries of proteins from fossils is, unfortunately, tempered by the realization that opportunities for such assays will necessarily remain extremely limited.

STANDARDIZATION OF DATA ANALYSES

The obsession with finding the "true" or "best" phylogenetic tree from given protein data has, in my opinion, carried one undesirable side effect: it has resulted in a vast profusion of different analyses which makes comparisons of results across studies quite difficult. For example, at least 14 different distance coefficients have been advocated for electrophoretic data (Nei, 1981). Although for a given data base these coefficients are usually highly correlated, they may differ considerably in absolute value. For some time, Nei's (1972) *D* coefficient won in popularity, but with the realization that *D* is not a metric, the search for a coefficient compatible with distance-Wagner analysis has been renewed. To this profusion of coefficients must be added an equally large number of protocols for distance matrix manipulation. As emphasized, qualitative data can also be analysed by many different algorithms. Given the general reservations about protein data in phylogenetic reconstruction, it often seems that researchers are attempting to extract more from the data than is really there.

Ironically, in the heated debate about best method of data analysis, perhaps the fundamental and most unique advantage of molecular genetic data in systematics has been largely lost from view. Unlike virtually any conventional systematic data (i.e. morphology, behaviour, etc.), molecular data readily provide "common yardsticks" for comparing *relative* levels of evolutionary divergence in phylogenetically distinct arrays of organisms. Suppose, for example, that one wishes to know whether a given taxonomic genus of birds is in any evolutionarily meaningful sense comparable to a particular genus of amphibians. This important issue seems seldom to have been attacked or even raised in traditional systematic literature (van Valen, 1973), no doubt because morphological criteria suggest no easy means of solution (but see Cherry *et al.*, 1982). However, it is a simple matter to choose for study analogous and often homologous cohorts of proteins in different phylads, and to compare relative levels of genetic divergence based on common molecular traits.

In other words, suppose a genus of birds contains species essentially identical to one another in electromorph frequencies at 30 loci, while at those same loci congeneric frogs are almost completely different from one another in electromorph composition. Even though these data would permit little clade delineation or opportunities for phylogenetic inference *within* either group, the contrast *between* the phylads is most informative. If these proteins were an unbiased and reliable sample of total genome divergence, the avian and amphibian genera would not be genetically equivalent. Yet, this conclusion would not have been apparent from morphological considerations alone (as is evidenced by the fact that both groups had been placed at the generic level of taxonomic recognition).

In an example of this comparative approach, Avise and Aquadro (1982) summarized the large literature on electrophoretically-assayed protein distances between species within 44 vertebrate genera. Despite the multivarious factors which might be expected to influence such distances (including both biological factors and differences in laboratory methodologies), some strong and consistent trends were apparent (Fig. 4). Avian congeners appear extremely conservative in magnitude of protein divergence while amphibian congeners exhibit huge genetic distances. Few reptilian genera have been assayed, but three of five genera exhibit a non-conservative pattern of divergence approaching that of amphibians. Fish and mammalian genera are highly variable in D, but generally fall intermediate in magnitude of protein divergence to birds and amphibians. These data suggest a conceptual framework for further comparative work (Avise, 1983). Perhaps avian congeners are younger than most non-avian congeners. Alternatively, protein evolution may have been decelerated in birds relative to other groups. To test these and related possibilities, it will be necessary to conduct similar comparative assays of other portions of the genome, as well as to re-examine fossil and other evidence for probable divergence times.

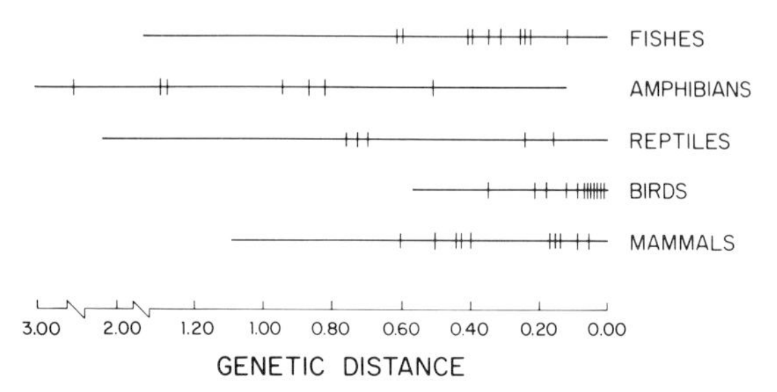

Fig. 4. A comparative summary of genetic distances (based on protein electrophoretic data) between species within each of 44 genera from the five vertebrate classes. Each vertical bar represents the mean genetic distance between all assayed pairs of species within a particular taxonomic genus. The horizontal lines indicate the total observed ranges of genetic distance between all pairs of assayed congeners within each vertebrate class (from Avise and Aquadro 1982).

Such approaches may not ultimately be limited to molecular data. Ironically, recent entrees of comparative thinking to morphological differentiation have come from molecular evolutionists (King and Wilson, 1975; Cherry *et al.*, 1978, 1982). For example, Cherry *et al.* (1978) conclude that from a "frog's perspective" the morphological difference between humans and chimpanzees is huge, while the genetic distance is very small. From these data, it is clear that protein divergence and organismal divergence can proceed at very different rates. I believe that such comparative approaches will provide some of the most exciting opportunities for any "new systematics".

SUMMARY

These are exciting but also somewhat frustrating times for molecular phylogeneticists. Molecular data of many sorts (including DNA sequences) are being generated at ever increasing rates. Concurrently, numerous methods of data analysis are being developed by various systematic camps. Recently, the algorithms have been produced faster than they can be interpreted, evaluated and assimilated into real data analysis. In addition, so many criteria exist for evaluating trees that no method is likely to be best for all. And the ultimate criterion for evaluation, the true phylogeny of an assemblage, may never be known with certainty. It is little wonder that the systematic controversies have been so fervid.

On the other hand, the debates have certainly served to focus thinking and enhance understanding of phylogenetic procedures. It is probable that the major conceptual principles of phylogenetic reasoning are now understood (this would not have been true 20 years ago). Most problems are now concerned with the appropriateness of various analyses to particular data. This involves consideration of whether the assumptions underlying the analysis are likely to hold for the characters being analysed. In the future, it is hoped that methods of data analysis and study of molecular mechanisms will *mutually* benefit from positive feedback to a greater extent than they have in the past.

There have been widely varied opinions about the value of protein and other molecular genetic data in systematics. For example, Selander (1982) concludes that "the continuing study of the structure of the gene at the most fundamental level will soon tell us more about the phylogenetic relationships of organisms than we have managed to learn in all the 173 years since Lamarck", while Straney (1981) was considerably more reserved: "This is not to say that genetic information is useless for phylogenetic inference . . . Genetic data are simply more obviously difficult to analyze phylogenetically than are other types of data". It is now apparent that many problems of phylogenetic inference apply with usual force to protein characters, and it is also clear that not all protein characters are equally informative to phylogeny reconstruction. Appropriate analyses of the more informative molecular character states will contribute substantially to understanding of phylogenetic relationships.

REFERENCES

Aquadro, C. F. and Avise, J. C. (1981). Genetic divergence between rodent species assessed by using two-dimensional electrophoresis. *Proc. Nat. Acad. Sci., U.S.A.* **78**, 3784–3788.

Aquadro, C. F. and Avise, J. C. (1982). Evolutionary genetics of birds VI. A. reexamination of protein divergence using varied electrophoretic conditions. *Evolution, Lancaster, Pa.* **36**, 1003–1019.

Aquadro, C. F. and Greenberg, B. D. (1983). Human mitochondrial DNA variation and evolution: analysis of nucleotide sequences from seven individuals. *Genetics* **103**, 287–312.

Avise, J. C. (1983). "A Molecular Perspective on Avian Speciation". (G. C. Clark and A. H. Brush, eds), *In* Ornithologists Union, Centennial volume in press. American Ornithologist's Union.

Avise, J. C. and Aquadro, C. F. (1982). A comparative summary of genetic distances in the vertebrates: patterns and correlations. *Evol. Biol.* **15**, 151–185.

Avise, J. C. and Kitto, G. B. (1973). Phosphoglucose isomerase gene duplication in the bony fishes: an evolutionary history. *Biochem. Genet.* **8**, 113–132.

Avise, J. C., Smith, M. H. and Selander, R. K. (1974). Biochemical polymorphism and systematics in the genus *Peromyscus*. VI. The *boylii* species group. *J. Mammal.* **55**, 751–763.

Avise, J. C., Giblin-Davidson, C., Laerm, J., Patton, J. C. and Lansman, R. A. (1979). Mitochondrial DNA clones and matriarchal phylogeny within and among geographic populations of the pocket gopher, *Geomys pinetis*. *Proc. Nat. Acad. Sci., U.S.A.* **76**, 6694–6698.

Avise, J. C., Patton, J. C. and Aquadro, C. F. (1980a). Evolutionary genetics of birds. I. Relationships among North American thrushes and allies. *Auk* **97**, 135–147.

Avise, J. C., Patton, J. C. and Aquadro, C. F. (1980b). Evolutionary genetics of birds. II. Conservative protein evolution in North American sparrows and relatives. *Syst. Zool.* **29**, 323–334.

Ayala, F. J. (1975). Genetic differentiation during the speciation process. *Evol. Biol.* **8**, 1–78.

Berlocher, S. H. (1981). A comparison of molecular and morphological data, and phenetic and cladistic methods, in the estimation of phylogeny in *Rhagoletis* (Diptera: Tephritidae). *In* "Application of Genetics and Cytology in Insect Systematics and Evolution" (M. W. Stock, ed.), pp. 1–31. University of Idaho Press, Idaho.

Bruce, E. J. and Ayala, F. J. (1979). Phylogenetic relationships between man and the apes: electrophoretic evidence. *Evolution, Lancaster, Pa.* **33**, 1040–1056.

Camin, J. H. and Sokal, R. R. (1965). A method for deducing branching sequences in phylogeny. *Evolution, Lancaster, Pa.* **19**, 311–326.

Carson, H. L. and Kaneshiro, K. Y. (1976). *Drosophila* of Hawaii: systematics and ecological genetics. *A. Rev. Ecol. Syst.* **7**, 311–345.

Cavalli-Sforza, L. L. and Edwards, A. W. F. (1967). Phylogenetic analysis: models and estimation procedures. *Evolution, Lancaster, Pa.* **21**, 550–570.

Cherry, L. M., Case, S. M. and Wilson, A. C. (1978). Frog perspective on the morphological difference between humans and chimpanzees. *Science, N.Y.* **200**, 209–211.

Cherry, L. M., Case, S. M., Kunkel, J. G., Wyles, J. S. and Wilson, A. C. (1982). Body shape metrics and organismal evolution. *Evolution, Lancaster, Pa.* **36**, 914–933.

Colless, D. H. (1980). Congruence between morphometric and allozyme data for *Menidia* species: a reappraisal. *Syst. Zool.* **29**, 288–299.

Coyne, J. A. (1976). Lack of genic similarity between two sibling species of *Drosophila* as revealed by varied techniques. *Genetics* **84**, 593–607.
deJong, R. (1980). Some tools for evolutionary and phylogenetic studies. *Z. Zool. Syst. EvolForsch.* **18**, 1–23.
Dobzhansky, T. (1970). "Genetics of the Evolutionary Process". Columbia University Press, New York.
Dobzhansky, T. H. and Eppling, C. (1944). "Contributions to the genetics, taxonomy and ecology of *Drosophila pseudoobscura* and its relatives." [Publ. 554, 1–183.] Carnegie Institute, Washington.
Estabrook, G. F. (1979). Some concepts for the estimation of evolutionary relationships in systematic botany. *Syst. Bot.* **3**, 146–158.
Farris, J. S. (1970). Methods for computing Wagner trees. *Syst. Zool.* **19**, 83–92.
Farris, J. S. (1971). The hypothesis of nonspecificity and taxonomic congruence. *A. Rev. Ecol. Syst.* **2**, 227–302.
Farris, J. S. (1972). Estimating phylogenetic trees from distance matrices. *Am. Nat.* **106**, 645–668.
Farris, J. S. (1973). On comparing the shapes of taxonomic trees. *Syst. Zoöl.* **22**, 50–54.
Farris, J. S. (1978). Inferring phylogenetic trees from chromosome inversion data. *Syst. Zool.* **27**, 275–284.
Felsenstein, J. (1978a). The number of evolutionary trees. *Syst. Zool.* **27**, 27–33.
Felsenstein, J. (1978b). Cases in which parsimony or compatibility methods will be positively misleading. *Syst. Zool.* **28**, 401–410.
Felsenstein, J. (1979). Alternative methods of phylogenetic inference and their interrelationship. *Syst. Zool.* **28**, 49–62.
Ferris, S. D. and Whitt, G. S. (1977). Loss of duplicate gene expression after polyploidization. *Nature, Lond.* **265**, 258–260.
Ferris, S. D. and Whitt, G. S. (1978). Phylogeny of tetraploid catastomid fishes based on the loss of duplicate gene expression. *Syst. Zool.* **27**, 189–206.
Ferris, S. D. and Whitt, G. S. (1979). Evolution of the differential regulation of duplicate genes after polyploidization. *J. Mol. Evol.* **12**, 267–317.
Fisher, S. E. and Whitt, G. S. (1978). Evolution of isozyme loci and their differential tissue expression. *J. Mol. Evol.* **12**, 25–55.
Fisher, S. E., Shaklee, J. B., Ferris, S. D. and Whitt, G. S. (1980). Evolution of five multilocus isozyme systems in the chordates. *Genetica* **52**, 73–85.
Fitch, W. M. (1971). Toward defining the course of evolution: minimum change for a specific tree topology. *Syst. Zool.* **20**, 406–416.
Fitch, W. M. (1975). Molecular evolutionary clocks. *In* "Molecular Evolution" (F. J. Ayala, ed.) pp. 160–178. Sinauer, Sunderland, Mass.
Fitch, W. M. and Margoliash, E. (1967). Construction of phylogenetic trees. *Science, N.Y.* **155**, 279–284.
Gottlieb, L. D. (1977). Electrophoretic evidence and plant systematics. *Ann. Mo. bot. Gdn.* **64**, 161–180.

Gottlieb. L. D. (1982). Conservation and duplication of isozymes in plants. *Science, N.Y.* **216**, 373–379.

Gottlieb, L. D. and Weeden, N. F. (1979). Gene duplication and phylogeny in *Clarkia. Evolution, Lancaster, Pa.* **33**, 1024–1039.

Hennig, W. (1966). "Phylogenetic Systematics". University of Illinois Press, Urbana.

Johnson, G. B. (1976). Hidden alleles at the α-glycerophosphate dehydrogenase locus in *Colias* butterflies. *Genetics* **83**, 149–167.

King, M. C. and Wilson, A. C. (1975). Evolution at two levels. Molecular similarities and biological differences between humans and chimpanzees. *Science, N.Y.* **188**, 107–116.

Klotz, L. C., Komar, N., Blanken, R. L. and Mitchell, R. M. (1979). Calculation of evolutionary trees from sequence data. *Proc. Nat. Acad. Sci., U.S.A.* **76**, 4516–4520.

Lansman, R. A., Shade, R. O., Shapira, J. F. and Avise, J. C. (1981). The use of restriction endonucleases to measure mitochondrial DNA sequence relatedness in natural populations. III. Techniques and potential applications. *J. Mol. Evol.* **17**, 214–226.

Lansman, R. A., Avise, J. C., Aquadro, C. F., Shapira, J. F. and Daniel, S. W. (1983). Extensive genetic variation in mitochondrial DNA's among geographic populations of the deer mouse, *Peromyscus maniculatus. Evolution, Lancaster, Pa.* **37**, 1–16.

LeQuesne, W. J. (1972). Further studies based on the uniquely derived character concept. *Syst. Zool.* **21**, 281–288.

Lowenstein, J. M. (1980). Species-specific proteins in fossils. *Naturwissenschaften* **67**, 343–346.

Lowenstein, J. M., Sarich, V. M. and Richardson, B. J. (1981). Albumin systematics of the extinct mammoth and Tasmanian wolf. *Nature, Lond.* **290**, 409–411.

Lewontin, R. C. (1974). "The Genetic Basis of Evolutionary Change". Columbia University Press, New York.

Li, W.-H. (1981). Simple method for constructing phylogenetic trees from distance matrices. *Proc. Nat. Acad. Sci., U.S.A.* **78**, 1085–1089.

Mayr, E. (1981). Biological classification: toward a synthesis of opposing methodologies. *Science, N.Y.* **214**, 510–516.

McDowell, R. E. and Prakash, S. (1976). Allelic heterogeneity within allozymes separated by electrophoresis in *Drosophila pseudoobscura. Proc. Nat. Acad. Sci., U.S.A* **73**, 4150–4153.

Mickevich, M. F. and Johnson, M. S. (1976). Congruence between morphological and allozyme data in evolutionary inference and character evolution. *Syst. Zool.* **25**, 260–270.

Mickevich, M. F. (1978). Taxonomic congruence. *Syst. Zool.* **27**, 143–158.

Nei, M. (1972). Genetic distance between populations. *Am. Nat.* **106**, 283–292.

Nei, M. (1975). "Molecular Population Genetics and Evolution". North Holland Publishing, Amsterdam.

Nei, M. (1978). The theory of genetic distance and evolution of human races. *Jap. J. Hum. Genet.* **23**, 341–369.
Nei, M. (1981). Genetic distance and molecular taxonomy. *In* "Problems in General Genetics" [Proc. XIV Int. Cong. Genet., Vol. 2.] (Y.P. Altukhov, ed.) pp. 7–22. MIR Publishers, Moscow.
Nei, M. and Roychoudhury, A. K. (1974). Sampling variances of heterozygosity and genetic distance. *Genetics* **76**, 379–390.
O'Farrell, P. H. (1975). High resolution two-dimensional electrophoresis of proteins. *J. Biol. Chem.* **250**, 4007–4021.
Ohno, S. (1970). "Evolution by Gene Duplication". Springer-Verlag, New York.
Parker, E. D., Jr., Burbanck, W. D., Burbanck, M. P. and Anderson, W. W. (1981). Genetic differentiation and speciation in the estuarine isopods *Cyathura polita* and *Cyathura burbancki*. *Estuaries* **4**, 213–219.
Patton, J. C. and Avise, J. C. (1983). An empirical evaluation of qualitative Hennigian analyses of protein electrophoretic data. *J. Mol. Evol.* (in press).
Popper, K. R. (1968). "The Logic of Scientific Discovery". Harper Torchbooks, New York.
Prager, E. M. and Wilson, A. C. (1976). Congruence of phylogenies derived from different proteins. *J. Mol. Evol.* **9**, 45–57.
Prager, E. M. and Wilson, A. C. (1978). Construction of phylogenetic trees for proteins and nucleic acids: empirical evaluation of alternative matrix methods. *J. Mol. Evol.* **11**, 129–142.
Prager, E. M., Wilson, A. C., Lowenstein, J. M. and Sarich, V. M. (1980). Mammoth albumin. *Science, N.Y.* **209**, 287–289.
Schopf, T. J. M. (1981). Evidence from findings of molecular biology with regard to the rapidity of genomic change: implications for species durations. *In* "Paleobotany, Paleoecology and Evolution". (K. J. Niklas, ed.), pp. 135–190. Praeger Publishers, New York.
Selander, R. K. (1976). Genic variation in natural populations. *In* "Molecular Evolution" (F. J. Ayala, ed.), pp. 21–45. Sinauer, Sunderland, Mass.
Selander, R. K. (1982). Phylogeny. *In* "Perspectives on Evolution" (R. Milkman, ed.), pp. 32–59. Sinauer, Sunderland, Mass.
Shaklee, J. B. and Whitt, G. S. (1981). Lactate dehydrogenase isozymes of gadiform fishes: divergent patterns of gene expression indicate a heterogeneous taxon. *Copeia* **1981**, 563–578.
Simpson, G. G. (1945). The principles of classification and a classification of mammals. *Bull. Am. Mus. Nat. Hist.* **85**, 1–350.
Singh, R. S., Lewontin, R. C. and Felton, A. A. (1976). Genetic heterogeneity within electrophoretic "alleles" of xanthine dehydrogenase in *Drosophila pseudoobscura*. *Genetics* **84**, 609–629.
Sneath, P. H. A. and Sokal, R. R. (1973). "Numerical Taxonomy". W. H. Freeman, San Francisco.
Stevens, P. F. (1980). Evolutionary polarity of character states. *A. Rev. Ecol. Syst.* **11**, 333–358.
Straney, D. O. (1981). The stream of heredity: genetics in the study of

phylogeny. *In* "Mammalian Population Genetics" (M. H. Smith and J. Joule, eds), pp. 100–138. University of Georgia Press, Athens.

Swofford, D. L. (1981). On the utility of the distance-Wagner procedure. *In* "Advances in Cladistics: proceedings of the 1st meeting of the Hennig Society 1981", (F. A. Funk and D. R. Brooks, eds), pp. 25–43. New York Bot. Garden, New York.

Templeton, A. R. (1983). Phylogenetic inference from restriction endonuclease cleavage site maps with particular reference to the evolution of humans and the apes. *Evolution, Lancaster, Pa.* **37**, 221–244.

Throckmorton, L. H. (1978). Molecular phylogenetics. *In* "Biosystematics in Agriculture" (J. A. Romberger, R. H. Foote, L. Knutson and P. L. Lentz, eds), pp. 221–239. Wiley, New York.

van Valen, L. (1973). Are categories in different phyla comparable? *Taxon* **22**, 333–374.

White, M. J. D. (1978). "Modes of Speciation". W. H. Freeman, San Francisco.

Whitt, G. S., Miller, E. T. and Shaklee, J. B. (1973). Developmental and biochemical genetics of lactate dehydrogenase isozymes in fishes. *In* "Genetics and Mutagenesis of Fish" (J. H. Schröder, ed.), pp. 243–276. Springer-Verlag, New York.

Whitt, G. S. (1981). Developmental genetics of fishes: isozymic analyses of differential gene expression. *Am. Zool.* **21**, 549–572.

Wiley, E. O. (1981). "Phylogenetics. The Theory and Practice of Phylogenetic Systematics." Wiley, New York.

Wilson, A. C., Carlson, S. S. and White, T. J. (1977). Biochemical evolution. *A. Rev. Biochem.* **46**, 573–639.

8 Enzyme Variation, Genetic Distance and Evolutionary Divergence in Relation to Levels of Taxonomic Separation

J. P. THORPE

Department of Marine Biology, University of Liverpool,
The Marine Biological Station, Port Erin,
Isle of Man, U.K.

Abstract: It is probable that there is a correlation between biochemical genetic diversity and taxonomic separation, especially since both are likely to be functions of evolutionary time. The hypothesis that the amount of biochemical divergence between conspecific populations, congeneric species and confamilial genera may be roughly similar across a wide range of taxa has some empirical support.

In the present work a large amount of eukaryote genetic distance data is analysed to look at typical levels of genetic differentiation within families, genera and species. Comparisons of major taxa (mammals, birds, reptiles, amphibians, fish, invertebrates, plants) confirm that there is generally reduced biochemical evolution in birds, but do not show any great differences between the other major taxa. It is tentatively suggested that where conventional methods are unable to resolve taxonomic problems an estimate of genetic distance may provide more objective and potentially useful data.

INTRODUCTION

The usefulness of studies of enzyme variation for the identification of species has long been known (Manwell and Baker, 1963) and since this time enzyme electrophoresis has been used to detect or confirm

Systematics Association Special Volume No. 24, "Protein Polymorphism: Adaptive and Taxonomic Significance", edited by G. S. Oxford and D. Rollinson, 1983, Academic Press, London and New York.

many cryptic species in a wide variety of organisms (e.g. Thorpe *et al.*, 1978a, c; Hedgecock, 1979; Mundy and Thorpe, 1979; Carter and Thorpe, 1981; Smith and Robertson, 1981).

Between sympatric populations, significant differences in allele frequencies at any locus will indicate the existence of a barrier to gene flow and the probability that two species are present. Various aspects of the uses of biochemical techniques, particularly enzyme electrophoresis, in systematics have been reviewed by several authors (Avise, 1974; Gottlieb, 1977; Throckmorton, 1977, 1978; Thorpe, 1979, 1982; Ferguson, 1980; Hurka, 1980; Rollinson, 1980; Funk and Brooks, 1981).

A more difficult and complex aspect of the use of biochemical genetics in taxonomy is the identification of putative species in cases where the populations concerned are not sympatric. Several workers have shown in various genera that the biochemical genetic similarity of populations decreases with increasing taxonomic diversity (e.g. Hubby and Throckmorton, 1968; Ayala and Tracey, 1974; Ayala *et al.*, 1974c; Avise *et al.*, 1975, 1977; Ayala, 1975; Avise, 1976). It has been suggested (Thorpe, 1979, 1982) that where morphological or other conventional criteria do not permit firm conclusions to be drawn as to the taxonomic status of a dubious "species" or morphotype, the level of biochemical divergence between populations may be used as an objective criterion. Similarly, levels of genetic variation between species could also possibly be used, in cases of dispute, as an indication of whether or not two species should be regarded as congeneric.

SYSTEMATIC DIVERGENCE AND GENETIC IDENTITY

Overall measures of electrophoretic similarity between populations or species, expressed as a single figure, can be given using any of the several published methods (for references see Thorpe, 1982) to estimate genetic similarity, genetic identity or genetic distance. The method in general use in population genetics is that of Nei (1972).

It has been proposed (Thorpe, 1979, 1982) that taxa at different levels of systematic divergence will have different probabilities of having any given value of genetic identity. Thorpe (1979) suggested that I value (Nei, 1972) between most pairs of conspecific populations would be above about 0·9, whilst between congeneric species most would be between about 0·25 and 0·85. It was predicted that between species of different genera many *I* values would fall below 0·3. Data available at the time that publication was written (1977) were inadequate to test distributions empirically. Although there was a fair body of published information on interspecific and intraspecific genetic distances, this was largely restricted to studies of vertebrates or *Drosophila*. Few estimates were available for other invertebrates or for plants and very few intergeneric distances had been published.

Thorpe (1982) was able to examine empirical distributions of Nei's genetic identity (*I*) with taxonomic diversity in data gathered from a wide range of groups of vertebrates, invertebrates and plants. These data showed overall distributions of probabilities of *I* within species, between congeneric species and between genera which were generally in good agreement with those previously predicted (Thorpe, 1979).

For the present work, an increased body of data has been employed to re-examine previous conclusions by using a larger number of estimates of *I* from an increased variety of taxa. Also, with the additional data this could be broken down so that distributions of *I* values with different levels of systematic diversity could be examined separately within various major taxa of eukaryotic organisms.

Data for this study do not include all available information on enzyme variation between populations or species. Apart from some studies which must inevitably have been overlooked, a number were deliberately excluded.

These fell into one or more of the following groups.

(1) Most publications where Nei's (1972) *I* or *D* values were not included by the authors. To have calculated *I* values from allele frequency data would have been very time consuming (*I* values were calculated where genetic distances (*D*) were given).

(2) Intergeneric genetic identity values involving different families (e.g. Ayala *et al.*, 1975; Bruce and Ayala, 1979). Such data would have to be analysed separately and there is too little for it to provide any useful results at present.

(3) Studies where earlier work had left doubt as to the taxonomic status of a population or species so that *I* values could not be unambiguously categorized as interspecific, intraspecific or intergeneric.

(4) *I* values based upon few loci were omitted for taxa where other estimates were available.

(5) For genera or species for which a great deal of data were already included, further (usually less comprehensive) pieces of work were omitted. This was in an attempt to avoid undue bias of the overall data towards groups which had been the subjects of the greatest research interest. In particular, not all of the many studies of *Drosophila* spp. were used.

(6) Cases where there is evidence that sudden speciation (e.g. chromosomal divergence) has occurred so that time-dependent biochemical divergence cannot be expected.

Also exluded from the overall data, although analysed separately, are the *I* values from the small number of studies on birds. As has already been pointed out (Avise *et al.*, 1980; Thorpe, 1982), these differ radically from those of any other major groups for which data are available.

The overall data for vertebrates (excluding birds), invertebrates and plants combined are summarized in Fig. 1. *I* values for each of the three levels of taxonomic separation considered in this study (*viz.* conspecific populations, congeneric species and confamilial genera) were broken down and classified into twenty 0·05 increments of *I*, from 0·00 to 1·00. From these data, the frequencies of *I* values in each 0·05 increment of *I* were obtained and, hence, the probabilities associated

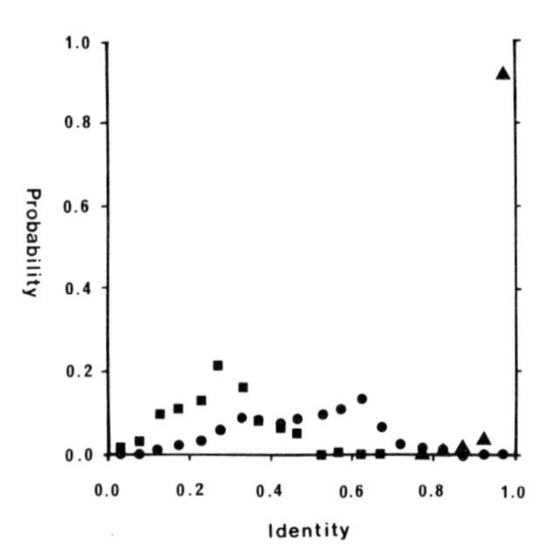

Fig. 1. Graph of probability against genetic identity (*I*; Nei, 1972) for conspecific populations (black triangles), congeneric species (black circles) and confamilial genera (black squares) for all plants and animals, excluding birds. For further explanation see text. Sources of data are indicated in the references.

with each increment. It is these probabilities which are plotted in Fig. 1. Approximate total numbers of estimates of *I* used for Fig. 1 are 7000 for conspecific populations, 900 for congeneric species and 160 for confamilial genera. Publications used as sources of data are too numerous to list in the text, but are shown in the references(*). Inevitably, reflecting availability, the data are biased towards vertebrates and towards species native to North America.

The general form of the three curves shown in Fig. 1 does not differ greatly from those previously derived from a smaller body of data (Thorpe, 1982). They demonstrate the same overall basic relationship between taxonomic separation and genetic identity. From these data, about 2% of estimates of *I* between conspecific populations fall below 0·9, whereas only about the same proportion of estimates between congeneric species exceed 0·85 (~ 0·5% above 0·9).

At the lower end of the range approximately 15% of *I* values between congeneric species, but nearly 80% of estimates between confamilial genera, fall below 0·35.

COMPARATIVE BIOCHEMICAL AND GENETIC DIVERGENCE IN VARIOUS MAJOR TAXA

In previous studies (e.g. Thorpe, 1982), no direct comparison has been made of genetic identity values at various levels of systematic divergence between organisms in different major taxa. How similar, for example, is the biochemical divergence found in a typical plant genus to that found within a genus of mammals? Until recently, it is doubtful whether available data were adequate for useful comparisons to be made. Even now, there are very few intergeneric *I* values and comparisons are necessarily limited to considerations of congeneric species and conspecific populations.

To look for possible differences between major groups, the available data were examined to see how they could best be divided. The extensive work on many species of vertebrate permits each class (mammals, birds, reptiles, amphibians and fish) to be examined separately. The bulk of invertebrate data are for arthropods but, since there are not enough for other phyla to be treated separately, all invertebrate *I* values have been pooled. Published genetic identities for lower plant species are almost unavailable and the vast majority of the data are for various angiosperms. Therefore, plants, like invertebrates, are treated as a single group.

The data have been examined, as previously, by looking at probabilities of *I* values falling into each 0·05 increment of *I* from 0·0 to 1·0. The interspecific *I* values were spread over a large range and the subdivision of the total data resulted in much reduced numbers in each increment and, consequently, large sampling errors. These could have been reduced by using larger (e.g. 0·1) increments of *I*, but it was considered preferable to use a running mean. To do this, the probability for each 0·05 increment of *I* was taken as the mean probability over that and the two immediately adjacent increments of *I*. The effect of this treatment on the overall data is shown in Fig. 2 which uses the same data as Fig. 1.

As mentioned above, it has already been pointed out (Avise *et al.*, 1980; Thorpe, 1982) that intergeneric, interspecific and conspecific *I* values for birds differ markedly from those of most other groups. This is clearly shown in Fig. 3 where the ranges of genetic identities for estimates between genera and between species of birds vary substantially from those shown for other groups (Figs 1 and 2).

To investigate possible differences in ranges of *I* values in other vertebrate classes, invertebrates and plants, probabilities against *I* values for congeneric species have been plotted separately for each major group (Fig. 4a, b, c, d, e, f, g). Probabilities in this case are running means calculated as outlined above.

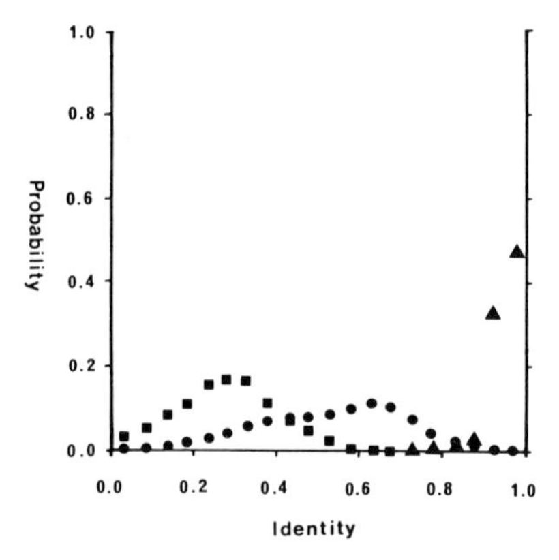

Fig. 2. As Fig. 1, but using running means to reduce effects of sampling errors (see text).

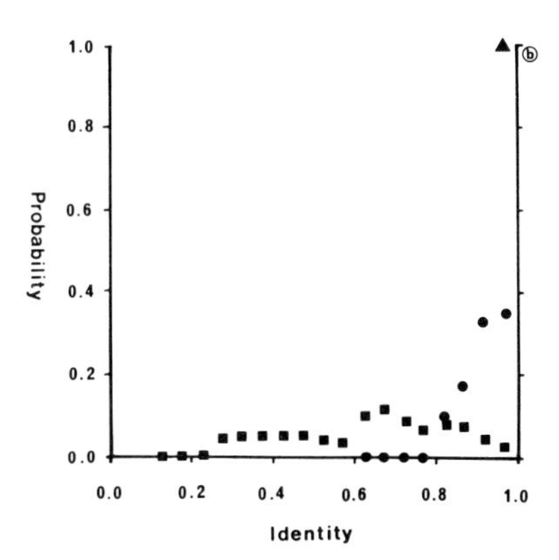

Fig. 3. As Fig. 1, but for bird data only.

Comparisons can also be made using available data on intraspecific variation (Fig. 5a, b, c, d, e, f, g). Here, there are more data and a smaller range than for interspecific comparisons and it is, therefore, not necessary to average probabilities beween increments.

DISCUSSION

The use of more extensive data to re-examine the relationship between systematic divergence and genetic distance only serves to confirm previous conclusions. Clearly, increased levels of taxonomic separation indicate generally lower levels of biochemical similarity. This is to be expected since both are likely to be functions of evolutionary time. The time-dependent evolutionary divergence of the structures of homologous protein molecules forms the basis of what has become known as the molecular clock hypothesis (see Fitch, 1973, 1976; Fitch and Langley, 1976; Wilson *et al.*, 1977; Carlson *et al.*, 1978; Thorpe, 1982). Some aspects of this hypothesis have been the subject of considerable controversy; in particular, the claims of some authors that biochemical divergence may be linearly related to evolutionary time (see Holmquist, 1972; Romero-Herrera *et al.*, 1978; Vawter *et al.*, 1980; and references above). Support for this suggestion is not universal (e.g. Jukes and Holmquist, 1972; Radinsky, 1978; Throckmorton, 1978; Lessios, 1979, 1981; Korey, 1981; Farris, 1981), but there is overwhelming evidence that, in general, following reproductive

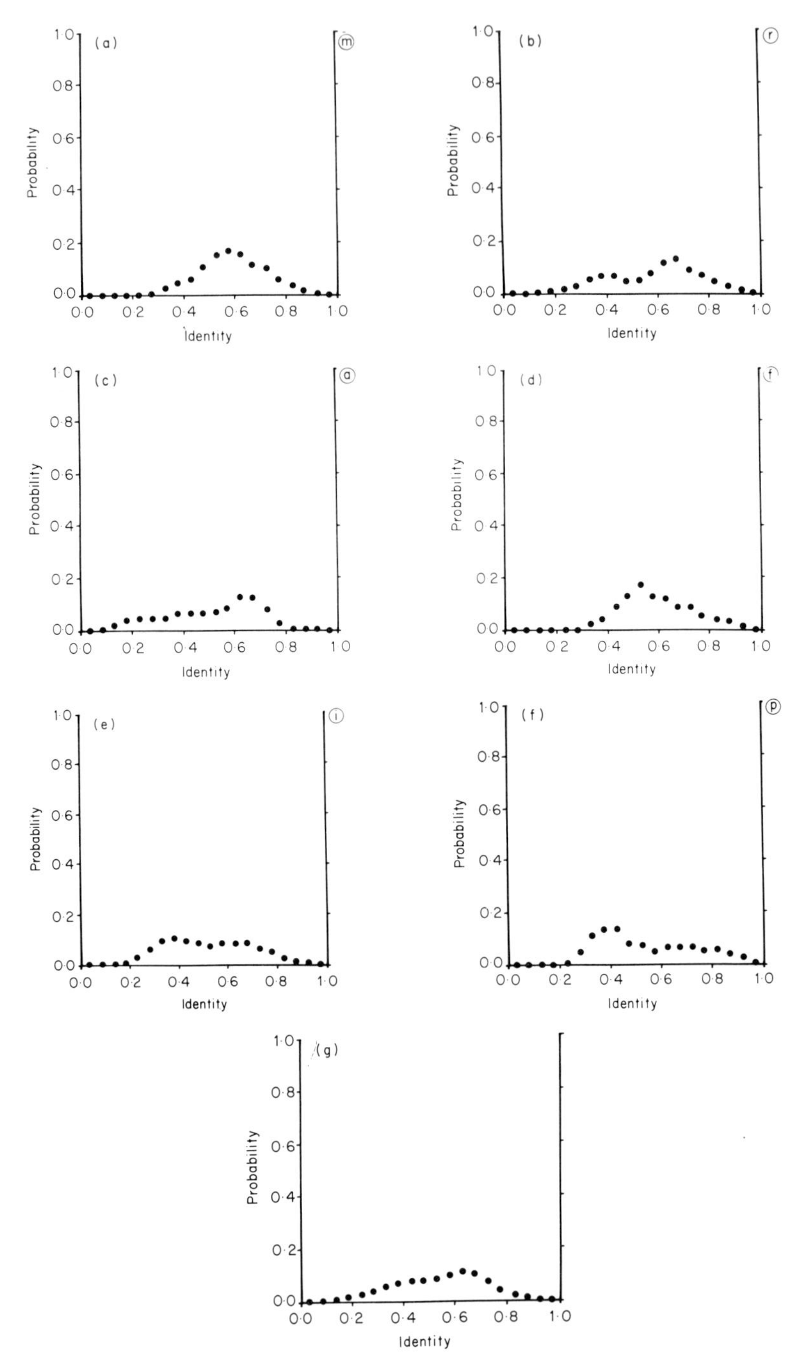

Fig. 4. Graphs of probability against genetic identity for congeneric species of (a) mammals, (b) reptiles, (c) amphibians, (d) fish, (e) invertebrates, (f) plants, (g) all these groups combined.

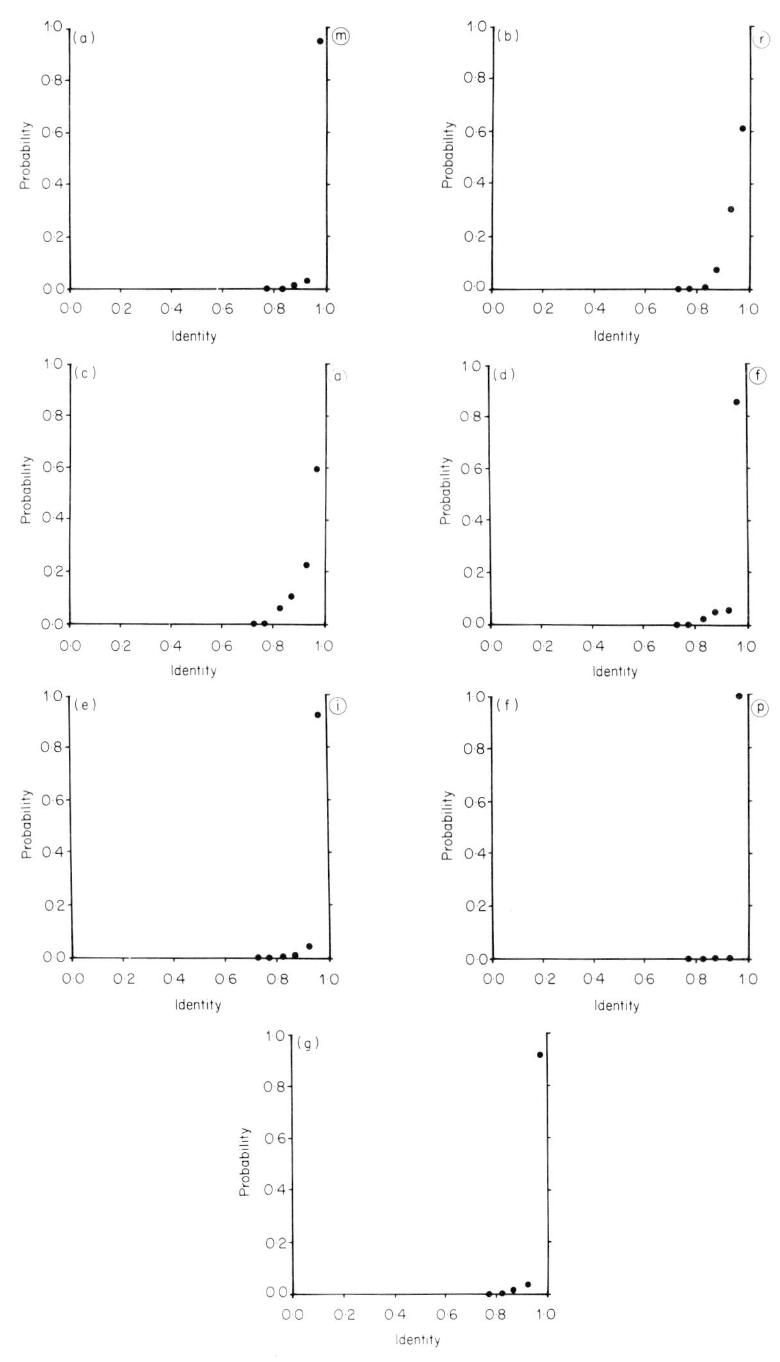

Fig. 5. Graphs of probability against genetic identity for conspecific populations of (a) mammals, (b) reptiles, (c) amphibians, (d) fish, (e) invertebrates, (f) plants, (g) all these groups combined.

isolation, the molecular structures of homologous proteins diverge, linearly or otherwise, with time. For purely taxonomic work, the precise linearity of Nei's (1972) genetic distance (*D*) with time is relatively unimportant. This is fortunate since it is improbable that *D* is better than approximately linear over even short periods of evolutionary time (see King, 1973; Nei and Chakraborty, 1973; Maxson and Maxson, 1979; Korey, 1981; Thorpe, 1982). A further problem is that various authors have proposed several different relationships of *D* to evolutionary time.

Despite these difficulties the suggestion that a *D* value of one is equivalent to about 18–20 million years (or about 35 albumin immunological distance units) is in approximate agreement with data from various taxa (Yang *et al.*, 1974; Wilson *et al.*, 1977; Sarich, 1977; Wyles and Gorman, 1980; Thorpe, 1982). This indicates that the minimum age for the divergence of most pairs of congeneric species ($I < 0{\cdot}85$) may be about three million years. On the same scale, the upper limit of diversity within a typical genus ($I \sim 0{\cdot}2$–$0{\cdot}3$) suggests a time of about 23–30 million years. However, the accuracy of these estimates is so doubtful that they cannot be considered better than highly speculative.

Probably of greater interest is the overall similarity of the amounts of genetic divergence found within genera and species of the six major taxa considered in Figs 4 and 5. These generally reinforce previous impressions (Thorpe, 1979, 1982), that most species or most genera in most major taxa (passerine birds are excluded) typically encompass roughly similar amounts of biochemical genetic diversity. The shortage of data and large sampling errors make it difficult to draw any firm conclusions but, given these large sampling errors, it is improbable that the range of *I* values found within species or within genera for any of the six taxa is significantly different from that for any other. Unfortunately, it is not a simple matter to make a statistical comparison since, within each taxon, all *I* values cannot be validly treated as independent samples. In practice, within, for example, a single genus, data frequently show a significant clustering of *I* values. Instances in which these occur are frequently those in which a genus can be divided into two groups of closely related species, and most comparisons of pairs of species between the two species groups give very similar *I* values. The reason for this is that those loci which vary between the closely related species within each group are largely those which also differ between the two groups of species. If a locus has no common alleles between two species groups, then further variation at this locus within either or both species groups will have no effect on *I* values between the two groups. The non-independence of *I* values is to be expected on theoretical grounds since it is known that rates of genetic divergence (i.e. amino acid substitution) differ radically between different proteins. (Skibinski and Ward, 1982). Hence, the faster evolving loci are the most likely to show differences between groups of species and are also most likely to evolve differences between species within a group. However, second or subsequent substitutions at a locus (back mutations are improbable) are unlikely to affect *I* values.

It has been suggested (Avise, this volume) that amphibian genera are larger (in terms of biochemical genetic diversity) than those of other vertebrate groups. This claim is based largely upon extensive data (some of it as yet unpublished) for the two genera *Hyla* (tree frogs) and *Plethodon* (salamanders). In the present study (which includes data from both of these genera), congeneric genetic identity values for Amphibia (Fig. 4c) do not appear to be dramatically different from the overall data (Fig. 4g) or from those for any other group of organisms. Given the necessarily subjective and often somewhat arbitrary nature of conventional taxonomy, particularly above the species level, great variation in the sizes of genera is to be expected. Therefore, if two, or even several, genera of Amphibia or any other group include an unusually wide range of genetic divergence, it is difficult to come to any firm conclusions as to whether these genera are typical. Even breaking down the total data into separate *I* values for analysis will not overcome uncertainties since, as mentioned above, *I* values between a single group of species cannot be validly treated as independent samples.

Overall in the present study, comparisons of *I* values between the various major taxa suggest that, in each case, the amount of biochemical divergence within a species or genus may be broadly similar in most vertebrates, invertebrates or plants. The comparability of taxa in different groups is open to debate (van Valen, 1973) but, where conventional studies leave taxonomic status in doubt, an estimate of genetic distance, from protein data, could provide an objective and potentially useful criterion. It is very doubtful whether it is desirable or even useful for the size of taxa, particularly genera, to be regulated in terms of gross biochemical diversity. If there is such a thing as a "natural" genus, it must, to be of practical use, surely be defined on morphological grounds with biochemical criteria only being employed where alternative methods have proved unsatisfactory.

REFERENCES

References used as sources of data for Figs 1–5 are indicated by an asterisk.*

Avise, J. C. (1974). Systematic value of electrophoretic data. *Syst. Zool.* **23**, 465–481.

Avise, J. C. (1976). Genetic differentiation during speciation. *In* "Molecular Evolution" (F. J. Ayala, ed.), pp. 106–122. Sinauer Associates, Sunderland, Massachusetts.

*Avise, J. C. and Smith, M. H. (1977). Gene frequency comparisons between sunfish (Centrarchidae) populations at various stages of evolutionary divergence. *Syst. Zool.* **26**, 319–335.

*Avise, J. C., Smith, M. H. and Selander, R. K. (1974). Biochemical polymorphism and systematics in the genus *Peromyscus*. VI. The *boylii* species group. *J. Mammal.* **55**, 751–763.

Avise, J. C., Smith, J. J. and Ayala, F. J. (1975). Adaptive differentiation with little genic change between two native California minnows. *Evolution, Lancaster, Pa.* **29**, 411–426.

*Avise, J. C., Patton, J. C. and Aquadro, C. F. (1980). Evolutionary genetics of birds. I. Relationships among North American thrushes and allies. *Auk* **97**, 135–147.

*Avise, J. C., Straney, D. O. and Smith, M. H. (1977). Biochemical genetics of sunfish. IV. Relationships of centrarchid genera. *Copeia* **1977**, 250–258.

Ayala, F. J. (1975). Genetic differentiation during the speciation process. *Evol. Biol.* **8**, 1–78.

*Ayala, F. J. and Tracey, M. L. (1974). Genetic differentiation within and between species of the *Drosophila willistoni* group. *Proc. Nat. Acad. Sci., U.S.A.* **71**, 999–1003.

*Ayala, F. J., Tracey, M. L., Barr, L. G. and Ehrenfeld, J. G. (1974a). Genetic and reproductive differentiation of the subspecies *Drosophila equinoxalis caribbensis*. *Evolution, Lancaster, Pa.* **28**, 24–41.

*Ayala, F. J., Tracey, M. L., Barr, L. G., McDonald, J. F. and Perez-Salas, S. (1974b). Genetic differentiation in five *Drosophila* species and the hypothesis of the selective neutrality of protein polymorphisms. *Genetics* **77**, 343–384.

*Ayala, F. J., Tracey, M. L., Hedgecock, D. and Richmond, R. C. (1974c). Genetic differentiation during the speciation process in *Drosophila*. *Evolution, Lancaster, Pa.* **28**, 576–592.

Ayala, F. J., Valentine, J. W., Hedgecock, D. and Barr, L. G. (1975). Deep sea asteroids: high genetic variability in a stable environment. *Evolution, Lancaster, Pa.*, **29**, 203–212.

*Babbel, G. R. and Selander, R. K. (1974). Genetic variability in edaphically restricted and widespread plant species. *Evolution, Lancaster, Pa.* **28**, 619–630.

*Baker, M. C. (1975). Song dialects and genetic differences in white-crowned sparrows (*Zonotrichia leucophrys*). *Evolution, Lancaster, Pa.* **29**, 226–241.

*Barrowclough, G. F. and Corbin, J. W. (1978). Genetic variation and differentiation in the Parulidae. *Auk* **95**, 691–702.

*Bellemin, J., Adest, G., Gorman, G. C. and Aleksiuk, M. (1978). Genetic uniformity in northern populations of *Thamnophis sirtalis* (Serpentes, Colubridae). *Copeia* **1978**, 150–151.

*Bezy, R. L., Gorman, G. C., Kim, Y. J. and Wright, J. W. (1977). Chromosomal and genetic divergence in the fossorial lizards of the family Annielidae. *Syst. Zool.* **26**, 57–71.

*Browne, R. A. (1977). Genetic variation in island and mainland populations of *Peromyscus leucopus*. *Am. Midl. Nat.* **97**, 1–9.

Bruce, E. J. and Ayala, F. J. (1979). Phylogenetic relationships between man and the apes: electrophoretic evidence. *Evolution, Lancaster, Pa.* **33**, 1040–1056.

*Buroker, N. E., Hershberger, W. K. and Chew, K. K. (1979a). Population genetics of the family Ostreidae. I. Intraspecific studies of *Crassostrea gigas* and *Saccostrea commercialis*. *Mar. Biol.* **54**, 157–169.

*Buroker, N. E., Hershberger, W. K. and Chew, K. K. (1979b). Population genetics of the family Ostreidae. II. Interspecific studies of the genera *Crassostrea* and *Saccostrea*. *Mar. Biol.* **54**, 171–184.

*Buth, D. G., Gorman, G. C. and Lieb, C. S. (1980). Genetic divergence between *Anolis carolinensis* and its Cuban progenitor *Anolis porcatus*. *J. Herpetol.* **14**, 279–284.

*Campbell, C. A. (1978). Genetic divergence between populations of *Thais lamellosa* (Gmellin). *In* "Marine Organisms: Genetics, Ecology and Evolution" (B. Battaglia and J. A. Beardmore, eds), pp. 150–170. Plenum Press, New York.

Carlson, S. S., Wilson, A. C. and Maxson, R. D. (1978). Do albumin clocks run on time? *Science, N.Y.* **200**, 1183–1185.

*Carter, M. A. and Thorpe, J. P. (1981). Reproductive, genetic and ecological evidence that *Actinia equina* var. *mesembryanthemum* and var. *fragacea* are not conspecific. *J. mar. biol. Assoc. U.K.* **61**, 79–94.

*Case, S. M. (1978a). Biochemical systematics of members of the genus *Rana* native to western North America. *Syst. Zool.* **27**, 299–311.

*Case, S. M. (1978b). Electrophoretic variation in two species of Ranid frogs, *Rana boylei* and *R. mucosa*. *Copeia* **1978**, 311–320.

*Case, S. M. and Wake, M. H. (1975). Electrophoretic patterns of certain proteins in caecilians (Amphibia: Gymnophiona). *Comp. Biochem. Physiol.* **52B**, 473–476.

★Case, S. M., Haneline, P. G. and Smith, M. F. (1975). Protein variation in several species of *Hyla*. *Syst. Zool.* **24**, 281–295.
★Chan, K. L. (1977). Enzyme polymorphism in Malayan rats of the subgenus *Rattus*. *Biochem. Syst. Ecol.* **5**, 161–168.
★Cheney, D. P. and Babbel, G. F. (1978). Biosystematic studies of the red algal genus *Eucheuma*. I. Electrophoretic variation among Florida populations. *Mar. Biol.* **47**, 251–264.
★Corbin, K. W., Sibley, C. G., Ferguson, A., Wilson, A. C., Brush, A. H. and Ahlquist, J. E. (1974). Genetic polymorphism in New Guinea starlings of the genus *Aplonis*. *Condor* **76**, 307–318.
★Corbin, K. W., Sibley, C. G. and Ferguson, A. (1979). Genic changes associated with the establishment of sympatry in orioles of the genus *Icterus*. *Evolution, Lancaster, Pa.* **33**, 624–633.
★Crawford, D. J. and Smith, E. B. (1982). Allozyme variation in *Coreopsis neucensoides* and *C. neucensis* (Compositae) a progenitor-derivative species pair. *Evolution, Lancaster, Pa.,* **36**, 379–386.
★Davis, B. J., De Martini, E. E. and McGee, K. (1981). Gene flow among populations of a teleost (painted greenling, *Oxylebius pictus*) from Puget Sound to Southern California. *Mar. Biol.* **65**, 17–23.
★Duncan, R. and Highton, R. (1979). Genetic relationships of the eastern large *Plethodon* of the Ouachita Mountains. *Copeia* **1979**, 95–110.
★Eisses, K. T., Van Dijk, H. and Van Delden, W. (1979). Genetic differentiation within the *melanogaster* species group of the genus *Drosophila* (Sophophora). *Evolution, Lancaster, Pa.* **33**, 1063–1068.
Farris, J. F. (1981). Distance data in phylogenetic analysis. *In* "Advances in Cladistics" (V. A. Funk and D. R. Brooks, eds), pp. 3–24. New York Botanic Garden, New York.
★Feder, J. H. (1979). Natural hybridization and genetic divergence between the toads *Bufo boreas* and *Bufo punctatus*. *Evolution, Lancaster, Pa.,* **33**, 1089–1097.
★Feder, J. H., Wurst, G. Z. and Wake, D. B. (1978). Genetic variation in western salamanders of the genus *Plethodon*, and the status of *Plethodon gordoni*. *Herpetologica* **34**, 64–69.
★Felley, J. D. and Smith, M. H. (1978). Phenotypic and genetic trends in bluegills of a single drainage. *Copeia* **1978**, 175–177.
Ferguson, A. (1980). "Biochemical Systematics and Evolution". Blackie, Glasgow.
★Fevolden, S. E. and Ayala, F. J. (1981). Enzyme polymorphism in antarctic krill (Euphasiacea); genetic variation between populations and species. *Sarsia* **66**, 167–181.
Fitch, W. M. (1973). Aspects of molecular evolution. *Ann. Rev. Genet.* **7**, 343–380.
Fitch, W. M. (1976). Molecular evolutionary clocks. *In* "Molecular Evolution" (F. J. Ayala, ed.), pp. 160–178. Sinauer Associates, Sunderland, Massachusetts.
Fitch, W. M. and Langley, C. H. (1976). Evolutionary rates in proteins: neutral

mutations and the molecular clock. *In* "Molecular Anthropology" (M. Goodman and R. E. Tashian, eds), pp. 197–219. Plenum Press, New York.

Funk, V. A. and Brooks, D. R. (eds) (1981). "Advances in Cladistics." New York Botanic Garden, New York.

*Galleguillos, R. A. (1981). "Genetic Relatedness in Marine Flatfish". Ph.D. thesis, University of Wales, Swansea.

*Galleguillos, R. A. and Ward, R. D. (1982). Genetic and morphological divergence between populations of the flatfish *Platichthys flesus* (L.) (Pleuronectidae). *Biol. J. Linn. Soc.* **17**, 395–408.

*Gill, P. D. (1981a). Enzyme variation in the grasshopper *Chorthippus brunneus* (Thunberg). *Biol. J. Linn. Soc.* **15**, 247–258.

*Gill, P. D. (1981b). Allozyme variation in sympatric populations of British grasshoppers. *Biol. J. Linn. Soc.* **16**, 83–91.

*Giuseffi, S., Kane, T. C. and Duggleby, W. F. (1978). Genetic variability in the Kentucky cave beetle *Neaphaenops tellkampfii* (Coleoptera: Carabidae) *Evolution, Lancaster, Pa.* **32**, 679–681.

*Gorman, G. C. and Kim, Y. J. (1975). Genetic variation and genetic distance among populations of *Anolis* lizards on two Lesser Antillean island banks. *Syst. Zool.* **24**, 369–373.

*Gorman, G. C. and Kim, Y. J. (1977). Genotypic evolution in the face of phenotypic conservativeness: *Abudefduf* (Pomacentridae) from the Atlantic and Pacific sides of Panama. *Copeia* **1977**, 694–697.

*Gorman, G. C., Buth, D. G., Soule, M. and Yang, S. Y. (1980). The relationships of the *Anolis cristatellus* species group: electrophoretic analysis. *J. Herpetol.* **14**, 269–278.

*Gorman, G. C., Kim, Y. J. and Rubinoff, R. (1976). Genetic relationships of three species of *Bathygobius* from the Atlantic and Pacific sides of Panama. *Copeia* **1976**, 361–364.

*Gorman, G. C., Soule, M., Yang, S. Y. and Nevo, E. (1975). Evolutionary genetics of insular adriatic lizards. *Evolution, Lancaster, Pa.* **29**, 52–71.

*Gottlieb, L. D. (1973a). Genetic differentiation, sympatric speciation and the origin of a diploid species of *Stephanomeria*. *Am. J. Bot.* **60**, 545–553.

*Gottlieb, L. D. (1973b). Enzyme differentiation and phylogeny in *Clarkia franciscana, C. rubicunda* and *C. amoena*. *Evolution, Lancaster, Pa.* **27**, 205–214.

*Gottlieb, L. D. (1974). Genetic confirmation of the origin of *Clarkia lingulata*. *Evolution, Lancaster, Pa.* **28**, 244–250.

Gottlieb. L. D. (1977). Electrophoretic evidence and plant systematics. *Ann. Mo. bot. Gdn.* **64**, 161–180.

*Graves, J. E. and Rosenblatt, R. H. (1980). Genetic relationships of the color morphs of the serranid fish *Hypoplectrus unicolor*. *Evolution, Lancaster, Pa.* **34**, 240–245.

*Graves, J. E. and Somero, G. N. (1982). Electrophoretic and functional enzymic evolution in four species of eastern Pacific barracudas from different thermal environments. *Evolution, Lancaster, Pa.* **36**, 97–106.

★Greenbaum, I. F. (1981). Genetic interactions between hybridizing cytotypes of the tent-making bat (*Uroderma bilobatum*). *Evolution, Lancaster, Pa.* **35**, 306–321.

★Greenbaum, I. F. and Baker, R. J. (1976). Evolutionary relationships in *Macrotus* (Mammalia: Chiroptera): biochemical variation and karyology. *Syst. Zool.* **25**, 15–25.

★Guries, R. P. and Ledig, F. T. (1982). Genetic diversity and population structure in pitch pine (*Pinus rigida* Mill.) *Evolution, Lancaster, Pa.* **36**, 387–402.

★Guttman, S. I., Wood, T. K. and Karlin, A. A. (1981). Genetic differentiation along host plant lines in the sympatric *Enchenopa binotata* Say complex (Homoptera: Membracidae). *Evolution, Lancaster, Pa.* **35**, 205–217.

★Halliday, R. B. (1981). Heterozygosity and genetic distance in sibling species of meat ants (*Iridomyrex purpureus* group). *Evolution, Lancaster, Pa.* **34**, 234–242.

★Handford, P. and Nottebohm, F. (1976). Allozymic and morphological variation in population samples of rufous-collared sparrow *Zonotrichia capensis* in relation to vocal dialects. *Evolution, Lancaster, Pa.* **30**, 802–817.

★Harrison, R. G. (1979). Speciation in North American field crickets: evidence from electrophoretic comparisons. *Evolution, Lancaster, Pa.* **33**, 1009–1023.

★Hauptli, H. and Jain, S. K. (1978). Biosystematics and agronomic potential of some weedy and cultivated amaranths. *Theor. Appl. Genet.* **52**, 177–185.

★Hedgecock, D. (1978). Population subdivision and genetic divergence in the red-bellied newt, *Taricha rivularis*. *Evolution, Lancaster, Pa.* **32**, 271–286.

★Hedgecock, D. (1979). Biochemical genetic variation and evidence of speciation in *Cthamalus* barnacles of the tropical eastern Pacific Ocean. *Mar. Biol.* **54**, 207–214.

★Hedgecock, D. and Ayala, F. J. (1974). Evolutionary divergence in the genus *Taricha* (Salamandridae). *Copeia* **1974**, 738–746.

Holmquist, R. (1972). Empirical support for a stochastic model of evolution. *J. Mol. Evol.* **1**, 211–222.

★Howard, J. H. and Wallace, R. L. (1981). Microgeographical variation of electrophoretic loci in populations of *Amblystoma macrodactylum columbianum* (Caudata: Amblystomatidae). *Copeia* **1981**, 466–471.

Hubby, J. L. and Throckmorton, L. H. (1968). Protein differences in *Drosophila*. IV. A study of sibling species. *Am. Nat.* **102**, 193–206.

Hurka, H. (1980). Enzymes as a taxonomic tool: a botanist's view. *In* "Chemosystematics: Principles and Practice" (F. A. Bisby, J. G. Vaughan and C. A. Wright, eds), pp. 103–121. Academic Press, London.

★Johnson, D. L. E. (1978). Genetic differentiation in two members of the *Drosophila athabasca* complex. *Evolution, Lancaster, Pa.* **32**, 798–811.

★Johnson, G. L. and Packard, R. L. (1975). Electrophoretic analysis of *Peromyscus comanche* Blair, with comments upon its systematic status. *Occ. Pap. Mus. Texas Tech. Univ.* **24**, 1–12.

*Johnson, W. E. and Selander, R. K. (1971). Protein variation and systematics in kangaroo rats (genus *Dipodomys*). *Syst. Zool.* **20**, 377–405.

*Johnson, W. E., Selander, R. K., Smith, M. H. and Kim, S. Y. (1972). Biochemical genetics of sibling species of the cotton rat (*Sigmodon*). *Univ. Texas Publ.* **7213**, 297–306.

Jukes, T. H. and Holmquist, R. (1972). Evolutionary clock: non-constancy of rate in different species. *Science, N.Y.* **177**, 530–532.

*Kim, Y. J., Gorman, G. C., Papenfuss, T. and Roychoudhury, A. K. (1976). Genetic relationships and genetic variation in the amphisbaenian genus *Bipes*. *Copeia* **1976**, 120–124.

King, J. L. (1973). The probability of electrophoretic identity of proteins as a function of amino acid divergence. *J. Mol. Evol.* **2**, 317–322.

*Kirkpatrick, M. and Selander, R. K. (1979). Genetics of speciation in lake whitefishes in the Allegash Basin. *Evolution, Lancaster, Pa.* **33**, 478–485.

Korey, K. A. (1981). Species number, generation length and the molecular clock. *Evolution, Lancaster, Pa.* **35**, 139–147.

*Kornfield, I. L., Ritte, U., Richler, C. and Wahrman, J. (1979). Biochemical and cytological differentiation among cichlid fishes of the Sea of Galilee. *Evolution, Lancaster, Pa.* **33**, 1–14.

*Larson, A. (1980). Paedomorphosis in relation to rates of morphological and molecular evolution in the salamander *Aneides flavipunctatus* (Amphibia: Plethodontidae). *Evolution, Lancaster, Pa.* **34**, 1–17.

*Larson, A. and Highton, R. (1978). Geographic protein variation and divergence in the salamanders of the *Plethodon welleri* group (Amphibia: Plethodontidae). *Syst. Zool.* **27**, 431–448.

*Larson, A., Wake, D. B., Maxson, L. H. and Highton, R. (1981). A molecular phylogenetic perspective on the origins of morphological novelties in the salamanders of the tribe Plethodontini (Amphibia: Plethodontidae). *Evolution, Lancaster, Pa.* **35**, 405–423.

Lessios, H. A. (1979). Use of Panamanian sea urchins to test the molecular clock. *Nature, Lond.* **280**, 599–601.

Lessios, H. A. (1981). Divergence in allopatry: molecular and morphological differentiation between sea urchins separated by the Isthmus of Panama. *Evolution, Lancaster, Pa.* **35**, 618–634.

*Levin, D. A. (1975). Interspecific hybridization, heterozygosity and gene exchange in *Phlox*. *Evolution, Lancaster, Pa.* **29**, 37–51.

*Levin, D. A. (1977). The organization of genetic variability in *Phlox drummondii*. *Evolution, Lancaster, Pa.* **31**, 477–494.

*Levin, D. A. (1978). Genetic variations in annual *Phlox*: self-compatible versus self-incompatible species. *Evolution, Lancaster, Pa.* **32**, 245–263.

*Levin, D. A. and Crepet, W. L. (1973). Genetic Variation in *Lycopodium lucidulum*: a phylogenetic relic. *Evolution, Lancaster, Pa.* **27**, 622–632.

*Levy, M. and Levin, D. A. (1975). Genic heterozygosity and variation in permanent translocation heterozygotes of the *Oenothera biennis* complex. *Genetics* **79**, 493–512.

*Lewis-Jones, L. J., Thorpe, J. P. and Wallis, G. P. (1982). Genetic divergence in four species of the genus *Raphanus*: implications for the ancestry of the domestic radish *R. sativus*. *Biol. J. Linn. Soc.* **18**, 35–48.

★Loudenslager, E. J. and Gall, G. A. E. (1980). Geographic patterns of protein variation and subspeciation in cutthroat trout *Salmo clarki*. *Syst. Zool.* **29**, 27–42.

★Manwell, C. and Baker, C. M. A. (1963). A sibling species of sea-cucumber discovered by starch gel electrophoresis. *Comp. Biochem. Physiol.* **10**, 39–53.

★Marcus, N. H. (1977). Genetic variation within and between geographically separated populations of the sea urchin, *Arbacia punctulata*. *Biol. Bull., Woods Hole* **153**, 560–576.

★Marincovic, D., Ayala, F. J. and Andjelkovic, M. (1978). Genetic polymorphism and phylogeny of *Drosophila subobscura*. *Evolution, Lancaster, Pa.* **32**, 164–173.

★Mashburn, S. J., Sharitz, R. R. and Smith, M. H. (1979). Genetic variation among *Typha* populations of the southeastern United States. *Evolution, Lancaster, Pa.* **32**, 681–685.

★Matsuoka, N. (1981). Phylogenetic relationships among five species of starfish of the genus *Asterina*: an electrophoretic study. *Comp. Biochem. Physiol.* **70B**, 739–743.

Maxson, L. R. and Maxson, R. D. (1979). Comparative albumin and biochemical evolution in plethodontid salamanders. *Evolution, Lancaster, Pa.* **33**, 1057–1062.

★McCommas, S. A. and Lester, L. J. (1980). Electrophoretic evaluation of the taxonomic status of two species of sea anemone. *Biochem. Syst. Ecol.* **8**, 289–292.

★Merritt, R. B., Rogers, J. F. and Kurz, B. J. (1978). Genic variability in the longnose dace *Rhinichthys cataractae*. *Evolution, Lancaster, Pa.* **32**, 116–124.

★Moyse, J., Thorpe, J. P. and Al-Hamadani, E. (1982). The status of *Littorina aestuarii* Jeffreys; an approach using morphology and biochemical genetics. *J. Conch.* **31**, 7–15.

★Mulley, J. C. and Latter, B. D. H. (1980). Genetic variation and evolutionary relationships within a group of thirteen species of penaeid prawns. *Evolution, Lancaster, Pa.* **34**, 904–916.

★Mundy, S. P. and Thorpe, J. P. (1979). Biochemical genetics and taxonomy in *Plumatella fungosa* and *P. repens* (Bryozoa: Phylactolaemata). *Freshwat. Biol.* **9**, 157–164.

Mundy, S. P. and Thorpe, J. P. (1980). Biochemical genetics and taxonomy in *Plumatella coralloides* and *P. fungosa*, and a key to the British and European Plumatellidae (Bryozoa: Phylactolaemata). *Freshwat. Biol.* **10**, 519–526.

★Murphy, R. W. and Ottley, J. R. (1980). A genetic evaluation of the leafnose snake, *Phyllorhyncus arenicolus*. *J. Herpetol.* **14**, 263–268.

★Murphy, R. W. and Papenfuss, T. J. (1980). Biochemical variation of *Phyllodactylus unctus* and *P. paucituberculatus*. *Biochem. Syst. Ecol.* **8**, 97–100.

Nei, M. (1972). Genetic distance between populations. *Am. Nat.* **106**, 283–292.

Nei, M. and Chakraborty, R. (1973). Genetic distance and electrophoretic identity of proteins between taxa. *J. Mol. Evol.* **2**, 323–328.
*Nicklas, N. L. and Hoffman, R. J. (1979). Genetic similarity between two morphologically similar species of polychaetes. *Mar. Biol.* **52**, 53–59.
*Nozawa, K., Shotake, T., Ohkura, Y. and Tanabe, Y. (1977). Genetic variation within and between species of Asian macaques. *Jap. J. Gen.* **52**, 15–30.
*Patton, J. C., Baker, R. J. and Avise, J. C. (1981). Phenetic and cladistic analyses of biochemical evolution in peromyscine rodents. *In* "Mammalian Population Genetics" (M. H. Smith and J. Joule, eds), pp. 288–308. University of Georgia Press, Athens, Georgia.
*Pierce, B. A., Mitton, J. B. and Rose, F. I. (1981). Allozyme variation among large, small and cannibal morphs of the tiger salamander inhabiting the Llano Estacado of West Texas. *Copeia* **1981**, 590–595.
*Post, T. J. and Uzzell, T. (1981). The relationships of *Rana sylvatica* and the monophyly of the *Rana boylii* group. *Syst. Zool.* **30**, 179–180.
*Prakash, S. (1977). Genetic divergence in closely related sibling species *Drosophila pseudoobscura, Drosophila persimilis* and *Drosophila miranda. Evolution, Lancaster, Pa.* **31**, 14–23.
Radinsky, L. (1978). Do albumin clocks run on time? *Science, N.Y.* **200**, 1182–1183.
Rollinson, D. (1980). Enzymes as a taxonomic tool: a zoologists view. *In* "Chemosystematics: Principles and Practice"(F. A. Bisby, J. G. Vaughan and C. A. Wright, eds), pp.123–146. Academic Press, London.
Romero-Herrera, A. E., Lehmann, H., Joysey K. A. and Friday, A. E. (1978). On the evolution of myoglobin. *Phil. Trans. R. Soc.* B **283**, 161–163.
*Roose, M. L. and Gottlieb, L. D. (1976). Genetic and biochemical consequences of polyploidy in *Tragopogon*. *Evolution, Lancaster, Pa.* **30**, 818–830.
*Ryman, N., Reuterwall, C., Nygren, K. and Nygren, T. (1980). Genetic variation and differentiation in Scandinavian moose (*Alces alces*). Are large mammals monomorphic? *Evolution, Lancaster, Pa.* **34**, 1037–1049.
*Salmon, M., Ferris, S. D., Johnston, D., Hyatt, G. and Whitt, G. S. (1979). Behavioural and biochemical evidence for species distinctiveness in the fiddler crabs *Uca speciosa* and *U. spinicarpa*. *Evolution, Lancaster, Pa.* **33**, 182–191.
Sarich, V. M. (1977). Rates, sample sizes and the neutrality hypothesis for electrophoresis in evolutionary studies. *Nature, Lond.* **265**, 24–28.
*Sattler, P. W. (1980). Genetic relationships among selected species of North american *Scaphiopus*. *Copeia* **1980**, 605–610.
*Sattler, P. W. and Guttman, S. I. (1976). An electrophoretic analysis of *Thamnophis sirtalis* from Western Ohio. *Copeia* **1976**, 352–356.
*Sbordoni, V., Allegrucci, G., Caccone, A., Cesaroni, D., Cobolli Sbordoni, M. and de Matthaeis, E. (1981). Genetic variability and divergence in cave populations of *Troglophilus cavicola* and *T. andreinii* (Orthoptera: Raphidophoridae). *Evolution, Lancaster, Pa.* **35**, 226–233.

*Schaal, B. A. and Smith, W. G. (1980). The apportionment of genetic variation within and between populations of *Desmodium nudiflorum*. *Evolution, Lancaster, Pa.* **34**, 241–221.

*Schoen, D. J. (1982). Genetic variation and the breeding system of *Gilia achilleifolia*. *Evolution, Lancaster, Pa.* **36**, 361–370.

*Schopf, T. J. M. and Murphy, L. S. (1973). Protein polymorphism of the seastars *Asterias forbesi* and *Asterias vulgaris*: an evolutionary paradigm for the Cape cod marine microcosm. *Biol. Bull., Woods Hole* **145**, 454–455.

*Schwaegerle, K. E. and Schaal, B. A. (1979). Genetic variability and founder effect in the pitcher plant *Sarracenia purpurea* L. *Evolution, Lancaster, Pa.* **33**, 1210–1218.

*Siebenaller, J. F. (1978). Genetic variation in deep-sea invertebrate populations: the bathyal gastropod *Bathybembix bairdii*. *Mar. Biol.* **47**, 265–275.

Skibinski, D. O. F. and Ward, R. D. (1982). Correlations between heterozygosity and evolutionary rate of proteins. *Nature, Lond.* **298**, 490–492.

*Skibinski, D. O. F., Cross, T. F. and Ahmad, M. (1980). Electrophoretic investigation of systematic relationships in the marine mussels *Modiolus modiolus* L., *Mytilus edulis* L. and *Mytilus galloprovincialis* Lmk. (Mollusca: Mytilidae). *Biol. J. Linn. Soc.* **13**, 65–74.

*Smith, J. K. and Zimmerman, E. G. (1976). Biochemical genetics and evolution of North American blackbirds, family Icteridae. *Comp. Biochem. Physiol.* **53B**, 319–324.

*Smith, P. J. and Robertson, D. A. (1981). Genetic evidence for two species of sprat (*Sprattus*) from New Zealand waters. *Mar. Biol.* **62**, 227–233.

*Straney, D. O., O'Farrell, M. J. and Smith, M. H. (1976). Biochemical genetics of *Myotis californicus* and *Pipistrellus hesperus* from southern Nevada. *Mammalia* **40**, 344–347.

*Tabachnick, W. J., Munstermann, L. E. and Powell, J. R. (1979). Genetic distinctness of related forms of *Aedes aegypti* in East Africa. *Evolution, Lancaster, Pa.* **33**, 287–295.

*Taylor, C. E. and Gorman, G. C. (1975). Population genetics of a 'colonising' lizard: natural selection for allozyme morphs in *Anolis grahami*. *Heredity* **35**, 241–247.

Thorpe, J. P. (1979). Enzyme variation and taxonomy: the estimation of sampling errors in measurements of interspecific genetic similarity. *Biol. J. Linn. Soc.* **11**, 369–386.

Thorpe, J. P. (1982). The molecular clock hypothesis: Biochemical evolution, genetic differentiation and systematics. *A. Rev. Ecol. Syst.* **13**, 139–168.

*Thorpe, J. P. and Mundy, S. P. (1980). Biochemical genetics and taxonomy in *Plumatella emarginata* and *P. repens* (Bryozoa: Phylactolaemata). *Freshwat. Biol.* **10**, 361–366.

*Thorpe, J. P., Beardmore, J. A. and Ryland, J. S. (1978a). Genetic evidence for cryptic speciation in the marine bryozoan *Alcyonidium gelatinosum*. *Mar. Biol.* **49**, 27–32.

⋆Thorpe, J. P. Beardmore, J. A. and Ryland, J. S. (1978b). Taxonomy, interspecific variation and genetic distance in the phylum Bryozoa. *In* "Marine Organisms: Genetics, Ecology and Evolution" (B. Battaglia and J. A. Beardmore, eds), pp. 425–445. Plenum Press, New York.

⋆Thorpe, J. P., Ryland, J. S. and Beardmore, J. A. (1978c). Genetic variation and biochemical systematics in the marine bryozoan *Alcyonidium mytili*. *Mar. Biol.* **49**, 343–350.

Throckmorton, L. H. (1977). *Drosophila* systematics and biochemical evolution. *A. Rev. Ecol. Syst.* **8**, 235–254.

Throckmorton, L. H. (1978). Molecular phylogenetics. *In* "Biosystematics in Agriculture" (J. A. Rhomberger, R. H. Foote, L. Knutson and P. L. Lentz, eds), pp. 221–239. Wiley, New York.

⋆Tilley, S. G., Merritt, R. B., Wu, B. and Highton, R. (1978). Genetic differentiation in salamanders of the *Desmognathus ochrophaeus* complex (Plethodontidae). *Evolution, Lancaster, Pa.* **32**, 93–115.

⋆Turner, J. R. G., Johnson, M. S. and Eanes, W. F. (1979). Contrasted modes of evolution in the same genome: Allozymes and adaptive change in *Heliconius*. *Proc. Nat. Acad. Sci., U.S.A.* **76**, 1924–1928.

van Valen, L. (1973). Are categories in different phyla comparable? *Taxon* **22**, 333–373.

Vawter, A. T., Rosenblatt, R. and Gorman, G. C. (1980). Genetic divergence among fishes of the eastern Pacific and Caribbean: support for the molecular clock. *Evolution, Lancaster, Pa.* **34**, 705–711.

⋆Vrigenhoek, R. C., Angus, R. A. and Schultz, R. J. (1977). Variation and heterozygosity in sexually versus clonally reproducing populations of *Poeciliopsis*. *Evolution, Lancaster, Pa.* **31**, 767–781.

⋆Wake, D. B., Maxson, L. R. and Wurst, G. Z. (1978). Genetic differentiation, albumin evolution and their biogeographic implications in plethodontid salamanders (genus *Hydromantes*) of California and southern Europe. *Evolution, Lancaster, Pa.* **32**, 529–539.

⋆Wake, D. B., Yang, S. Y. and Papenfuss, T. J. (1980). Natural hybridization and its evolutionary implications in Guatemalan lethodontid salamanders of the genus *Bolitoglossa*. *Herpetologica* **36**, 335–345.

⋆Wallis, G. P. and Beardmore, J. A. (1980). Genetic evidence for naturally occurring fertile hybrids between two goby species, *Pomatoschistus minutus* and *P. lozanoi* (Pisces, Gobiidae). *Mar. Ecol. Prog. Ser.* **3**, 309–315.

⋆Ward, P. A. (1980). Genetic variation and population differentiation in the *Rhytidoponera impressa* group, a species complex of ponerine ants (Hymenoptera: Formicidae). *Evolution, Lancaster, Pa.* **34**, 1060–1076.

⋆Ward, R. D. and Galleguillos, R. A. (1978). Protein variation in the plaice, dab and flounder, and their genetic relationships. *In* "Marine Organisms: Genetics, Ecology and Evolution" (B. Battaglia and J. A. Beardmore, eds), pp. 71–94. Academic Press, New York.

⋆Webster, T. P. (1975). An electrophoretic comparison of the hispaniolan lizards *Anolis cybotes* and *A. marcanoi*. *Breviora* **431**, 1–8.

★Webster, T. P., Selander, R. K. and Yang, S. Y. (1972). Genetic variability and similarity in the *Anolis* lizards of Bimini. *Evolution, Lancaster, Pa.* **26**, 523–535.

Wilson, A. C., Carlson, S. S. and White, T. J. (1977). Biochemical evolution. *A. Rev. Biochem.* **46**, 573–639.

★Winans, G. A. (1980). Geographic variation in the milkfish *Chanos chanos*. I. Biochemical evidence. *Evolution, Lancaster, Pa.* **34**, 558–574.

★Wyles, J. and Gorman, G. C. (1980). The albumin immunological and Nei electrophoretic distance correlation: a calibration for the saurian genus *Anolis* (Iguanidae). *Copeia* **1980**, 66–71.

★Yang, S. Y., Soule, M. and Gorman, G. C. (1974). *Anolis* lizards of the Eastern Caribbean: a case study in evolution. I. Genetic relationships, phylogeny and colonisation sequence of the *roquet* group. *Syst. Zool.* **23**, 387–399.

★Zimmerman, E. G. and Wotten, M. C. (1981). Allozymic variation and natural hybridization in sculpins, *Cottus confuses* and *Cottus cognatus*. *Biochem. Syst. Ecol.* **9**, 341–346.

★Zimmerman, E. G., Hart, B. J. and Kilpatrick, C. W. (1975). Biochemical genetics of the *boylii* and *truei* groups of the rodent genus *Peromyscus*. *Comp. Biochem. Physiol.* **52B**, 541–545.

9 | Taxonomy and Evolution in the Brine Shrimp *Artemia*

J. A. BEARDMORE and F. A. ABREU-GROBOIS

Department of Genetics, University College of Wales, Swansea SA2 8PP, U.K.

Abstract: Electrophoretic, cytological and anatomical investigation of about 80 populations of *Artemia* has produced results confirming that *A. salina, A. franciscana, A. persimilis* and *A. urmiana*, though morphologically very similar, are well differentiated species. The population from Lake Mono although genetically very similar to *A. franciscana* is considered to merit specific status as *A. monica*. Parthogenetic populations are numerous and diverse and the parthenogenetic mode seems likely to have been derived from a common ancestor with *A. urmiana* possibly originating after the drying of the Mediterranean Basin about 5My B.P.

INTRODUCTION

The brine shrimp *Artemia* is an Anostracan crustacean with a striking physiological tolerance of salinity, some strains being able to live in saturated brine concentrations about one order of magnitude higher than water in the open sea. *Artemia* is morphologically relatively invariable and is widely used for experiments in chemistry, nutrition and toxicology often with the assumption that all material belongs to a single species *A. salina*. Details of the life history of *Artemia* can be found in Sorgeloos (1980).

Bowen and her colleagues (Bowen, 1964; Clark and Bowen, 1976; Bowen *et al.*, 1980) carried out experiments to determine the extent to

Systematics Association Special Volume No. 24, "Protein Polymorphism: Adaptive and Taxonomic Significance", edited by G. S. Oxford and D. Rollinson, 1983, Academic Press, London and New York.

which populations from different regions could be crossed and from the results obtained considered that there are five bisexual species. The existence of parthenogenetic strains was demonstrated long ago (Artom, 1911).

The work described here is part of a programme designed to examine a large number of populations electrophoretically, cytologically and anatomically to enable the construction of a taxonomy for the available species of bisexual and parthenogenetic *Artemia* and to relate this to their evolution. A more detailed account of much of the earlier part of the work is given in Abreu-Grobois and Beardmore (1983).

MATERIALS AND METHODS

Some 80 geographic sites distributed over four continents were sampled by means of collections of the highly resistant cysts formed, often under unfavourable conditions, in natural populations. The methods for electrophoresis of adults are detailed in Abreu-Grobois and Beardmore (1980). Electrophoretic analysis was of whole animal homogenates and was based on 23 loci with sample sizes usually in the range of 40–50.

Methods of culture of *Artemia* are given in Abreu-Grobois and Beardmore (1983) and methods for cytological analysis were based on those of Barigozzi and Baratelli-Zambruni (1981).

1. *Cytology*

The presence of chromocentres in nuclei (Table I) is limited to New World populations and those populations resulting from transplantation of such populations. The number of chromocentres varies from population to population, those of the North American mainland having higher numbers of chromocentres per nucleus than geographically marginal populations in the Caribbean and Mexico.

The basic haploid number of chromosomes is 21 and all populations except that of Buenos Aires (BA) have a multiple of this figure. The BA population is recognized as *Artemia persimilis* and is characterized by $2n = 44$. This chromosome number probably involves duplication of one chromosome pair though at present different chromosome pairs cannot be sufficiently well distinguished from one another for

Table I. Results of chromosome and chromocentre studies in a sample of the populations assayed.

Population	Chromosome No.	Chromocentres per nucleus
Rockhampton, Australia[a]	42	17·10
Bahia Salinas, Puerto Rico	42	7·25
Great Salt Lake, U.S.A. (1977)	42	17·00
Macau, Brazil[a]	42	16·90
Mono Lake, California	42	16·65
San Francisco Bay, California	42	17·00
Yavaros, Mexico	42	3·90
Buenos Aires, Argentina	44	3·03
Barbarena, Spain	42	0
Salina di Stato, Boetto, Sardinia	42	0
Urmia, Iran	42	0
Lavalduc, France	42	0
San Felix, Spain	42	0
Eilat, Israel	63	0
Alcochete, Portugal	84	0
Izmir, Turkey	105	0

[a] Populations derived by transplantation from San Francisco Bay.

this to be established with certainty. However, as this species has 2 *Pgi* loci the hypothesis of duplication is not without support.

2. *Anatomical characters*

The anatomical paired characters used were the form of the furca, the presence of a pair of lower thoracic protruberances on the female, shape of the male clasper knob and the presence of penis spines. These have all been studied in the Argentine, North American and European populations by several workers but the Mono Lake population had not been fully studied previously and the Lake Urmia populations had been examined only for anatomical characters by Günther in 1900. The findings of our study accord with those of earlier work and the results are summarized in Table II, for the five species distinguished by Clark and Bowen (1976).

Table II. Cytological and anatomical characters in bisexual species of *Artemia*.

				Anatomical characters			
Species	Distribution	2*n*	Chromocentres[c, d]	Furca	♀ l.t.p	♂ clasper knob	Penis spine
A. salina[a, b] (or *tunisiana*)	Europe and N. Africa	42	absent	two lobed many setae	absent	sub-conical	absent
A. franciscana[a, b] (or *gracilis*)	N. America and Caribbean	42	+ + + + +	two lobed many setae	present	subspherical	present
A. monica[f]	Mono Lake, California	42	+ + +	two lobed many setae	present	subspherical	present
A. persimilis[a, b]	Argentina	42	+	rudimentary few setae	present	subspherical	present
A. urmiana[f]	Lake Urmia, Iran	42	absent	rudimentary few setae	present	subspherical	present

l.t.p. = lower thoracic protuberances.
Column 4: + = 1–3 per nucleus, + + (Caribbean + Mexico) = 3–8, + + + (N. America) = 14–22.
[a] Barigozzi (1974); [b] Amat (1980); [c] Abreu-Grobois and Beardmore (1983); [d] Barigozzi and Baratelli-Zambruni (1981); [e] Günther (1900); [f] this study.

3. *Electrophoretic data*

These data were used to estimate the values of P, H_e, Nei's D and G_{ST} statistics, and to generate cluster dendrograms by the UPGMA method in Wishart (1979) and Wagner trees (Swofford, 1982). Table III shows the values of H_e, and G_{ST} for the five bisexual species and the values for gene diversity and proportion of loci heterozygous for a sample of the parthenogenetic populations. Figure 1 gives the dendrogram of genetic distance for many of the populations. Those omitted from the dendrogram in the interests of clarity do not materially alter the pattern when they are included in the analysis. Figures 2 and 3 show the complete Wagner trees for the Eurasian bisexual and parthenogenetic populations and for the bisexual species respectively.

Table III. Estimates of the proportion of polymorphic loci, $\bar{H}_e$ and G_{ST} for the five bisexual species (only one population is known for each of *A. monica, A. persimilis* and *A. urmiana*) and of gene diversity for some parthenogenetic populations.

Species	P′	P″	$\bar{H}_e$	G_{ST}
A. salina	0·48	0·245	0·092 (0·041)	0·117 (± 0·118)
A. franciscana	0·83	0·251	0·086 (0·033)	0·237 (± 0·045)
A. monica	–	0·391	0·177 (0·052)	–
A. persimilis	–	0·375	0·129 (0·037)	–
A. urmiana	–	0·478	0·135 (0·036)	–
Lavalduc (2*n*)	–	0·17	0·065 (±0·036)	–
Israel (3*n*)	–	0·22	0·099 (±0·040)	–
Alcochete (4*n*)	–	0·48	0·217 (±0·051)	–
Izmir (5*n*)	–	0·43	0·188 (±0·099)	–

For bisexual species P′ = proportion of loci polymorphic in total over all populations, P″ = mean proportion of loci polymorphic; both at the 0·05 level. For parthenogenetic species P″ = proportion of heterozygous loci.

DISCUSSION

The anatomical and cytological distinctions between *A. salina, A. franciscana, A. persimilis* and *A. urmiana* are clear and specific distinction is confirmed by genetic distance data. Indeed the genetic distances are relatively large by comparison with many other cases of sibling species

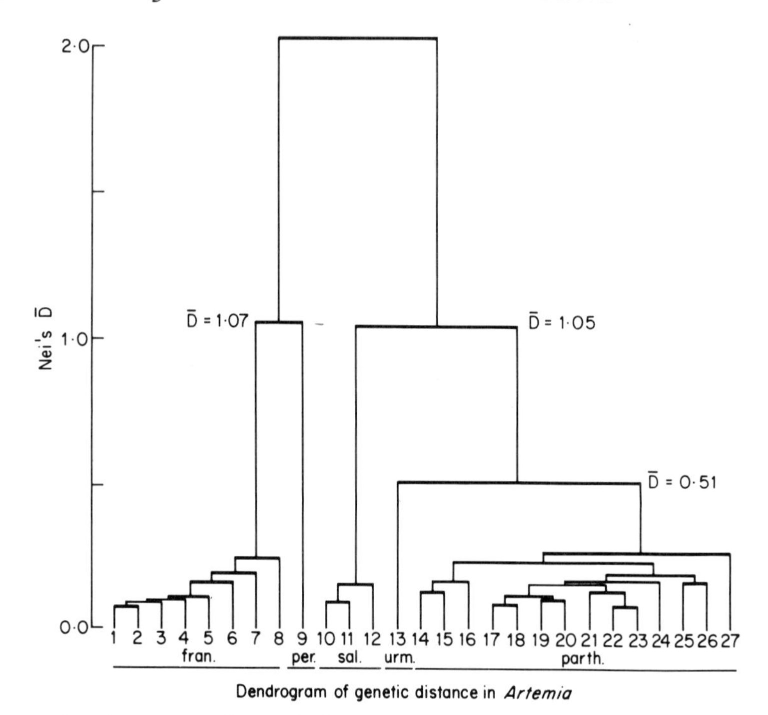

Fig. 1. Dendrogram of genetic distances in *Artemia*. 1: SFB; 2: GSL; 3: PA; 4: YAV; 5: Caribbean; 6: JES; 7: GP + RP; 8: BS; 9: BA; 10: BAR; 11: LAR; 12: SPO; 13: URM; 14: Crimea + Altai + SG; 15: Karedhi Lake + Caspian; 16: SPO(2*n*); 17: AL; 18: TIEN; 19: CAD + CAL; 20: SF; 21: BP(2*n*); 22: LAV + SE + MS; 23: EIL; 24: LT; 25: TUT; 26: 4*n* + 5*n*; 27: IZM. (Key to abbreviations in Figs 2 and 3.)

in the literature (Avise, 1976). The Mono lake population is not clearly differentiated from the populations of *A. franciscana* on the North American mainland, typical 'D' values being 0·06–0·09. However, the Mono lake population is unusual in several respects. The lake has a high level of carbonate ion and Mono shrimp cannot live in waters with high chloride levels. Conversely animals taken from saline water do not survive in Mono water. There is, thus, an effective ecological barrier to gene exchange and this led Bowen *et al*. (1980) to accord this population the specific status of *A. monica*. The degree of genetic differentiation (as expressed by the D statistic) of *A. monica* from typical *A. franciscana* populations is very small and considerably smaller than values commonly associated with between species dif-

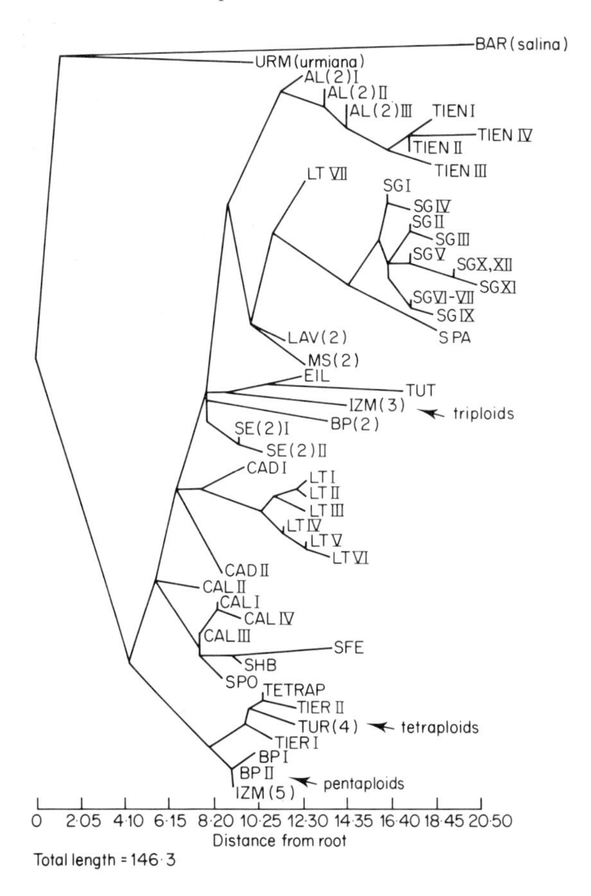

Fig. 2. Wagner tree of Eurasian bisexual and parthenogenetic *Artemia* (AL: Alcochete, Port; BP: Burgas Pomerije, Bulg; CAD: Cadiz, Spain; CAL: Calpe, Spain; EIL: Eilat, Israel; IZM: Ismir, Turkey; LT: Lake Techirgiol, Romania; MS: Margherita de Savoia, Italy; SE: Sete, France; SFE: San Felix, Spain; SG: Salin de Giraud, France; TIER: Tierzo, Spain.

ferences. However, in some cases specific distinction appears to depend upon a very small number of genes as in some lacewings where barriers to gene exchange appear to be controlled by as few as two or three loci (Tauber and Tauber, 1977; Tauber *et al.*, 1977). The Mono lake population is characterized by a very high level of genetic variability and lies on the flight paths of migratory bird species. The opportunities for considerable and frequent input of genes from other populations are thus, in principle, abundant. Whether the high level of

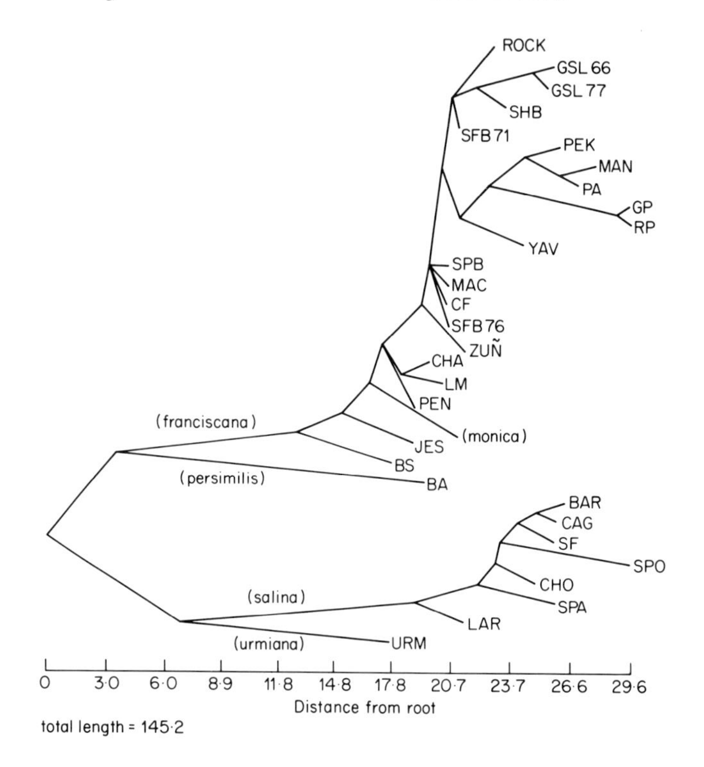

Fig. 3. Wagner tree of bisexual *Artemia*. BA: Buenos Aires; BAR: Barbarena, Spain; BS: Bahia Salina; Puerto Rico; CAG: Cagliari, Italy; CHA: Chaplin Lake, Canada; CHO: Chott, Tunisia; CF: Cabo Frio, Brazil; GP: Green Pond, USA; GSL: Gt. Salt Lake, U.S.A.; JES: Jesse Lake, U.S.A.; LAR: Larnaca, Cyprus; LM: L. Manitou Lake, Canada; MAC: Macau, Brazil; MAN: Manaure, Columbia; PA: Punta Araya, Venezuela; PEK: Pekelmeer, Bonaire; RP: Red Pond, U.S.A.; SFB: San Francisco Bay, U.S.A.; SPA: San Pablo, Spain; SPO: Santa Pola, Spain; URM: Lake Urmia, Iran; ZUN: Zuni Lake, U.S.A.

genetic variability owes much to this input or whether it derives mainly from the process of adaptation to unusual and, for *Artemia*, difficult ecological conditions in the lake is an open question. However, *Artemia* in general display high levels of genetic variability and this may suggest that there is a general effect of difficult ecological conditions promoting the retention of heterozygosity in populations as suggested by Parsons (1973) and Smith (1974). On the basis of present evidence the specific distinction of *A. monica* on ecological grounds appears to be justified.

Both the dendrogram and Wagner tree representations give a very satisfactory clustering of populations. In particular the individual polyploid types tend to cluster together though the diploid parthenogenetic populations are more widely scattered. The European bisexual populations are close together and the North American, Central American and Caribbean populations reasonably so though there is more differentiation visible in this latter group as already noted with respect to chromocentre number. Figures 4 and 5 are, respectively, a skeletal dendrogram and a skeletal Wagner tree incorporating, for the sake of clarity and comparison, a sample of populations from all bisexual species and some of the parthenogenetic types. However, the omission of some populations does not materially alter the form of either representation.

A comparison of the pattern of relatedness offered by the skeletal Wagner tree with the skeletal dendrogram is instructive particularly as some criticisms have been made of the lack of agreement using the two methods. In this case a very satisfactory concordance appears to exist, the only obvious difference between the two diagrams being the place

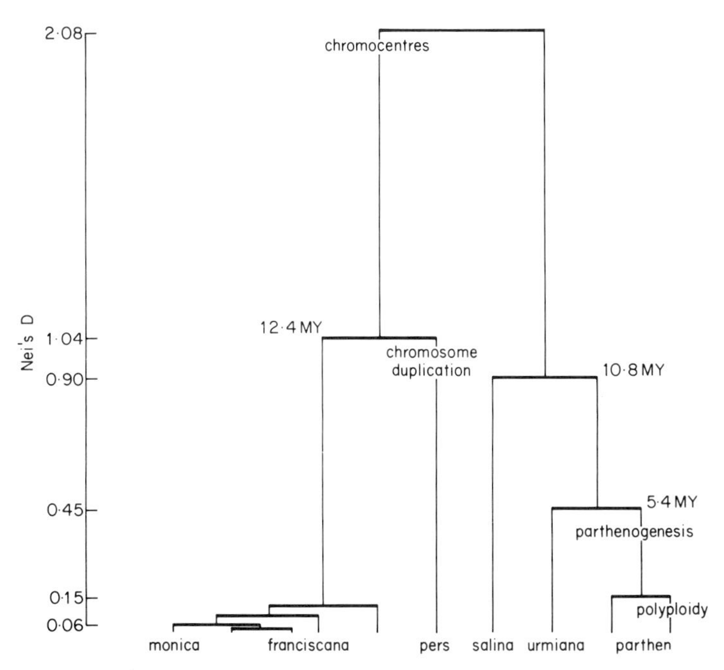

Fig. 4. Genetic distance and evolution in *Artemia*.

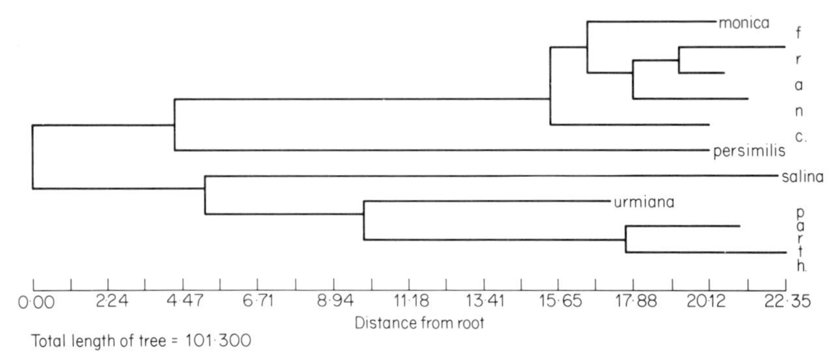

Fig. 5. Computer generated Wagner tree of evolution in *Artemia*.

of the population from Lake Mono which, as has already been pointed out, is unusual.

Using the Nei (1972) and Sarich (1977; as modified by Carlson *et al.*, 1978) approaches to evolutionary change it is possible to represent a possible evolutionary pattern of the genus *Artemia* based on genetic distance as seen in Fig. 4. The times indicated are millions of years B.P. and are the mean of the Nei and Sarich conversions from genetic distance. This figure shows that the division leading to the European and American lines occurred about 25My ago and that the origin of the South American species *A. persimilis* occurred some 12My later, after the evolution of nuclear chromocentres had taken place. The significance of chromocentres is unclear, they are considered to represent regions of AT repetitiveness of unknown function. Chromosome duplication of one pair of chromosomes seems likely to have occurred at some time thereafter in the line leading to *A. persimilis*.

At about 11My ago the lines leading to typical *A. salina* and *S. urmiana* diverged while the evolution of parthenogenesis and polyploidy would seem to have taken place at roughly 5·4 to 1·7 My ago respectively. It seems likely from these data that the origin of parthenogenesis is monophyletic and relates closely to the line which led to *A. urmiana*.

The centre of origin of *Artemia* is probably the Mediterranean region where according to geological theory a dry period was experienced about 6×10^6 years ago. This would have produced many opportunities for colonization by parthenogenetic forms and it is interesting to see that the figure for the time since the origin of parthenogenesis in Fig. 4 fits quite well with the geological evidence.

REFERENCES

Abreu-Grobois, F. A. and Beardmore, J. A. (1980). Genetical characterisation of *Artemia* populations: An electrophoretic approach. *In* "The Brine Shrimp *Artemia*". (G. Persoone, P. Sorgeloos, O. Roels and E. Jaspers, eds), pp. 133–146. Universa Press, Wetteren, Belgium.

Abreu-Grobois, F. A. and Beardmore, J. A. (1983). Genetic differentiation and speciation in the brine shrimp *Artemia*. *In* "Mechanisms of Speciation" (C. Barigozzi, ed.), pp. 345–376. Liss, New York.

Amat, F. (1980). Differentiation in *Artemia* strains from Spain. *In* "The Brine Shrimp *Artemia*" (G. Persoone, P. Sorgeloos, O. Roels and E. Jaspers, eds), pp. 19–39. Universa Press, Wetteren, Belgium.

Artom, C. (1911). Analisi comparativa della sostanza cromatica pelle mitosi di segmentazione dell' uovo dell' *Artemia sessuota* di cagliari (univalens) e dell' uovo dell' *Artemia partenogenetica* di Capodistria (bivalens). *Arch. f. Zell.* **7**, 277–295.

Avise, J. C. (1976). Genetic differentiation during speciation. *In* "Molecular Evolution" (F. J. Ayala, ed.), pp. 106–122. Sinauer Associates, Sunderland, Massachusetts.

Barigozzi, C. (1974). *Artemia*. A survey of its significance in genetic problems. *Evol. Biol.* **7**, 221–252.

Barigozzi, C. and Baratelli-Zambruni L. (1981). Presence and absence of chromocentres in populations of *Artemia*. Communication to "Academia dei Lincei, Rome.

Bowen, S. T. (1964). The genetics of *Artemia salina* IV. Hybridization of wild populations with mutant stocks. *Biol. Bull., Woods Hole* **126**, 333–344.

Bowen, S. T., Davis, M. L., Fenster, S. R. and Lindwall, G. A. (1980). Sibling species of *Artemia*. *In* "The Brine Shrimp *Artemia*" (G. Persoone, P. Sorgeloos, O. Roels and E. Jaspers, eds), pp. 155–167. Universa Press, Wetteren, Belgium.

Carlson, S. S., Wilson, A. C. and Maxson, R. D. (1978). Do albumin clocks run on time? A reply. *Science, N.Y.* **200**, 1183–1185.

Clark, L. S. and Bowen, S. T. (1976). The genetics of *Artemia salina* VII. Reproductive isolation. *J. Hered.* **67**, 385–388.

Günther, R. T. (1900). Contributions to the natural history of Lake Urmia, N. W. Persia, and its neighbourhood. *Zool. J. Linn. Soc.* **27**, 345–453.

Nei, M. (1972). Genetic distance between populations. *Am. Nat.* **106**, 283–292.

Parsons, P. A. (1973). "Behavioural and ecological genetics: A study in *Drosophila*." Clarendon Press, Oxford.

Sarich, V. M. (1977). Rates, sample sizes and the neutrality hypothesis for electrophoresis in evolutionary studies. *Nature, Lond.* **265**, 24–28.

Smith, M. H., Garten, C. T. and Ramsay, P. E. (1975). Genetic heterozygosity and population dynamics in small mammals. *In* "Isozymes 4,

Genetics and Evolution" (C. L. Markert, ed.), pp. 85–102. Academic Press, New York.
Sorgeloos, P. (1980). Life history of the brine shrimp *Artemia*. *In* "The Brine Shrimp *Artemia*" (G. Persoone, P. Sorgeloos, O. Roels and E. Jaspers, eds), pp. 19–23. Universa Press, Wetteren, Belgium.
Swofford, D. L. (1982). "Wagner Procedure Program (Version 3)." Department of Genetics and Development, University of Illinois, Urbana.
Tauber, C. A. and Tauber, M. J. (1977). Sympatric speciation based on allelic changes at three loci: evidence from natural populations in two habitats. *Science, N.Y.* **194**, 1298–1299.
Tauber, C. A., Tauber, M. J. and Tauber, J. R. (1977). Two genes control seasonal isolation in sibling species. *Science, N.Y.* **197**, 592–593.
Wishart, D. (1979). "Clustan IB Computer Programs." Computing Laboratory, University of St. Andrews.

10 Biochemical Systematics and Genetic Variation in Flatfish of the Family Pleuronectidae

R. D. WARD and R. A. GALLEGUILLOS†*

Department of Genetics, University College of Swansea, Singleton Park, Swansea, SA2 8PP, U.K.

Abstract: Six British species of the family Pleuronectidae were screened for electrophoretically detectable protein variation, and their systematic relationships thereby assessed. In addition, three subspecies of the flounder (*Platichthys flesus*) were screened and the validity of their subspecific ranks confirmed. Levels of differentiation between populations of the plaice (*Pleuronectes platessa*) and between populations within subspecies of the flounder were very low. Divergence time estimates derived from genetic distance values appear reasonable (where one unit of distance (Nei) is equal to 19×10^6 years). The overall mean level of heterozygosity in the Pleuronectidae is significantly higher than in non-pleuronectid fishes.

INTRODUCTION

During the last ten to 15 years, as evidenced by this volume, studies in systematics and evolutionary biology have been revitalized by the application of techniques from molecular biology. One of the most significant advances has been the use of protein electrophoresis to

Present addresses: * Department of Human Sciences, Loughborough University, Loughborough, Leicestershire LE11 3TU, U.K.

† Department of Biology and Marine Technology, Catholic University, Cafsilla 127, Talcahuano, Chile.

Systematics Association Special Volume No. 24, "Protein Polymorphism: Adaptive and Taxonomic Significance", edited by G. S. Oxford and D. Rollinson, 1983, Academic Press, London and New York.

estimate levels of genetic similarity between pairs of populations or related species (e.g. Ayala, 1975). This paper describes an electrophoretic investigation into species relationships and levels of genetic variability in members of the flatfish family Pleuronectidae.

Pleuronectidae and other flatfish belong to the order Pleuronectiformes (or Heterosomata), an order comprising six families, 117 genera and about 500 extant species (Nelson, 1976). Very young flatfish are bilaterally symmetrical, but early in their development complex modifications of body structure occur, particularly in the head region where one eye migrates to the other side leaving one side blind. The upper side (eyeside) is, in general, pigmented, whereas the lower side is usually white. We have examined six of the seven British members of the Pleuronectidae, the one not examined being the halibut, *Hippoglossus hippoglossus*. Interpopulation differentiation has been examined in *Pleuronectes platessa* and in *Platichthys flesus* and, in addition, subspecies relationships in the *P. flesus* group have been investigated.

MATERIALS AND METHODS

The species examined and sample localities are given in Table I. *Platichthys flesus italicus* and *P. flesus luscus* are restricted to the Adriatic and Black Sea regions respectively, the other species are widely distributed throughout the north-east Atlantic.

Starch gel electrophoresis was used to compare allozyme mobilities. The following proteins were screened, although not all proteins were typed in all species: alcohol dehydrogenase, octanol dehydrogenase, xanthine dehydrogenase, α glycerophosphate dehydrogenase (2 loci), sorbitol dehydrogenase, lactate dehydrogenase (3 loci), malate dehydrogenase (2 loci), malic enzyme, isocitrate dehydrogenase (2 loci), 6 phosphogluconate dehydrogenase, glyceraldehyde-3-phosphate dehydrogenase (2 loci), "nothing" dehydrogenase, superoxide dismutase, glutamate oxalate transaminase (2 loci), creatine kinase (3 loci), adenylate kinase, phosphoglucomutase (2 loci), esterase (5 loci, including *Est-D*, specific for methylumbelliferyl esters), alkaline and acid phosphatases, leucine aminopeptidase, aminopeptidase (3 loci), adenosine deaminase, glyoxylase, phosphoglucose isomerase (2 loci) aconitase, muscle protein (4 loci) and haemoglobin (2 loci). Hom-

Table I. Sample localities and genetic variability in the species assayed.

	Nloci	$\bar{H}_o$	P	$\bar{N}$ind
Pleuronectes platessa (plaice)				
(i) Carmarthen Bay, Bristol Channel	46	0·102	0·48	678·8
(ii) N.E. Irish Sea (off Anglesey)	35	0·100	0·51	105·6
(iii) Danish Belt Sea	29	0·078	0·28	34·8
Platichthys flesus flesus (flounder)				
(i) Carmarthen Bay, Bristol Channel	36	0·074	0·33	49·3
(ii) River Thames estuary, London	37	0·085	0·40	38·0
(iii) River Tamar estuary, Plymouth	36	0·092	0·39	38·1
(iv) Danish Belt Sea	28	0·108	0·39	39·1
Platichthys flesus italicus				
(i) N. Adriatic Sea – I	36	0·035	0·33	63·2
(ii) N. Adriatic Sea – II	37	0·036	0·24	19·5
Platichthys flesus luscus				
(i) Istanbul (Sea of Marmara?)	37	0·036	0·11	6·0
(ii) Constanta, Roumania. Black Sea	34	0·016	0·09	7·0
Microstomus kitt (lemon sole)				
English Channel, off Plymouth	38	0·057	0·18	58·5
Hippoglossoides platessoides (long rough dab)				
N.E. Atlantic (off west coast of Scotland)	35	0·083	0·40	47·4
Glyptocephalus cynoglossus (witch)				
N.E. Atlantic (off west coast of Scotland)	31	0·063	0·29	21·9
Limanda limanda (dab)				
Carmarthen Bay, Bristol Channel	36	0·066	0·22	68·3

Nloci = number of loci scored, $\bar{H}_o$ = mean observed heterozygosity per locus, P = proportion of loci polymorphic (frequency of most common allele $\leq$ 0·99), $\bar{N}$ind = mean number of individuals scored per locus.

ologies between alleles were ascertained by side by side sample running, alleles being numbered with respect to the commonest allele in the Bristol Channel population of *P. platessa*, designated 100. Further details are given in Ward and Beardmore (1977) and, with alterations in some locus designations, Galleguillos and Ward (1982).

Genetic identities (I), distances (D), and standard errors of distances between population pairs were estimated using the methods of Nei (1971, 1972). Measures of genetic distance are commonly used as estimators of divergence times of two lineages, and, although this molecular clock hypothesis is controversial (e.g. Wilson *et al.*, 1977; Goodman *et al.*, 1976), empirical evidence from a number of fish species separated by the Panamanian isthmus (Vawter *et al.*, 1980) indicates that given a reasonably large random sample of proteins, a Nei distance of one is equivalent to about 19 My of divergence. Such a calibration had earlier been proposed by Sarich (1977) and Carlson *et al.* (1978). We use this calibration in the present paper, although we recognise that the derived time estimates must be subject to large standard errors. Even if the molecular clock hypothesis is accepted, it is clear that different proteins evolve at different rates (Wilson *et al.*, 1977; Sarich, 1977; Skibinski and Ward, 1982). We have not attempted any corrections of time based upon the particular mix of proteins used.

RESULTS

Mean observed heterozygosities per locus, proportion of polymorphic loci, numbers of loci, and mean numbers of individuals scored per locus, are given in Table I. The allele frequencies of the most polymorphic loci in each taxa are given in Table II; these loci would be those most profitably investigated in any subsequent investigations into species specific population structure. Additional information is available in Ward and Beardmore (1977), Ward and Galleguillos (1978) and Galleguillos and Ward (1982).

1. *Intraspecific relationships*

a. *Pleuronectes platessa* The three populations assayed had pairwise I values of 0·999–1·0 and corresponding D values of 0·0009–0·0 with 23, 26 and 34 loci used in the three pairwise comparisons. The apparently decreased heterozygosity of the Danish population (Table I) is an artefact caused by the non-typing of certain highly polymorphic (liver specific) loci in this sample, notably *Sdh, Idh-1* and α*Gpdh-2*.

b. *Platichthys flesus* Comparisons between these populations are described in detail in Galleguillos and Ward (1982). Intrasubspecies I values range from 0·997–0·999, with corresponding D values of 0·003–0·001. These I and D values were derived from comparisons

Table II. Allele frequencies and designations for those loci with observed heterozygosities equal to or greater than 0·20.

Species	Locus	Allele frequencies
P. platessa[a]	*αGpdh-1*	*135* 0·055, *113* 0·002, *100* 0·872, *73* 0·066, *65* 0·001, *58* 0·001, *50*, 0·003
	αGpdh-2	*143* 0·002, *131*, 0·012, *115*, 0·352, *100* 0·599, *79* 0·036
	Sdh	*172* 0·003, *137* 0·342, *100* 0·655.
	Mdh-2	*126* 0·005, *100* 0·876, *77* 0·118
	Idh-1	*131* 0·002, *118* 0·102, *100* 0·512, *81* 0·374, *74* 0·009
	Got-2	*116* 0·167, *100* 0·463, *84* 0·358, *66* 0·011
	Pgm-1	*228* 0·012, *186* 0·006, *171* 0·002, *148* 0·002, *140* 0·360, *100* 0·611, *66* 0·005, *59* 0·002
	Pgm-2	*118* 0·036, *110* 0·259, *100* 0·652, *93* 0·045
	Ada	*117* 0·002, *108* 0·110, *100* 0·740, *96* 0·013 *91* 0·128, *80* 0·007
Pl. f. flesus[b]	*αGpdh-1*	*45* 0·437, *40* 0·563
	Got-2	*110* 0·138, *100* 0·812, *90* 0·038, *83* 0·012
	Pgm-1	*140* 0·881, *100* 0·109, *85* 0·010
	Est-4	*100* 0·853, *97* 0·088, *92*, 0·059
	Ada	*100* 0·750, *91* 0·250
	Pgi-2	*110* 0·760, *100* 0·240
	Aco-2	*100* 0 125, *95* 0·573, *88* 0·292, *80* 0·010
Pl. f. italicus[c]	*αGpdh-2*	*100* 0·600, *79* 0·390, *77* 0·010
Pl. f. luscus[c]	*Odh*	*100* 0·143, *75* 0·837
	Got-1	*110* 0·833, *79* 0·167
	Pgm-1	*186* 0·036, *140* 0·888, *100* 0·077
M. kitt	*Sdh*	*180* 0·737, *137* 0·263
	Idh-1	*90* 0·627, *45*, 0·373
	Pgm-1	*190* 0·050, *140* 0·883, *100* 0·058, *90* 0·009
	Ada	*75* 0·033, *50* 0·250, *40* 0·517, *27* 0·133, *20* 0·058, *13* 0·009
	Pgi-2	*93* 0·225, *80* 0·717, *70* 0·058
H. platessoides	*Mdh-2*	*100* 0·063, *77* 0·479, *50* 0·458
	G3pdh-1	*330* 0·610, *183* 0·390
	Pgm-1	*346* 0·080, *284* 0·870, *253* 0·010, *228* 0·030, *200* 0·010
	Ada	*98* 0·143, *84* 0·785, *70* 0·072
	Pgi-1	*580* 0·010, *550* 0·200, *470* 0·710, *290* 0·080
C. cynoglossus	*αGpdh-1*	*43* 0·100, *40* 0·260, *20*, 0·540, *6*, 0·100
	Ada	*84* 0·075, *70* 0·525, *60* 0·275, *50* 0·125
	Pgi-2	*90* 0·036, *84* 0·071, *80* 0·903
L. limanda	*Pgm-1*	*228* 0·024, *160* 0·695, *120* 0·281
	Est-2	*121* 0·050, *115* 0·100, *103* 0·450, *102* 0·290, *100* 0·090, *97* 0·020
	Ada	*118* 0·017, *112* 0·542, *104* 0·342, *98* 0·090, *84* 0·009
	Pgi-2	*104* 0·246, *100* 0·746, *90* 0·008

[a] Carmarthen Bay population. [b] Thames River population.
[c] Mean of the two samples of each subspecies.

utilizing from 27 to 36 loci. However, since information was available in each subspecies on allele frequencies at 38 loci (in one or more samples), and since there is very little observed genetic heterogeneity between populations within subspecies, each subspecies may be allocated mean unweighted allele frequencies for each of these 38 loci. These frequencies may then be used to estimate intersubspecific relationships, and the resulting *I*/*D* matrix and UPGMA dendrogram (Sneath and Sokal, 1973) are given in Table III and Fig. 1. It can be seen

Table III. Genetic identities above diagonal and distances (S.E.) below diagonal of three subspecies of *Platichthys flesus*. Allele frequencies used are the mean unweighted values per subspecies at 38 loci. The bottom row gives the mean expected heterozygosity per locus per subspecies.

Subspecies	*flesus*	*italicus*	*luscus*
Pl. flesus flesus	–	0·877	0·891
Pl. flesus italicus	0·131 (0·061)	–	0·870
Pl. flesus luscus	0·115 (0·057)	0·139 (0·063)	–
Mean heterozygosity per locus	0·095	0·036	0·039

that the extent of genetic differentiation between subspecies is much greater than that within subspecies, confirming the reality of the three subspecific taxa. Furthermore, at some loci different subspecies are fixed or nearly fixed for alternative alleles (e.g. subspecies *flesus* for $Ck\text{-}M^{100}$, $Ck\text{-}1^{95}$ and Sdh^{90}, *italicus* for $Ck\text{-}M^{92}$, $Ck\text{-}1^{80}$ and Sdh^{99}, *luscus* for $Ck\text{-}M^{100}$, $Ck\text{-}1^{95}$ and Sdh^{99}). Levels of genetic variation in subspecies *flesus* approximate those in *P. platessa*, whereas those in *italicus* and *luscus* are substantially lower (Tables I and III).

2. *Interspecific relationships*

A matrix of genetic identity and distance values between the six British pleuronectid species assayed is given in Table IV. In deriving this matrix, the Carmarthen Bay populations of plaice and flounder were used. The resulting UPGMA dendrogram is given in Fig. 2a. The paired species comparisons used in drawing this dendrogram are based on varying numbers of loci (23–34), which may have had the

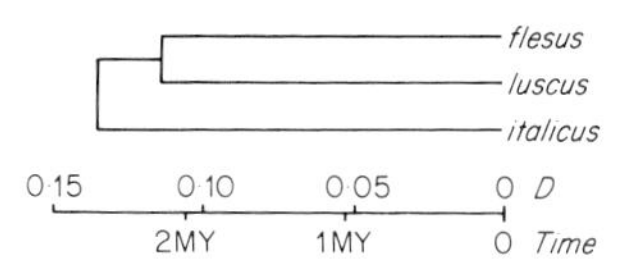

Fig. 1. UPGMA dendrogram showing relationships among the *Platichthys flesus* subspecies. 38 loci used.

effect of slightly distorting species relationships (since, for example, different proteins have different evolutionary rates). This problem has been overcome by deriving a second matrix and dendrogram (Fig. 2b) based only on the 22 loci typed in all species. In fact, it can be seen that the two dendrograms are very similar, the only substantial change being in the position of the witch (*G. cynoglossus*), which is no longer clustered with the two species of dab.

DISCUSSION

1. *Levels of genetic variation in pleuronectidae*

A summary of observations on genetic variability in Pleuronectidae and other fish species is given in Table V. It is clear that levels of

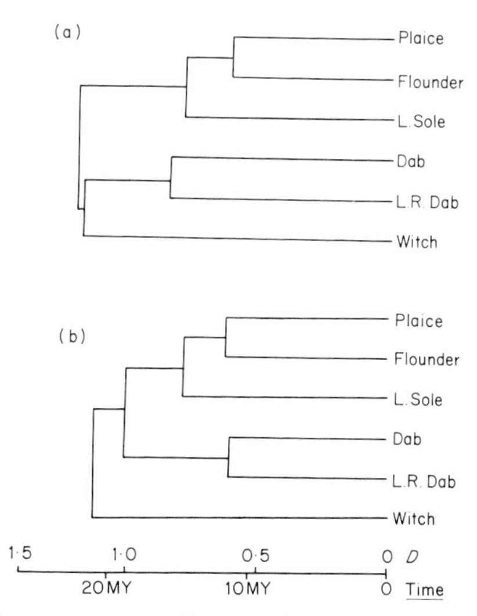

Fig. 2. UPGMA dendrograms drawn from interspecies genetic distance matrices. (a) Number of loci per per paired comparison 23–34. (b) all pairwise comparisons based on 22 common loci.

Table IV. Genetic identities (and number of loci scored) above diagonal and genetic distance (S.E.) below diagonal of six species of Pleuronectidae.

	P. platessa	*Pl. flesus*	*M. kitt*	*H. platessoides*	*G. cynoglossus*	*L. limanda*
P. platessa		0·539 (34)	0·441 (27)	0·327 (28)	0·340 (26)	0·371 (34)
Pl. flesus	0·618 (0·159)		0·464 (27)	0·236 (28)	0·337 (25)	0·318 (34)
M. kitt	0·819 (0·217)	0·767 (0·207)		0·321 (25)	0·254 (23)	0·264 (27)
H. platessoides	1·116 (0·271)	1·443 (0·340)	1·137 (0·291)		0·254 (25)	0·427 (28)
G. cynoglossus	1·079 (0·273)	1·088 (0·275)	1·370 (0·357)	0·370 (0·343)		0·379 (26)
L. limanda	0·991 (0·223)	1·147 (0·251)	1·332 (0·321)	0·852 (0·219)	0·969 (0·251)	

Table V. Genetic variation in Pleuronectidae compared with other fish species.

	No. of species	Mean heterozygosity per locus per species	Mean number of loci scored per species
Pleuronectidae			
This survey	6	0·076	37·2
Johnson and Utter (1976)	7	0·047	17·3
Fujio and Kato (1979)	15	0·082	21·6
Fairbairn (1981)	1	0·058	16
Other fish			
Johnson and Utter (1976)	8	0·016	16·6
Nevo (1979)	57	0·051	21·2
Fujio and Kato (1979)	26	0·046	18·0

variation in Pleuronectidae are, in general, higher than for other fishes. The somewhat lower levels of variation observed by Johnson and Utter (1976) for both groups of fishes are probably attributable to the high proportion of non-enzymatic proteins used in their tests: such proteins generally show low degrees of variability (see, for example, Skibinski and Ward, 1982). Consistent with this assumption, the single species common to both their survey and that of Fujio and Kato (1979), the flatfish *Platichthys stellatus*, was found to have mean heterozygosities per locus of 0·047 and 0·120, respectively. If Johnson and Utter's results are excluded, then the mean heterozygosity per locus of 22 species of Pleuronectidae is 0·079, and for the 83 non-pleuronectids 0·049. This difference is statistically significant (using arcsine transformed values and a t-test, $P < 0{\cdot}001$). The frequency distribution of mean heterozygosity per species is given in Fig. 3. The reality of this distinction is given further weight by the observation that *within* Johnson and Utter's data set, the pleuronectids are more variable than other fishes.

This phenomenon has been noted before (Johnson and Utter, 1976; Ward and Galleguillos, 1978; Fujio and Kato, 1979), but its significance is still not clear. It cannot simply be that Pleuronectidae are marine species, since non-pleuronectid marine species show reduced variability. It may be that pleuronectids have, on average, larger effective

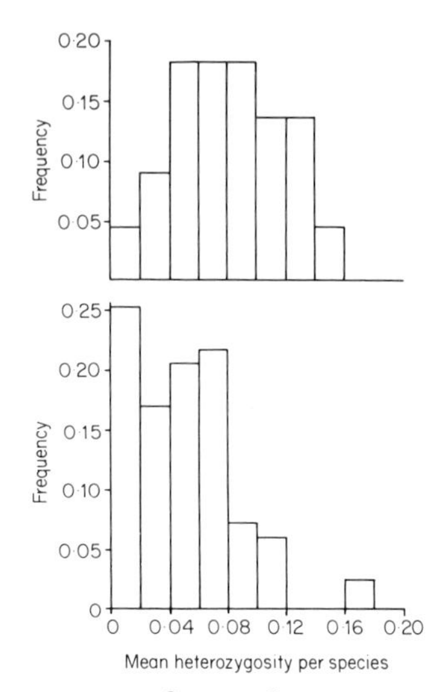

Fig. 3. Frequency histograms of mean heterozygosities per species. Upper distribution is of Pleuronectidae (n = 22), lower distribution is of non-pleuronectid fishes (n = 83).

population sizes than the majority of other fishes so far screened, and that this is reflected in increased heterozygosity. This is in line with neutral theory (Kimura and Ohta, 1971; Nei, 1975), but difficulties met in attempting to estimate population size make it hard to test rigorously such a hypothesis. Yet it is certainly true that pleuronectids have very large population sizes, and consistent with such a causal relationship is our observation that the *italicus* and *luscus* subspecies of *Platichthys flesus* are substantially less variable than the Atlantic subspecies *flesus*. Subspecies *flesus* may realistically be assumed to have a much larger population size than either *italicus* or *luscus* with their very restricted geographic ranges. However, one cannot, of course, rule out the possibility that the high levels of variation in pleuronectids are an adaptive consequence of some particular feature of their life histories: for example, perhaps flatfish, with bottom dwelling adults and pelagic eggs and larvae, are in their lifetime subject to a more spatially and temporally variable environment than most other fishes. In general, attempts to discriminate between neutral and selection models simply by considering such types of patterns are going to give unsatisfactory and equivocal conclusions. Far more knowledge is needed about aspects of population structure and ecology of the species being compared.

2. Genetic differentiation within and between species

A summary of overall levels of differentiation is provided in Table VI. These levels are similar to those recorded in other animal groups (e.g. Ayala, 1975).

Genetic differentiation between populations of the same species (*Pleuronectes platessa*) or subspecies (*Platichthys flesus*) is very limited Such a result can in principle be explained through a spectrum of possibilities ranging from uniform selection pressures operating in distinct populations to migration rates between populations of sufficient magnitude to maintain effectively panmictic mating, the allozymes themselves being selectively neutral. Some migration of these two species between spawning populations is known to occur (Macer, 1972; Dando, pers. comm.).

Table VI. Genetic differentiation in the Pleuronectidae.

		Genetic identity		Genetic distance	
	N	$\bar{I}$	range	$\bar{D}$	range
Populations	9	0·99	1–0·99	0·01	0–0·01
Subspecies	3	0·88	0·89–0·87	0·13	0·12–0·14
Species[a]	1	0·54	–	0·62	–
Genera	14	0·34	0·46–0·24	1·11	0·77–1·44

[a]Classifying *P. platessa* and *Pl. flesus* as conspecific (see text).

Differentiation between subspecies of *P. flesus*, with a mean distance value of 0·128, is far greater than that between populations within subspecies. Geological evidence, described in more detail in Galleguillos and Ward (1982), indicates that about 5·5 My ago, the Mediterranean and Black Seas were more or less dry basins with some freshwater or brackish lakes, and were refilled from the Atlantic some 5 My ago. The first immigrants were tropical species, but as the climate cooled, northern temperate species (such as the flounder) moved south and spread into the Mediterranean perhaps some 4 My ago. By this time, the Black Sea had separated from the Mediterranean and contact was only re-established some 10 000 years ago. This scenario indicates that the *P. flesus* subspecific radiation must have occurred within the last 4 My. Using the molecular clock calibration already referred to in the Materials and Methods section, a *D* value of 0·128 places the radiation at around 2·4 My ago (with a standard error of about 1 My). Thus, this molecular clock, with all its doubts and uncertainties, is at least consistent with the geological evidence and, if accepted, implies that subspecies *luscus* evolved prior to the reconnection of the Black Sea with the Mediterranean, possibly in the Aegean basin.

Alternatively, it could be argued that attempts to date cladogenic events from molecular phylogenies are suspect and misleading. It might, for example, be suggested that *luscus* evolved in the Black Sea in the last 10 000 years from an *italicus* or *flesus*-like ancestor. This would entail a high rate of gene substitution at a number of loci. Of course, it might be true that the observed allozyme differences (*and* similarities) between subspecies are selectively maintained. Then it would have to be argued that environmental pressures on *flesus* and *luscus* are similar to each other and different from those on *italicus* with respect to products of the *Ck-M* and *Ck-1* loci, but that pressures on *flesus* and *italicus* are similar and different from those on *luscus* with respect to the *Sdh* locus. A further corollary of a selectionist explanation must be that environmental differences operating between *flesus*, *italicus* and *luscus* subspecies are on average very much greater than those operating between populations of subspecies *flesus*, which show very little genetic differentiation yet are well separated geographically.

It is clear from the electrophoretic data that *flesus, italicus* and *luscus* represent three valid subspecies, and thus the contention of Tortonese (1971) that *italicus* and *luscus* are synonyms is incorrect. The latter two subspecies are no more closely related to each other than either is to *flesus*.

The genetic and morphological similarities between *P. platessa* and *Platichthys flesus*, together with the facts that in the Danish Belt Sea where breeding areas overlap, hybrids are produced which develop normally (Sick *et al.*, 1963; our unpublished observations), and that experimentally produced hybrid males are fertile (Ubisch, 1951), suggest that *Platichthys flesus* should be returned to the genus *Pleuronectes*.

The dendrograms of species relationships in British Pleuronectidae show that the plaice, flounder and lemon sole form one grouping, with the dab and long rough dab in the other group. The position of the witch is equivocal. This radiation appears to have taken place some

11–22 My ago (with a standard error of between 3–5 My), in the Miocene epoch. Unfortunately, fossil remains of Pleuronectidae are few and do not allow internal checking of the molecular clock. However, it is known that the related families Bothidae and Soleidae existed in the Middle Eocene (38–54 My ago), and there is evidence that a species similar to *P. platessa* existed in the lower Oligocene, 25–36 My ago (Woodward, 1901; Norman, 1934). In conclusion, the molecular clock calibration proposed by Sarich (1977) and Carlson *et al.* (1978), and supported empirically by Vawter *et al.* (1980), can be tested and in principle rejected if found to be a generally poor indicator of elapsed time. The data presented here, both for subspecific differentiation in *Platichthys flesus* and, less powerfully, for species differentiation in Pleuronectidae, do not enable us to reject such a clock.

REFERENCES

Ayala, F. J. (1975). Genetic differentiation during the speciation process. *Evol. Biol.* **8**, 1–78.

Carlson, S. S., Wilson, A. C. and Maxson, R. D. (1978). Do albumin clocks run on time? A reply. *Science, N.Y.* **200**, 1183–1185.

Fairbairn, D. J. (1981). Biochemical genetic analysis of population differentiation in Greenland halibut (*Rheinhardtius hippoglossoides*) from the northwest Atlantic, Gulf of St. Lawrence, and Bering Sea. *Can. J. Fish. Aquat. Sci.* **38**, 669–677.

Fujio, Y. and Kato, Y. (1979). Genetic variation in fish populations. *Bull. Jap. Soc. Sci. Fish.* **45**, 1169–1178.

Galleguillos, R. A. and Ward, R. D. (1982). Genetic and morphological divergence between populations of the flatfish *Platichthys flesus* (L.) (Pleuronectidae). *Biol. J. Linn. Soc.* **17**, 395–408.

Goodman, M., Tashian, R. E. and Tashian, J. H. (1976) (eds). "Molecular Anthropology, Genes and Proteins in the Evolutionary Ascent of the Primates". Plenum Press, New York and London.

Johnson, A. G. and Utter, F. M. (1976). Electrophoretic variation in intertidal and subtidal organisms in Puget Sound, Washington. *Anim. Blood Grps. biochem. Genet.* **7**, 3–14.

Kimura, M. and Ohta, T. (1971). Protein polymorphism as a phase of molecular evolution. *Nature* **229**, 467–469.

Macer, M. (1972). The movements of tagged adult plaice in the Irish Sea. *Fish. Investig., Lond., ser. 2* **27** (6), 1–41.

Nei, M. (1971). Interspecific gene differences and evolutionary time estimated from electrophoretic data on protein identity. *Am. Nat.* **105**, 385–398.

Nei, M. (1972). Genetic distance between populations. *Am. Nat.* **106**, 283–292.

Nei, M. (1975). "Molecular Population Genetics and Evolution". North-Holland Publishing Company, Amsterdam.

Nelson, J. S. (1976). "Fishes of the World". Wiley, London.

Nevo, E. (1979). Genetic variation in natural populations: patterns and theory. *Theor. Pop. Biol.* **13**, 121–177.

Norman, J. R. (1934). "A Systematic Monograph of the Flatfishes (Heterosomata). Volume 1. Psettodidae, Bothidae, Pleuronectidae". British Museum, London.

Sarich, V. M. (1977). Rates, sample sizes and the neutrality hypothesis for electrophoresis in evolutionary studies. *Nature, Lond.* **265**, 24–28.

Sick, K., Frydenberg, O. and Nielsen, J. T. (1963). Haemoglobin patterns of plaice, flounder and their natural and artificial hybrids. *Nature, Lond.* **198**, 411–412.

Skibinski, D. O. F. and Ward, R. D. (1982). Correlations between heterozygosity and evolutionary rate of proteins. *Nature, Lond.* **298**, 490–493.

Sneath, P. H. A. and Sokal, R. R. (1973). "Numerical Taxonomy: the Principles and Practise of Numerical Classification". Freeman, San Francisco.

Tortonese, E. (1971). I pesci Pleuronettiformi delle coste Romene del Mar Nero in relazione alle forme affini viventi nel Mediterraneo. *Annali Mus. civ. Stor. nat. Giacomo Doria* **78**, 322–352.

Ubisch, L. V. (1951). Untersuchungen über Pleuronektiden. II. Ambikoloration, inversion and bilateralität. *Wilhelm Roux Arch. EntwMech. Org.* **145**, 1–61.

Vawter, A. T., Rosenblatt, R. and Gorman, G. C. (1980). Genetic divergence among fishes of the eastern Pacific and the Caribbean: support for the molecular clock. *Evolution, Lancaster, Pa.* **34**, 705–711.

Ward, R. D. and Beardmore, J. A. (1977). Protein variation in the plaice, *Pleuronectes platessa* L. *Genet. Res.* **30**, 45–62.

Ward, R. D. and Galleguillos, R. A. (1978). Protein variation in the plaice, dab and flounder, and their genetic relationships. *In* "Marine Organisms: Genetics, Ecology and Evolution" (B. Battaglia and J. A. Beardmore, eds), pp. 71–93. Plenum Press, New York and London.

Wilson, A. C., Carlson, S. S. and White, T. J. (1977). Biochemical evolution. *A. Rev. Biochem.* **46**, 573–639.

Woodward, A. S. (1901). "Catalogue of Fossil Fishes in the British Museum of Natural History, Part IV". British Museum, London.

11 | Taxonomic and Evolutionary Inferences from Electrophoretic Studies of Various Animal Groups

L. BULLINI

Institute of Genetics, University of Rome, Città Universitaria, 00185 Rome, Italy

Abstract: A number of cases are illustrated, showing different applications of multilocus electrophoretic techniques to the study of genetic differentiation, systematics and speciation of various animal groups, such as Ascarid worms, stick insects, mosquitoes, snails and Plethodontid salamanders.

The following topics are examined: (1) detection of sibling species both sympatric and allopatric; (2) testing for reproductive isolation between allopatric populations by means of release experiments; (3) comparison of allozyme and chromosome differentiation in the genus *Anopheles*; and (4) study of speciation by hybridization in stick insects and snails. The evolutionary and taxonomic significance of the reported data is briefly discussed.

INTRODUCTION

The use of electrophoretic techniques in the study of genetic variation of both animals and plants has been providing data of critical importance to evolutionary biologists and taxonomists (for extensive reviews, see Avise, 1975; Ayala, 1975; Bullini and Sbordoni, 1980; Ferguson, 1980). In this paper, a number of cases are reported, showing different applications of electrophoretic data to the study of genetic differentiation, systematics and speciation of various animal groups.

Systematics Association Special Volume No. 24, "Protein Polymorphism: Adaptive and Taxonomic Significance", edited by G. S. Oxford and D. Rollinson, 1983, Academic Press, London and New York.

DETECTION OF SIBLING SPECIES IN ASCARID WORMS

The most important application of multilocus electrophoretic techniques to species-level taxonomy concerns the detection of sibling species. Good examples are provided by the ascarid worms *Ascaris, Parascaris* and *Anisakis*.

The electrophoretic study of the "univalens" and 'bivalens" forms of *P. equorum* has shown that they are, in fact, two distinct species: *P. equorum* and *P. univalens*, having a very high genetic distance (Nei's $D = 1 \cdot 96$). A number of natural hybrids has been detected by electrophoresis (one male and 18 females out of more than 2000 adult specimens tested), but they do not allow gene flow between the two species (Biocca *et al.*, 1978; Bullini *et al.*, 1978, 1981). *P. equorum* and *P. univalens* are the product of a very ancient process of speciation which affected a large part of the genome; a simple process of tetraploidization is ruled out both by the chromosome structure, which is completely different in the two species, and by the electrophoretic data. The extreme morphological similarity between *P. univalens* and *P. equorum*, which are genetically so much differentiated, is apparently due to their endoparasitic life in the same habitat (equine intestine).

The genetic distance between the man and pig ascarids, *A. lumbricoides* and *A. suum*, is much lower: $D = 0 \cdot 27$, indicating a more recent speciation process (Nascetti *et al.*, 1979; Bullini *et al.*, 1981).

A similar value of genetic distance ($D = 0 \cdot 33$) was found between two sibling species recently discovered in the genus *Anisakis, A. simplex A* and *A. simplex B* (Nascetti *et al.*, 1981). These two species have different intermediate hosts, consisting of fishes of the genera *Trachurus, Micromesistius* and *Merluccius* for *A. simplex A* and of the genus *Scomber* for *A. simplex B*.

As pointed out above, the wide range of genetic distance (*D* from $0 \cdot 3$ to 2) found between Ascarid sibling species appears related to their time of evolutionary divergence, and not to different modes of speciation – geographic speciation being the most likely hypothesis for all the cases reported (Bullini, 1982).

Single-host ascarids such as *Ascaris, Parascaris, Neoascaris, Toxocara, Toxascaris* and *Ascaridia*, show a very low mean heterozygosity, ranging from 2 to 8%. On the contrary, in the *Anisakis* species, which need three or more different hosts both homothermous and heterothermous to develop, mean heterozygosity ranges from 15 to 20%. The higher genetic variability of multiple-host ascarid species is in contrast with the higher probability of bottle-neck events; it can be accounted for as being related with the adaptation to a much more time-varying environment.

SIBLING SPECIES OF PLETHODONTID SALAMANDERS AND THEIR PHYLOGENETIC RELATIONSHIPS

Another group of organisms in which a number of sibling species have been detected by electrophoresis is that of plethodontid salamanders (Yanev, 1978, 1980; Larson and Highton, 1978; Highton, 1979; Wake, 1981).

Only two European members of this family had been recognized *Hydromantes italicus* from extreme southeastern France, northern and western Italy, and *H. genei* from Sardinia, both with a number of geographic races. The electrophoretic study of various populations from continental Italy and Sardinia has provided a quite different picture (Lanza *et al.*, 1982). At least six distinct taxa are recognized, provisionally indicated as follows: *H. italicus, H. ambrosii, H. genei, H. flavus, H. imperialis* and *H. sp.*. Furthermore, the eastern Sardinian species *H. flavus, H. imperialis* and *H. sp.*, previously considered as belonging to *H. genei*, appear to be more related to the continental species *H. italicus* and *H. ambrosii* (Fig. 1). This last point is also supported by recent karyological observations indicating that both continental and eastern Sardinian species show heterogametic males, whereas no heterochromosomes were found in *H. genei* (Nardi *et al.*, unpublished).

Both Ascarid worms and Plethodontid salamanders provide instances of unrelated degrees of morphological and electrophoretic divergence. In the reported examples, as in many others, this discordance appears due to the use for taxonomic purposes of an excessively low number of morphological characters, often reflecting phenomena of adaptive convergence and/or parallelism. In these cases, electrophoretic data appear to be more reliable than the "standard" morphological ones in establishing phylogenetic relationships among taxa.

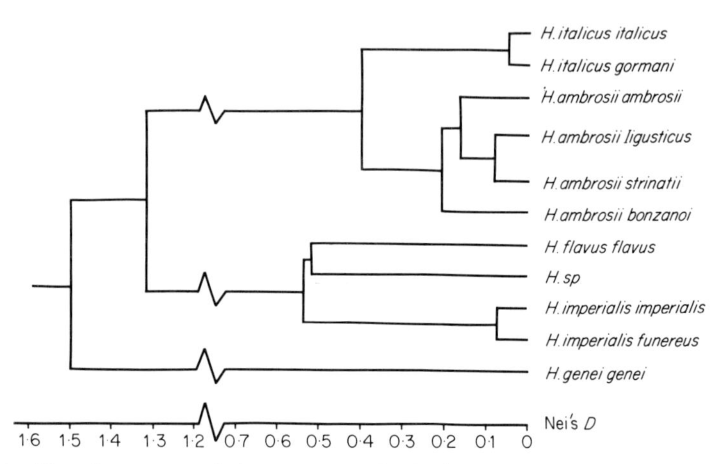

Fig. 1. Dendrogram of the presumed phylogenetic relationships among European taxa of the genus *Hydromantes* (UPGMA cluster analysis).

GENETIC DISTANCE AND REPRODUCTIVE ISOLATION BETWEEN ALLOPATRIC POPULATIONS

The separation of sympatric or partially sympatric sibling species by means of multilocus electrophoretic techniques is generally easy and fully reliable, the only exception being speciating taxa, not yet differentiated at the structural genes level. Also, in the case of allopatric populations of morphologically similar organisms, the electrophoretic approach provides essential information, such as the values of average genetic distance between taxa, when obtained on a sufficiently large and representative sample of loci. These values are often considerably higher between distinct species than between geographic races (Ayala, 1975; Ferguson, 1980). As suggested by Highton (Wake, 1981), the buildup of genetic distance is a better predictor than conventional morphology of whether two allopatric populations will interbreed upon recontact. Another more direct and effective approach to the problem of the taxonomic status of allopatric populations is represented by release experiments testing their reproductive isolation: electrophoretic variants offer good natural markers to identify the populations made artificially sympatric, and their possible hybrids. This approach was applied for the first time in the study of Mediterranean mosquitoes of the *Aedes mariae* complex (Coluzzi and Bullini,

1971; Coluzzi *et al.*, 1976; Bullini and Coluzzi, 1980, 1982). This includes three allopatric forms (*mariae, zammitii* and *phoeniciae*) slightly differentiated morphologically and showing various degrees of hybrid sterility. A series of release experiments, involving populations of the three forms monomorphic for alternative *Pgm* alleles, showed an almost complete absence of hybrids (Table I), and no evidence of introgression. These results indicated the existence of efficient pre-mating reproductive isolation mechanisms (RIMs), which are not apparent in the laboratory, and demonstrated that *Ae. mariae, Ae. zammitii* and *Ae. phoeniciae* are good species.

Release experiments using biochemical markers were also performed to test the possible role of stenogamy in preventing gene flow between *Culex pipiens pipiens* and *C. pipiens molestus* (Urbanelli *et al.*, 1981; Bullini, 1982). We observed in the release areas homogeneous nuptial swarms consisting of *molestus* males near the ground, and of *pipiens* males at a height of 2–3m, near the foliage of trees. Very few hybrids were found between the two forms (Table II), while they are obtained in laboratory crosses in the expected numbers. The stenogamy of *C. pipiens molestus*, therefore, represents a very efficient premating RIM. A certain amount of gene flow between the two forms, preventing their complete reproductive isolation, is, however, allowed by "intermediate" populations, having both stenogamous and eurygamous individuals. Accordingly, a taxonomic status of semispecies has been proposed for *C. pipiens pipiens* and *C. pipiens molestus* (Bullini *et al.*, 1982).

ALLOZYME AND CHROMOSOME DIFFERENTIATION IN SIBLING SPECIES OF THE GENUS *ANOPHELES*

The presence of good polytene chromosomes in the genus *Anopheles* allowed us to compare the genetic differentiation found at the gene and at the chromosome level. The results obtained indicate a similar overall picture of phylogenetic relationships with the two approaches (see for instance Figs 2 and 3); however, no correlation exists between values of genetic distance and extent of chromosomal rearrangements (Bullini and Coluzzi, 1978). Relatively high values of genetic distance can be found between homosequential species, like *An. labranchiae* and *An. atroparvus* ($D = 0{\cdot}25$), while lower values of D are found to coexist

Table I. Absolute frequencies of *Pgm* genotypes in samples collected at different times after the release dates: 4 July 1970 (*Ae. mariae, Ae. zammitii*), 6 June 1971 and 10 June 1973 (*Ae. mariae, Ae. zammitii* and *Ae. phoeniciae*).

Date of sampling	*Pgm* genotypes AA	AB	AC	BB	BC	CC	Total tested
15 July 1970	166	–		48			214
19 July 1970	103	–		9			112
22 July 1970	105	–		18			123
26 July 1970	112	1		32			145
2 August 1970	93	3		70			166
9 August 1970	60	–		77			137
16 August 1970	155	1		144			300
23 August 1970	55	–		62			117
30 August 1970	235	2		87			324
6 September 1970	29	–		166			195
13 September 1970	27	–		180			207
20 September 1970	112	1		56			169
27 September 1970	118	2		64			184
4 October 1970	42	–		89			131
11 October 1970	2	–		10			12
20 June 1971	83	1	–	109	1	105	299
4 July 1971	162	1	–	145	–	116	424
16 July 1971	117	–	1	161	1	108	388
28 July 1971	96	–	–	116	1	86	299
9 August 1971	61	–	1	81	–	50	193
21 August 1971	71	1	1	94	2	62	231
5 September 1971	50	1	–	65	1	28	145
19 September 1971	54	1	–	90	–	60	205
24 June 1973	161	1	–	99	2	106	369
8 July 1973	121	–	–	214	1	127	463
20 July 1973	141	1	1	228	1	128	500
2 August 1973	86	1	–	174	2	122	385
18 August 1973	47	–	1	73	1	52	174
1 September 1973	42	1	–	121	–	35	199

Ae. zammitii: Pgm^A/Pgm^A; *Ae. mariae*: Pgm^B/Pgm^B; *Ae. phoeniciae*: Pgm^C/Pgm^C.

Table II. Absolute frequencies of *Pgm* genotypes in samples collected at regular intervals after the release (18 June 1973) of the $Pgm^{92/92}$ strain of *C. pipiens molestus* in a biotope inhabited by *C. pipiens pipiens*.

Date of sampling	*Pgm* genotypes 92/92	92/★	★/★	Total sampled	92/★ relative frequency
28 June	152	1	217	370	0·003
5 July	121	3	261	385	0·008
12 July	116	4	218	338	0·018
19 July	104	3	197	304	0·010
26 July	83	6	171	260	0·025
2 August	122	2	192	316	0·006
9 August	91	4	142	237	0·018
16 August	89	4	169	262	0·015
23 August	101	7	177	285	0·025
30 August	77	4	201	282	0·014
6 September	81	4	268	353	0·011
13 September	93	3	245	341	0·009
20 September	89	4	227	320	0·013
27 September	78	2	192	272	0·007
4 October	69	1	175	245	0·004
11 October	85	1	222	308	0·003
18 October	79	2	256	337	0·006
25 October	64	1	213	278	0·004
1 November	33	–	258	291	0·000
8 November	6	1	212	219	0·005
15 November	–	–	143	143	0·000

In the *C.p.pipiens* population Pgm^{92} was found only as a rare variant (frequency about 0·002).
★ represents a cumulation of all the *Pgm* alleles observed, excluding *92*.

with fairly high levels of chromosomal divergence, as between *An. gambiae* and *An. arabiensis* (D = 0·13). When the species of the *An. gambiae* and *An. maculipennis* complexes are compared, a higher number of chromosomal rearrangements is found among the former (Fig. 4), whereas gene differentiation is higher in the latter (Cianchi *et al.*, 1982). This can be correlated with differences in the speciation processes. In the *An. maculipennis* complex, speciation occurred over a longer period of time and in areas with greater opportunities for geographic isolation. This kind of speciation does not necessarily have to be accompanied by chromosomal rearrangements. On the contrary, in the speciation process of the *An. gambiae* complex, which

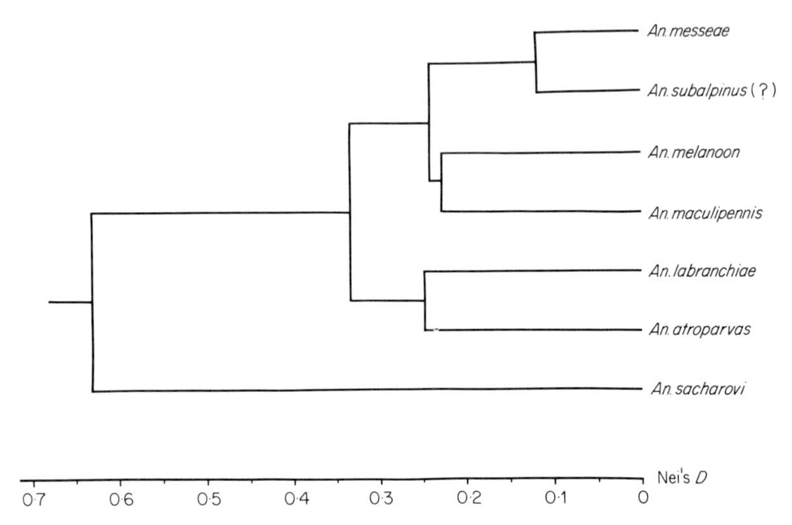

Fig. 2. Dendrogram of the presumed phylogenetic relationships among the Italian species of the *Anopheles maculipennis* complex (UPGMA cluster analysis).

occurred in the Afrotropical region in areas with fewer opportunities for geographic isolation, chromosomal rerrangements seem to have played a fundamental role in favouring the preservation of coadapted multilocus complexes selected in marginal niches. This appears to be confirmed by the recent discovery of a number of sympatric populations of *An. gambiae* (*sensu strictu*) showing partial reproductive isola-

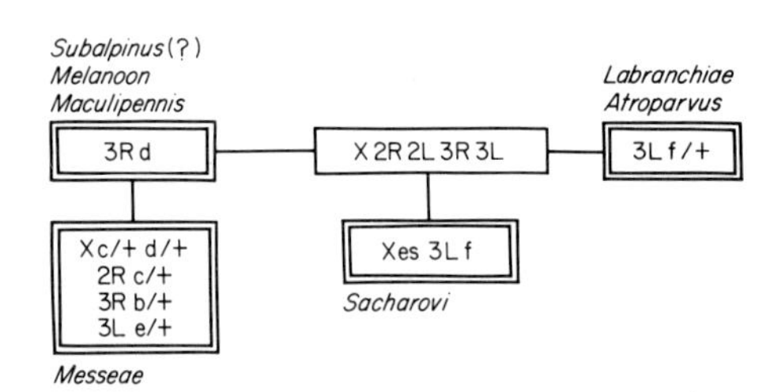

Fig. 3. Polytene chromosome relationships among the Italian species of the *Anopheles maculipennis* complex. Read the inversion formula for each species additively by following the lines leading from the standard chromosome sequence for the five polytene arms (X, 2R, 2L, 3R and 3L). Letters appearing singly represent fixed inversions, while polymorphic inversions are indicated with the symbol of heterozygosity (a/+).

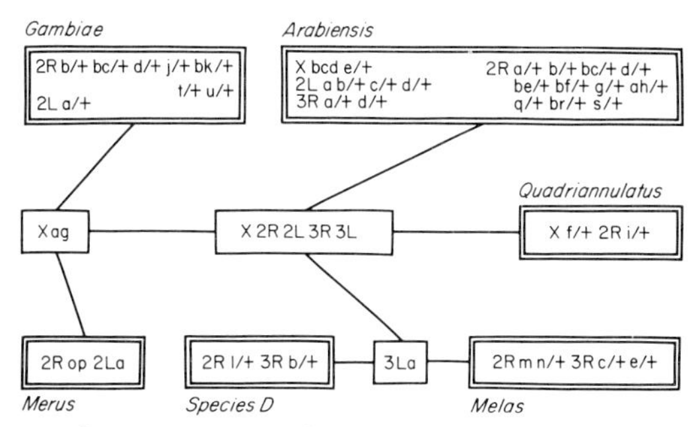

Fig. 4. Polytene chromosome relationships among the six species of the *Anopheles gambiae* complex (from Coluzzi *et al.*, 1979). (For explanations see Fig. 3).

tion, a very low differentiation at the gene level ($D = 0{\cdot}01$), and remarkably differentiated chromosomal rearrangements (Bryan *et al.*, 1982; Cianchi *et al.*, 1982; and unpublished data).

The following model has been suggested (Coluzzi, 1982). Consider a large population and a smaller one, living in an ecologically differentiated environment, the building up of an appropriate gene complex in the marginal population is broken down whenever the central population expands its range and overlaps it. If, by mutation, an inversion occurs in the marginal population, which protects the coadapted gene complex against such break-down, this inversion will be selected for and eventually get fixed in the marginal population. In the proposed model, inversions partially substitute geographic isolation in the evolution of populations. The coadapted gene complexes could also act as the first step in the setting up of premating barriers which, once fully evolved, will isolate the whole gene pools of the populations. Premating isolation mechanisms are very efficient in the species of both the *An. gambiae* and *An. maculipennis* complexes, while various degrees of hybrid sterility are shown by interspecific experimental crosses.

SPECIES OF HYBRID ORIGIN

Before the "allozyme revolution", reliable evidence in favour of a hybrid origin of some species was available only when its ancestors had differentiated karyotypes, and their haploid complements were both recognizable in the diploid hybrid. Also in these cases, however, the picture could be complicated by chromosome rearrangements occurring after hybridization. Now, hybridization between species may easily be detected by electrophoresis, since the hybrids are heterozygous for the loci diagnostic between the parental species. Using this approach, we discovered a number of cases of speciation by hybridization in both stick insects and snails, some of which are reported here.

Bacillus whitei is a diploid thelytokous species widespread in central and southern Sicily, whose hybrid origin was recently demonstrated on the basis of both allozyme and chromosome studies (Nascetti and Bullini, 1982a; Nascetti *et al.*, 1982). The bisexual ancestors of *B. whitei* are *B. rossius* and *B. grandii*. These two species are differentiated at many loci and their alleles coexist in *B. whitei* (Table III). Therefore, the mean heterozygosity per locus of *B. whitei* is very high: 67%. A second case of speciation by hybridization in stick insects is that of *Bacillus atticus* (Nascetti and Bullini, 1982b; Bullini, 1982). This species is also diploid and reproduces by thelytokous parthenogenesis. One of *B. atticus* bisexual ancestors is *B. grandii*, the second one is still unknown. Other parthenogenetic stick insects whose hybrid origin was revealed by electrophoresis are *Clonopsis gallica, Leptynia hispanica* and *Carausius morosus* (Nascetti and Bullini, 1982b; and unpublished data).

A hybrid origin was also recently evidenced by means of allozyme studies in the *Isidora truncata* complex (Nascetti and Bullini, 1980). These snails are tetraploid and reproduce by self-fertilization. The analysis of samples from southern Europe, Africa and the Middle East, showed that about ¼ of the 24 loci tested were fixed in the heterozygous condition. Possible ancestors of *I. truncata* appear to be *I. natalensis* and *I. tropica*, both diploid, showing many of the alleles found in the hybrid species.

CONCLUSIONS

The rapidly expanding application of multilocus electrophoretic techniques to the analysis of systematic and evolutionary problems is transforming taxonomic work. The researches summarized in the present paper appear representative of this situation. The building up

Table III. Genotype frequencies of 20 enzyme-loci in *B. rossius*, *B. grandii* and *B. whitei* from Sicily.

Loci	Genotype	*B. rossius* (Patti)	*B. grandii* (Noto)	*B. whitei* (Noto)	Loci	Genotype	*B. rossius* (Patti)	*B. grandii* (Noto)	*B. whitei* (Noto)
α-Gpdh	100/100	1·00	1·00	1·00	*Hbdh*	82/82	–	1·00	–
Sdh	100/100	0·10	–	–		82/100	–	–	1·00
	100/106	0·30	–	–		100/100	1·00	–	–
	106/106	0·60	–	–	*Sod*	82/82	–	1·00	–
	106/115	–	–	1·00		82/100	–	–	1·00
	115/115	–	1·00	–		100/100	1·00	–	–
Ldh	100/100	1·00	1·00	1·00	*Got-1*	92/92	–	1·00	–
Mdh-1	100/100	1·00	–	–		92/100	–	–	1·00
	100/107	–	–	0·96		100/100	1·00	–	–
	100/120	–	–	0·04					
	107/107	–	0·61	–	*Got-2*	100/100	1·00	1·00	1·00
	107/120	–	0·34	–	*Hk-1*	100/100	1·00	–	–
	120/120	–	0·05	–		100/104	–	–	1·00
Mdh-2	100/100	1·00	–	–		104/104	–	1·00	–
	100/105	–	–	1·00	*Hk-2*	100/100	1·00	1·00	1·00
	105/105	–	1·00	–	*Adk*	100/100	0·09	–	–
Idh-1	90/90	–	0·02	–		100/105	0·42	–	0·10
	90/100	–	0·12	0·05		105/105	0·49	1·00	0·90
	100/100	1·00	0·86	0·95	*Pgm*	100/100	0·90	–	–
Idh-2	92/92	–	1·00	–		100/105	0·10	–	1·00
	92/100	–	–	1·00		105/105	–	1·00	–
	100/100	1·00	–	–	*Ald*	100/100	1·00	1·00	1·00
6-Pgdh	100/100	1·00	–	–	*Mpi*	100/100	1·00	–	–
	100/106	–	–	1·00		100/110	–	–	1·00
	106/106	–	1·00	–		110/110	–	1·00	–
G3pdh	100/100	1·00	–	–	*Pgi*	73/73	–	1·00	–
	100/115	–	–	1·00		73/100	–		1·00
	115/115	–	1·00	–		100/100	1·00	–	–

of a systematic arrangement more consistent with the genetic relationships of organisms is expected as a major result of this new approach.

REFERENCES

Avise, J. C. (1975). Systematic value of electrophoretic data. *Syst. Zoo.* **23** 465–481.

Ayala, F. J. (1975). Genetic differentiation during the speciation process. *Evol. Biol.* **8**, 1–78.

Biocca, E., Nascetti, G., Iori, A., Costantini, R. and Bullini, L. (1978). Descrizione di *Parascaris univalens* (Hertwig, 1890) parassita degli equini e suo differenziamento da *Parascaris equorum* (Goeze, 1782). *Acc. Naz. Lincei, Rend. Cl. Sc. Fis. Mat. e Nat.* **65**, 133–140.

Bryan, J. H., Di Deco, M. A., Petrarca, V. and Coluzzi, M. (1982). Chromosomal polymorphism in *Anopheles gambiae* s.s. in The Gambia, West Africa. *Genetica* **59**, 167–176.

Bullini, L. (1982). Genetic, Ecological, and Ethological Aspects of the Speciation Process. *In* "Mechanisms of Speciation" (C. Barigozzi, ed.), pp. 242–264. Liss, New York.

Bullini, L. and Coluzzi, M. (1978). Applied and theoretical significance of electrophoretic studies in mosquitoes (Diptera: Culicidae). *Parassitologia* **20**, 7–21.

Bullini, L. and Coluzzi, M. (1980). Ethological mechanisms of reproductive isolation in *Culex pipiens* and *Aedes mariae* complexes (Diptera-Culicidae). *Monitore Zool. Ital,* n.s. **14**, 95–96.

Bullini, L. and Coluzzi, M. (1982). Evolutionary and taxonomic inferences of electrophoretic studies in mosquitoes. *In* "Recent Developments in the Genetics of Insect Disease Vectors" (W. W. M. Steiner, W. J. Tabachnick, K. S. Rai and S. Narang, eds) pp. 465–482. Stipes Publishing, Champaign, Illinois.

Bullini, L. and Sbordoni, V. (1980). Electrophoretic studies of gene enzyme systems: microevolutionary processes and phylogenetic inference. *Boll. Zool.* **47** (suppl.), 95–112.

Bullini, L., Nascetti, G., Ciafrè, S., Rumore, F. and Biocca, E. (1978). Ricerche cariologiche ed elettroforetiche su *Parascaris univalens* (Hertwig, 1890) e *Parascaris equorum* (Goeze, 1782). *Acc. Naz. Lincei, Rend. Cl. Sc. Fis. Mat. e Nat.* **65**, 151–156.

Bullini, L., Nascetti, G. and Grappelli, C. (1981). Nuovi dati sulla divergenza e sulla variabilità genetica delle specie gemelle *Ascaris lumbricoides – A. suum* e *Parascaris univalens – P. equorum. Parassitologia* **23**, 139–142.

Bullini, L., Urbanelli, S., Cianchi, R. and Coluzzi, M. (1982). Genetic divergence and taxonomy of European *Culex pipiens. Proc. V Int. Congr. Parasitol.* Toronto, 358.

Cianchi, R., Sabatini, A., Petrarca, V., Di Deco, M. A., Coluzzi, M. and

Bullini, L. (1982). Differenziamento genico e cromosomico nei complessi *Anopheles maculipennis* e *An. gambiae*. *Boll. Zool.* **49** (suppl.), 37–38.
Coluzzi, M. (1982). Spatial Distribution of Chromosomal Inversion and Speciation in Anopheline Mosquitoes. *In* "Mechanisms of Speciation" (C. Barigozzi, ed.), pp. 143–153. Liss, New York.
Coluzzi, M. and Bullini, L. (1971). Enzyme variants as markers in the study of precopulatory isolating mechanisms. *Nature, Lond.* **231**, 455–456.
Coluzzi, M., Bianchi Bullini, A. P. and Bullini, L. (1976). Speciazione nel complesso *mariae* del genere *Aedes* (Diptera, Culicidae). *Atti Ass. Genet. Ital.* **21**, 218–233.
Coluzzi, M., Sabatini, A., Petrarca, V., Di Deco, M. A. (1979). Chromosomal differentiation and adaptation to human environments in the *Anopheles gambiae* complex. *Trans. R. Soc. trop. Med. Hyg.* **73**, 483–497.
Ferguson, A. (1980). "Biochemical Systematics and Evolution". Blackie, Glasgow.
Highton, R. (1979). A new cryptic species of salamander of the genus *Plethodon* from the southeastern United States (Amphibia: Plethodontidae). *Brimleyana* **1**, 31–36.
Lanza, B., Nascetti, G. and Bullini, L. (1982). Tassonomia biochimica del genere *Hydromantes*. *Boll. Zool.* **49** (suppl.), 103.
Larson, A. and Highton, R. (1978). Geographic protein variation and divergence in the salamanders of the *Plethodon welleri* group (Amphibia, Plethodontidae). *Syst. Zool.* **27**, 431–448.
Nascetti, G. and Bullini, L. (1980). Genetic differentiation in the *Mandahlbarthia truncata* complex (Gastropoda Planorbidae). *Parassitologia* **22**, 269–274.
Nascetti, G. and Bullini, L. (1982a). *Bacillus grandii* n.sp. and *B. whitei* n.sp.: two new stick insects from Sicily (Cheleutoptera, Bacillidae). *Boll. Ist. Ent. Univ. Bologna* **36**, 245–258.
Nascetti, G. and Bullini, L. (1982b). Differenziamento genetico e speciazione in fasmidi dei generi *Bacillus* e *Clonopsis* (Cheleutoptera, Bacillidae). *Atti XII Congr. Naz. Entomol.*, Roma (in press).
Nascetti, G., Grappelli, C. and Bullini, L. (1979). Ricerche sul differenziamento genetico di *Ascaris lumbricoides* e *Ascaris suum*. *Acc. Naz. Lincei, Rend. Cl. Sc. Fis. Mat. e. Nat.* **67**, 457–465.
Nascetti, G., Paggi, L., Orecchia, P., Mattiucci, S. and Bullini, L. (1981). Divergenza genetica in popolazioni del genere *Anisakis* del Mediterraneo. *Parassitologia* **23**, 208–210.
Nascetti, G., Bianchi Bullini, A. P. and Bullini, L. (1982). Ricerche elettroforetiche e cariologiche su un fasmide partenogenetico di origine ibrida, *Bacillus whitei* e sui suoi progenitori bisessuati, *B. rossius* e *B. grandii*. *Boll. Zool.* **49**, (suppl.), 1330.
Urbanelli, S., Cianchi, R., Petrarca, V., Sabatinelli, G., Coluzzi, M. and Bullini, L. (1981). Adaptation to the urban environment in the mosquito *Culex pipiens* (Diptera, Culicidae). *In* "Ecologia" (A. Moroni, O. Ravera and A. Anelli, eds), pp. 305–316. Zara, Parma.

Wake, D. B. (1981). The Application of Allozyme Evidence to Problems in the Evolution of Morphology. *In* "Evolution Today", Proceedings of the II Intern. Congr. of Systematics and Evolutionary Biology (G.G.E. Scudder and J. L. Reveal, eds), pp. 257–270.
Yanev, K. P. (1978). "Evolutionary Studies of the Plethodontid Salamander Genus *Batrachoseps*." Ph.D. thesis, University of California, Berkeley.
Yanev, K. P. (1980). Biogeography and distribution of three parapatric salamander species in coastal and border land California. *In* "The California islands: Proceedings of a Multidisciplinary Symposium" (D. M. Power, ed.), pp. 531–554. Santa Barbara Museum of Natural History, Santa Barbara.

12 Relative Roles of Molecular Genetics, Anatomy, Morphometrics and Ecology in Assessing Relationships Among North American Unionidae (Bivalvia)

*G. M. DAVIS**

Academy of Natural Sciences of Philadelphia,
19th and the Parkway, Philadelphia,
Pennsylvania 19103, U.S.A.

Abstract: Protein variation and the ability to analyse it has added an important class of evidence for assessing relationships among molluscan individuals, species, genera and higher taxa. Others are those of comparative anatomy and conchology. The degree to which there is congruence among data sets has been used to demonstrate the relative value of each data set for taxonomic purposes. Ranking data sets from greatest to least value for assessing relationships yields: immunological and allozyme data > comparative anatomy > conchology.

I demonstrate the value of allozymes for distinguishing sympatric sibling species of the genus *Uniomerus* as has been done before for species of *Elliptio*. *Uniomerus* has been considered a monospecific genus. However, at least three species have been found on the basis of allozymes, yet only two are discernible using traditional conchology. Major classifications have been erected on the basis of conchology. Multivariate analyses of allozyme data and shell morphometric data are used to demonstrate that convergent evolution of shell character-states often masks genetic and anatomical differences between species, genera and higher taxa.

Congruence of different protein data sets has uncovered parallel evolution in some anatomical life history characters of unionids. In discarding such charac-

* This work was supported by a U.S. National Science Foundation grant DEB-01550, and in part by DEB 8118963.

Systematics Association Special Volume No. 24, "Protein Polymorphism: Adaptive and Taxonomic Significance", edited by G. S. Oxford and D. Rollinson, 1983, Academic Press, London and New York.

ters, anatomical and molecular data sets are brought into harmony for discerning major clades. However, the extent of divergence of the Anodontinae from the Margaritiferinae and Ambleminae was not recognized until protein data sets were produced.

Ecological and life history data are also essential for discerning species in the unionid radiation where genetic and anatomical variances are low.

INTRODUCTION

Assessments of relationships among organisms that lead to one's grouping of taxa into clades, and the establishment of classifications, continue as in the past to be made predominantly from anatomical data bases. In malacology, systematics has been dominated by conchology. There have been insufficient detailed anatomical studies to pursue modern cladistic analyses for most groups within any superfamily.

Establishment of clades, i.e. monophyletic lineages, depends on identifying convergences (Cain and Harrison, 1960). However, *a priori* (Cain, 1964, 1982) and *a posteriori*, convergence is probably the most underestimated phenomenon in systematic analyses (Davis, 1979, 1981a; Davis and Fuller, 1981; Davis and Silva, 1983). In particular, extreme problems are caused by conchological convergences (Davis, 1979; Davis and Silva, 1983).

Reliance on a single kind of data base for taxonomic purposes may lead to considerable error. There is no *a priori* way of knowing for a little-studied group of organisms which type of data may best serve to assess relationships. Convergence in anatomical features may go undetected. For this reason an integrated approach to systematics has long been advocated (Davis, 1973, 1979). Congruence among different data bases may reinforce a conclusion about cladistic relationships, but lack of congruence may point out errors in interpretation, convergences or other problems. However, the assumption is made that the probability is vanishingly small that there can be parallel convergence in different complex data sets.

Generalities drawn in this paper are based on the study of freshwater clams of the family Unionidae. Unionids comprise one of the most diverse macroinvertebrate radiations in freshwater. In North America alone there are about 50 nominal genera with over 225 species and subspecies currently recognized (Davis and Fuller, 1981). Of concern here is the subfamily Ambleminae, tribes Pleurobemini and Amblemini endemic to North America.

Shell characters alone have been the basis for unionid classification by some (e.g. Modell, 1942, 1949, 1964). Others rejected conchological data but relied on anatomical and reproduction characters (Heard and Guckert, 1971). Through analyses of protein variation it has been shown that there are both shell and reproduction character convergences (Davis and Fuller, 1981). With the elimination of convergent characters there is considerable congruence among anatomical and protein data sets (Davis and Fuller, 1981; Davis *et al.*, 1981). It is also evident that certain shell characters are associated with a given clade, e.g. pustulate or chevron-shaped pustulate sculpture on the shell disc is characteristic of many species of Amblemini (Davis and Fuller, 1981). The implications from these studies are that analyses of protein variation are of considerable use in unionid taxonomy and are perhaps of greater use than anatomical or conchological data alone.

Within unionids there are considerable problems in establishing the limits of species and genera. These problems stem, in part, from the apparent lack of qualitative anatomical characters, considerable inter-population shell phenotypic variability, uncertainty as to the extent of environmentally induced shell phenotypic variance, the tradition of describing taxa solely on the basis of the shell, and the uncritical placement of numerous nominal species into synonymy because of the general shell resemblance.

The primary purpose of this paper is to examine the relative values of protein variation, the shell, and comparative anatomy for assessing unionid relationships. The congruence of allozyme and shell morphometric data is examined. The study was prompted by finding sympatric sibling species in what was supposed to be a population of *Uniomerus tetralasmus* (Say), a monotypic genus (Johnson, 1970). After electrophoretic results showed that there were two distinctly different genotypes in this so-called population, re-examination of individually-numbered shells revealed two slightly different classes of shell phenotype.

A further purpose of this paper is to examine the relationships of *Uniomerus* and *Quincuncina* to other genera of the Ambleminae, tribes Amblemini and Pleurobemini. These relationships were not sufficiently clear on the basis of immunoelectrophoretic data because of insufficient information for *Quincuncina* (Davis and Fuller, 1981). On the basis of shell disc sculpture and overall shape, *Quincuncina* is most closely related to *Quadrula* of the Amblemini yet shares many anatomical characters with *Fusconaia* of the Pleurobemini. *Uniomerus* shares more characters with *Pleurobema* and *Elliptio* of the Pleurobemini (Davis and Fuller, 1981). Allozyme data resolve the relationships.

MATERIALS AND METHODS

1. *Taxa studied*

Eight species of six genera were studied. Names, classification, Academy of Natural Sciences (ANSP) catalog numbers, localities, and number of individuals electrophoresed are given in Table I. *Lampsilis teres* (Lt), tribe Lampsilini, is the outgroup comparator for taxa of the Amblemini and Pleurobemini. *Uniomerus declivis* and *U. excultus* are the sympatric species found in Mosquito Creek. Choice of specific names and reasons for these choices are given in the Appendix.

2. *Shell morphometrics*

Representative shells of the eight species are shown in Fig. 1. Ten shells of each population were scored for 39 characters, except where fewer than ten shells were available (Table II, Figs 2, 3). The largest shells were chosen when more than ten shells were available. Measurements were to determine shape parameters as determined by departure from a rectangle (Fig. 2). Shells were oriented on tracing paper such that a line drawn through the anterior and posterior adductor muscle scars (not figured) was parallel to the dorsal and ventral lines of the rectangle enclosing the shell (Fig. 2). The ligament was scraped away from the shell; the shell was traced on the tracing paper after proper alignment and most measurements were made from the tracing. Multivariate analyses were done with all 39 characters and then with only the first 25 characters. The two analyses were done to determine the extent to which shell characters other than measurements contributed towards distinguishing between taxa.

Table I. Species studied; catalogue numbers, localities, numbers of specimens electrophoresed.

Subfamily Ambleminae	Code	ANSP	Locality	No.
Tribe Pleurobemini				
1. *Elliptio complanata* (Lightfoot)	Ec^{8-2}	352824	Swartswood Lake, Sussex Co., New Jersey 41° 05′ N; 74° 50′ W	27
2. *Fusconaia flava* (Raf.)	Ff^{3}	350060	Mississippi River, 2 mi. SW of Stoddard, Vernon Co., Wisconsin. 43° 37′ N; 91° 02′ W	16
3. *Uniomerus carolinianus* (Bosc)	Uc	350750 stream bank 350751 stream centre	Mosquito Creek, Gadsden Co., Florida 30° 41′ N; 84° 51′ W	5
4. *Uniomerus declivis* (Say)	Ud	350750 350751	as in 3 above	3
5. *Uniomerus excultus* (Conrad)	Ue	353133	Buckhead Creek, Magnolia Springs State Park, Jenkins Co., Georgia. 32° 49′ N; 81° 55′ W	18
Tribe Amblemini				
6. *Quincuncina infucata* (Conrad)	Qi^{3}	348844	Ochlockonee River, Leon Co., Florida 30° 30′ N; 84° 28′ W	25
7. *Quadrula quadrula* (Raf.)	Qq^{1}	350057	as in 2 above	20
Tribe Lampsilini				
8. *Lampsilis teres* (Raf.)	Lt^{3}	348872	Apalachicola River, Gadsden Co., Florida 30° 40′ N; 84° 53′ W	24

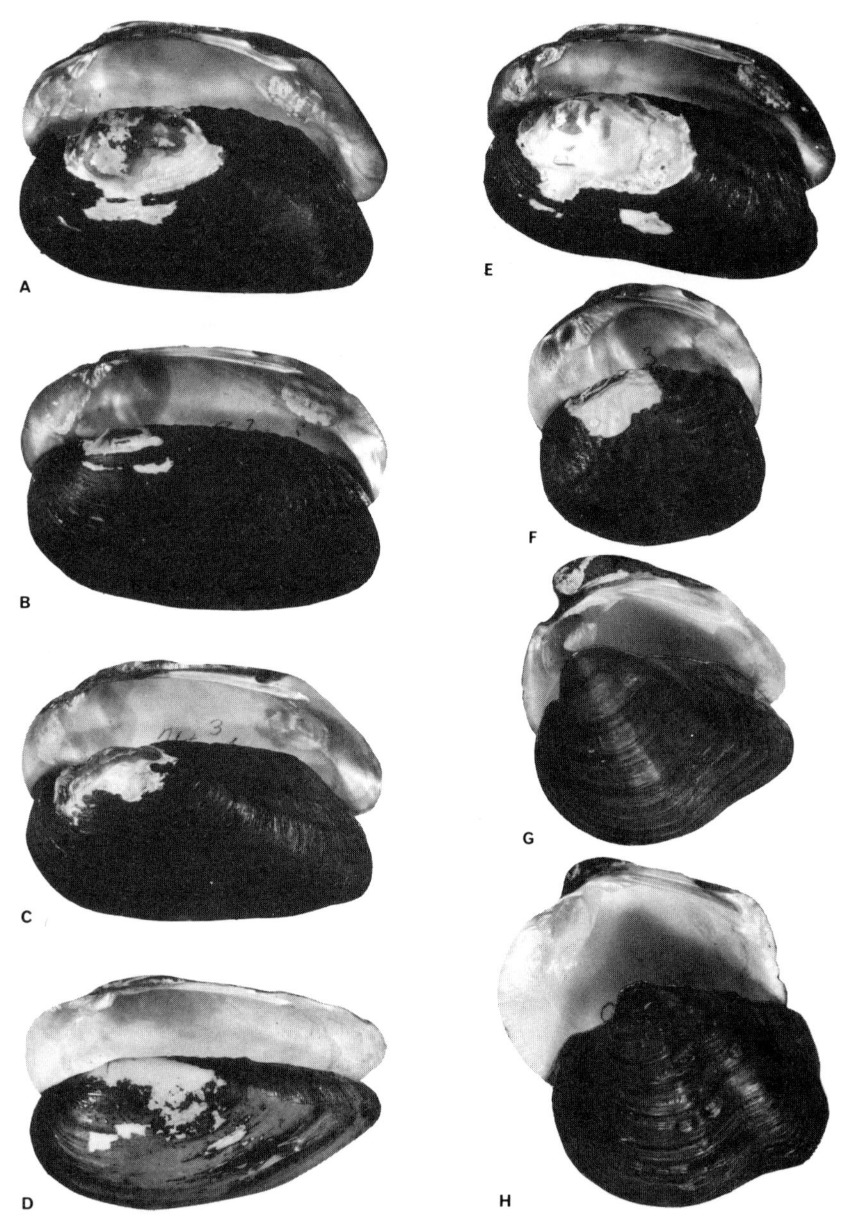

Fig. 1. Representative shells of the eight species studied. A, *Uniomerus declivis*, L = 79·5 mm, B, *Elliptio complanata*, L = 75·5 mm; C, *U. excultus*, L = 99·7 mm; D, *Lampsilis teres*, L = 112·5 mm; E, *U. carolinianus*, L = 83·0 mm; F, *Quincuncina infucata*, L = 41·0 mm; G, *Fusconaia flava*, L = 72·0 mm; H, *Quadrula quadrula*, L = 84·5 mm.

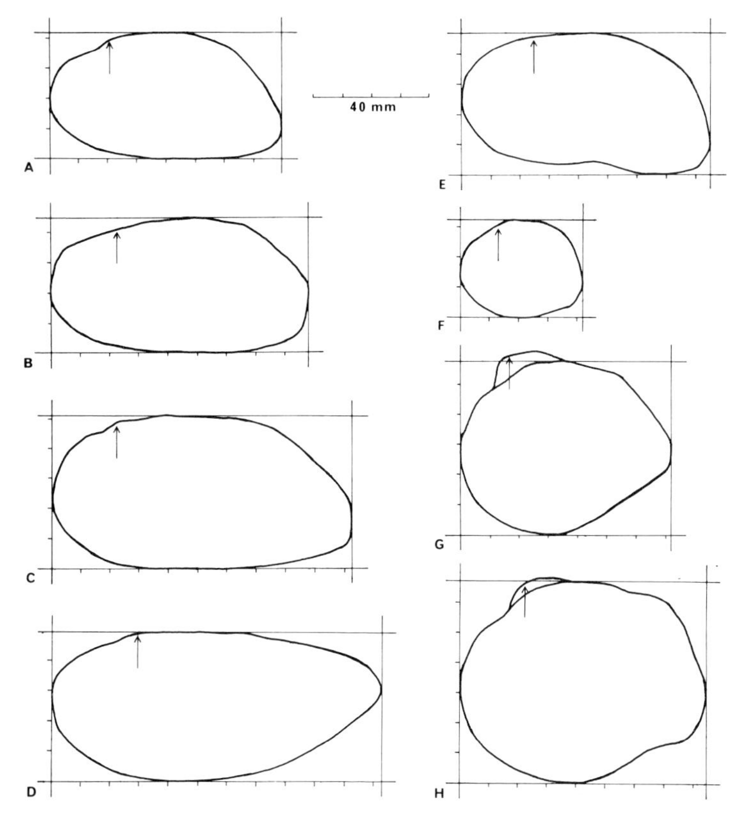

Fig. 2. Outlines of representative shells aligned on graph paper to facilitate measurements. See text for details, Fig. 1 caption for species names.

3. Anatomy

Anatomical characters brought into the Discussion are tabulated in Davis and Fuller (1981).

4. Allozymes

Electrophoretic procedures involving horizontal starch gels are those of Davis *et al.* (1981). The study involves 14 loci studied by Davis *et al.* (1981). Nei's genetic distance was calculated as were average individual heterozygosity (H) and polymorphism (P).

Table II. Shell characters used for morphometric analysis.

	Character	
1.	Shell length; standard malacological measure	(1,1)[a]
2.	Shell height; standard malacological measure	(2,2)
3.	Shell width; standard malacological measure	(3,3)
	Height measurements; departure from rectangle	
4.	1/2 distance from umbo to anterior end; ventral margin	(4,4)
5.	1/2 distance from umbo to anterior end; dorsal margin	(5,5)
6.	Ventral anterior tangent	(6,6)
7.	Dorsal anterior tangent	(7,7)
8.	3/5 distance from umbo to posterior end; ventral	(8,8)
9.	4/5 distance from umbo to posterior end; ventral	(9,9)
10.	Ventral posterior tangent	(10,10)
11.	Dorsal posterior tangent	(11,11)
12.	Posterior vertical contact	(12,12)
13.	Anterior vertical contact	(13,13)
14.	Dorsal departure from posterior vertical congruence	(23,22)
15.	Dorsal departure from anterior vertical congruence	(24,23)
	Length measurements	
16.	Dorsal congruence	(14,14)
17.	Ventral congruence	(15,15)
18.	Dorsal posterior departure	(16,16)
19.	Ventral posterior departure	(17,17)
20.	Dorsal anterior departure	(18,18)
21.	Ventral anterior departure	(19,19)
22.	Distance from umbo to anterior end	(20,20)
23.	Distance between centres of adductor muscle scars (posterior and anterior)	(21,21)
24.	Length of the lateral tooth	(26,25)
	Angle	
25.	Posterior angle	(25,24)
	Additional characters	
26.	Length from centre of ventral concavity to posterior end; scored "0" if no concavity	(22)
[a]27.	Posterior slope swelling. 0 = normally rounded as in *Elliptio complanata*; 1 = slightly swollen as in *Quincuncina infucata*; 2 = greatly swollen as in *Fusconaia flava*.	(27)
28.	Posterior slope sulcus. 0 = none as in *E. complanata*; 1 = one sulcus as in *Fusconaia flava*; 2 = two sulci as in *Quadrula quadrula*.	(28)
29.	Posterior slope; number of discernible raised lines; 0 = no lines; 1 = 1; 2 = 2; 3 = 3 lines.	(29)

Table II *continued*

[a]30. Inside nacre colour intensity. 0 = white; 1 = not white, slight colour; 2 = brightly coloured.	(30)
[a]31. Compression of valves. 0, none; 1, slight; 2, pronounced. This is the relative amount the sides of the shells – near centre – are depressed inward from expected contour. 60% = 0, 40% = 1 in *E. complanata*. 100% = 1 in *F. flava*; 100% = 2 in *Quadrula quadrula*; 100% = 0 in *Quincuncina infucata*; 100% = 0 in *Lampsilis teres, Uniomerus excultus*; 66% = 0, 33% = 1 in *U. declivis*; 100% = 1 in *U. carolinianus*.	(31)
[a]32. Beak cavity depth. 0, shallow; 1, moderate; 2, deep.	(33)
[a]33. Concavity of ventral margin. 0, none; 1, slight; 2, pronounced.	(32)
[a]34. Posterior slope ridge. 0, none; 1, slight; 2, pronounced.	(34)
35. Sculpture on shell disc. 0, none; 1, scattered pustules; 2, chevron-shaped sculpture.	(35)
36. Posterior fluting. 0 = none (all taxa except *Q. infucata*); 1 = with fluting.	(36)
37. Beak height. 0, not above hinge line; or mm above hinge line.	(37)
38. Pseudocardinal size; 0, slight; 1, massive.	(38)
39. Nacre colour. 0, white; 1, white-blue; 2, yellow-peach; 3, copper; 4, purple.	(39)

Measurements in mm. Refer to Fig. 3. [a] First number refers to character No. in Table III. Second number refers to character No. in Table IV. Numbers discussed in text are those to the left of the Table.
Subjectivity of scoring these character-states is emphasized.

5. *Analyses of data*

NT-SYS programs (Rohlf *et al.*, 1972) were used to analyse shell and allozyme data. For shell data, UPGMA was used to generate distance coefficients. There were 68 OTUs and 39 characters in the first matrix, 68 × 25 in the second. Principal component analysis (PCA), characters correlations and ordination of PCA projections of OTU's were done. The Prim Network and Subsets were used. For allozyme data, a matrix of Nei's genetic distances was the starting point for multidimensional scaling with the production of ordination diagrams. The Prim Network was used with Subsets.

Non-parametric statistics were used to compare immunology and allozyme data sets by comparing distances along the Prim Networks using Spearman Rank Correlation (Siegel, 1956).

Immunoelectrophoretic data are from Davis and Fuller (1981);

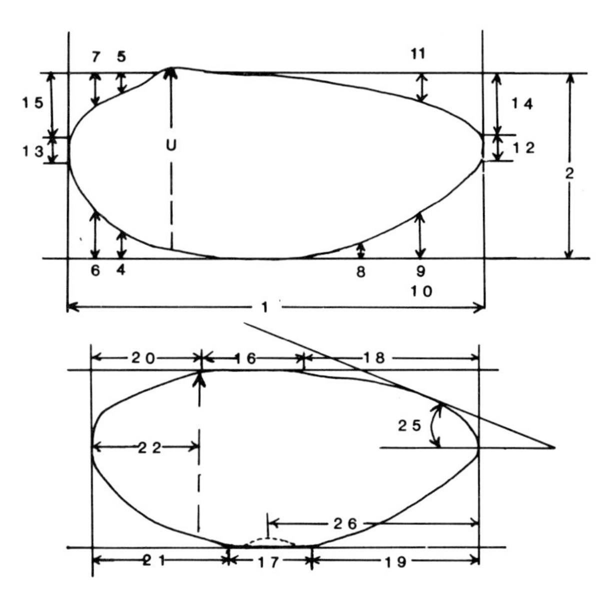

Fig. 3. Schematic diagrams showing how measurements were made. Numbers refer to characters listed in Table II.

allozyme data are from Davis *et al.* (1981) and this paper. The following seven species were used: *Margaritifera margaritifera, Anodonta cataracta, Lampsilis teres, Fusconaia flava, Elliptio complanata, Quincuncina infucata* and *Quadrula quadrula*. *Uniomerus* could not be included because of insufficient immunological data for this species, and because the population used in the immunological study may have been considerably different from the populations used in this study.

The Nei distances between species along the Prim Network were ranked (N-1 comparisons) and then corresponding immunological distances for the same pair-wise comparisons were recorded. Then, the immunological distances between species were ranked and the corresponding Nei distances were added. Significance was determined for each of the two rankings.

RESULTS

1. Shell characters

In the first analysis, the matrix correlation for taxonomic distances was 0·871. Eigen values dropped below 1·0 after six factors were identified accounting for 86% of the variance. Interpretation of factor loading (Table III) is done with analysis of the ordination diagrams (Fig. 4). Matrix correlation for distances × PCA projections on 3-dimensional space was 0·963. In Fig. 4 the first factor accounts for 34·77% of the variance, the second for 28·12% (total of 62·89%).

In the analyses of the reduced data set, the matrix correlation for taxonomic distances was 0·762. In PCA, eigen values dropped below 1·0 after four factors were identified accounting for 82% of the variance (Table IV). Matrix correlation for distances × PCA projections on 3-dimensional space was 0·961. In Fig. 5, the first factor accounts for 41·65% of the variance, the second for 22·03% (total of 63·68%).

In both analyses the following characters were so highly correlated (0·90 to 0·95) as to suggest reducing the character number by 6 characters. Shell length alone will suffice if characters 23, 24 and 18 are eliminated; shell height (No. 2) instead of shell height and shell width (No. 3); character 5 instead of both 5 and 7; character 8 instead of both 8 and 9.

Examining Figs 4 and 5, the results are quite similar. The main points are: *Fusconaia flava* clusters with *Quadrula quadrula* as one would expect examining the shapes of these taxa in Fig. 1. *Quincuncina infucata* is distinctly separated, and conchologically distantly related to *Fusconaia* and *Quadrula*. *Lampsilis teres* is clearly discerned with the help of the Prim Network. While individuals of *U. excultus* form a more or less cohesive group, there are no separate clusters for *Elliptio complanata, U. carolinianus* and *U. declivis*.

In comparing these two ordination diagrams, use of measurements alone (2nd analysis, 25 characters) is associated with a much cleaner separation of species, specifically the separation of *U. excultus* and *L. teres* (Fig. 5). Following along the Prim Network, on both diagrams not all individuals of the *Elliptio, U. declivis, U. carolinianus* groupings are connected with another of the appropriate species, i.e. there are misclassifications (Table V). *U. carolinianus* has a misclassification frequency of only 0·20 when all characters are included; this is increased with the reduced character set. Some shells of *E. complanata* and *U. declivis* are misidentified; some *U. carolinianus* look like *U. declivis*. Shells of *U. declivis* are easily confused with the other three species. Use of multi-dimensional scaling (MDS) yields the same mismatches for species of *Uniomerus* and *Elliptio complanata* as are seen in column 2, Table V. However, these three taxa are all connected

Table III. Factor loading on the first 6 components (% variance = 86·2). Matrix of 68 OTUs × 39 characters.

Character No. (Nos in parentheses in Table II)	Factors 1	2	3	4	5	6
1	0·282	−0·938	0·155	−0·021	0·071	−0·031
2	0·933	−0·301	−0·037	−0·097	0·093	−0·038
3	0·952	−0·107	0·033	−0·021	−0·071	0·035
4	0·837	−0·254	0·106	0·029	−0·033	0·061
5	0·777	0·244	−0·269	−0·308	−0·019	−0·081
6	0·755	−0·345	0·178	−0·119	0·211	−0·045
7	0·805	0·038	−0·130	−0·353	0·020	−0·087
8	0·719	0·143	0·560	0·128	−0·058	0·163
9	0·681	0·016	0·676	0·097	−0·031	0·091
10	0·762	0·097	0·483	0·206	0·197	−0·150
11	0·347	−0·261	−0·414	0·095	−0·177	−0·551
12	0·236	−0·103	−0·402	−0·639	0·024	0·353
13	0·373	−0·059	−0·186	0·141	−0·488	−0·196
14	0·110	−0·709	0·457	−0·049	0·062	−0·303
15	−0·249	−0·561	−0·299	−0·064	−0·168	−0·465
16	0·194	−0·874	0·177	0·015	−0·059	0·175
17	0·457	−0·592	0·388	−0·197	0·027	0·172
18	0·448	−0·647	−0·416	−0·029	0·246	0·021
19	0·188	−0·766	0·053	0·267	0·249	0·072
20	0·176	−0·838	0·284	0·070	0·332	−0·083
21	0·255	−0·946	0·130	−0·006	0·046	−0·026
22	0·196	−0·167	−0·646	0·489	0·124	0·208
23	0·633	−0·241	−0·656	−0·138	0·054	−0·117
24	0·828	−0·084	−0·253	−0·329	0·090	0·065
25	0·214	0·684	−0·559	−0·071	0·262	−0·134
26	0·309	−0·933	0·015	−0·096	0·058	−0·020
27	0·681	0·717	0·071	0·001	−0·065	0·011
28	0·877	0·381	0·011	0·102	0·168	−0·094
29	−0·004	−0·632	−0·114	−0·084	−0·605	0·184
30	−0·244	−0·434	−0·656	0·131	0·094	0·206
31	0·774	0·180	−0·383	0·266	0·210	−0·025
32	0·578	0·168	−0·392	0·516	0·099	0·177
33	0·681	0·717	0·071	0·001	−0·065	0·011
34	0·905	0·265	−0·165	−0·101	−0·174	0·070
35	−0·168	0·742	0·135	−0·168	0·482	−0·107
36	−0·527	0·658	0·138	−0·223	0·228	−0·005
37	0·774	0·442	0·021	0·115	−0·297	0·081
38	0·866	0·441	0·016	0·088	−0·152	0·012
39	−0·476	−0·627	−0·462	−0·021	0·090	0·129
% of variance	34·8	28·1	11·6	4·5	4·2	3·0

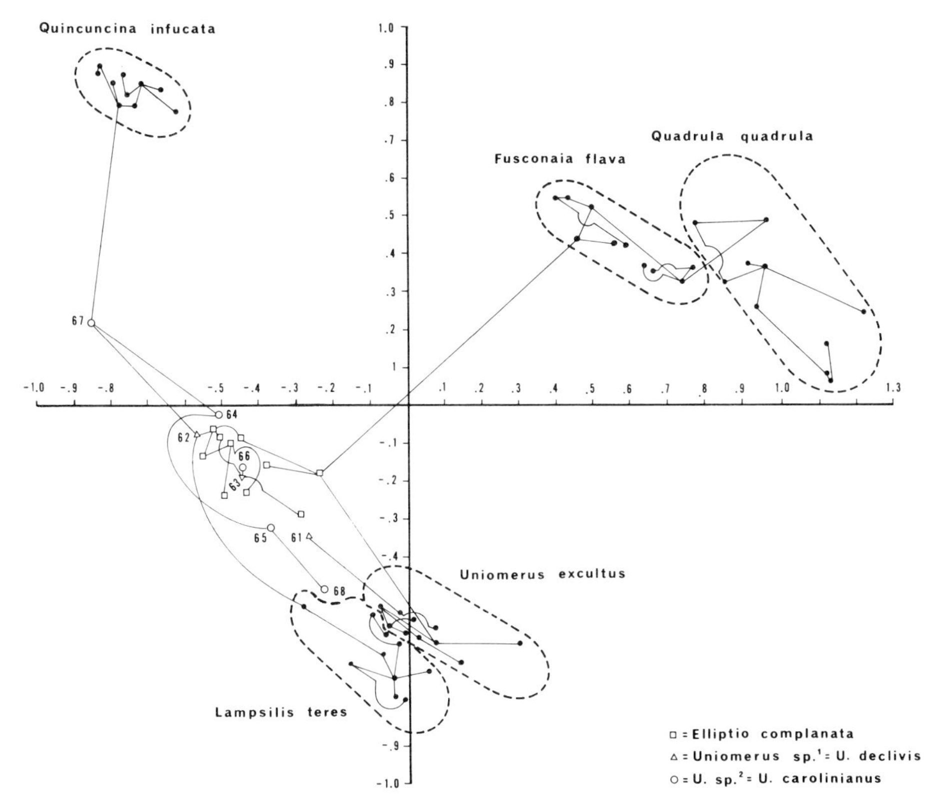

Fig. 4. Ordination diagram of OTUs on first two factors; with Prim Network. The dotted lines are hand-drawn around dots representing individuals of one species. From matrix of 68 OTUs, 39 shell characters.

among themselves and thus no taxa link to *U. excultus* along the MST from within the linkages of the 10 individuals of *U. excultus* as seen in Fig. 5 (one *E. complanata*, one *U. declivis*). While MDS separates *U. excultus* as an entirely separate group with only one linkage to another species, use of MDS results in mismatches between *Quadrula quadrula* and *Fusconaia flava* (Table V).

In the ordination diagrams, separation along the first axis is 1) overall size, small to large (left to right; excluding length in Fig. 4; including length in Fig. 5). Height measurements that indicate shape changes (characters 4–9, Table II) load highly on this axis; also dorsal departure from anterior and posterior congruences. Posterior slope

Table IV. Factor loading on the first four components (% variance = 81·8). Matrix of 68 OTUs × 25 characters.

Character No. (second number in parentheses in Table II)	Factors 1	2	3	4
1	0·835	− 0·542	0·052	− 0·016
2	0·909	0·387	0·061	0·006
3	0·788	0·512	− 0·073	0·054
4	0·802	0·332	− 0·119	0·045
5	0·436	0·785	0·207	0·016
6	0·829	0·197	− 0·115	− 0·131
7	0·601	0·624	0·133	− 0·064
8	0·502	0·445	− 0·669	0·037
9	0·570	0·295	− 0·723	0·030
10	0·569	0·422	− 0·555	0·146
11	0·382	0·096	0·463	0·542
12	0·227	0·224	0·497	− 0·621
13	0·290	0·249	0·154	0·641
14	0·586	− 0·549	− 0·234	0·058
15	0·130	− 0·535	0·540	0·213
16	0·724	− 0·547	− 0·006	− 0·033
17	0·780	− 0·183	− 0·277	− 0·175
18	0·728	− 0·111	0·506	− 0·076
19	0·636	− 0·486	0·050	0·029
20	0·716	− 0·585	− 0·083	− 0·073
21	0·815	− 0·561	0·074	0·001
22	0·578	0·376	0·657	0·072
23	0·679	0·555	0·257	− 0·199
24	− 0·315	0·745	0·395	− 0·052
25	0·840	− 0·481	0·196	− 0·044
% of variance	41·7	22·0	13·0	5·1

ridge (No. 34), beak height (No. 37), and pseudocardinal size (No. 38) also load highly on this axis, i.e. > 0·650 (Fig. 4 only). In Fig. 4, taxa to the right of axis 2 have increasingly prominent posterior slope ridges. To the far right are taxa with massive pseudocardinals and prominent beaks; to the far left, taxa with shell fluting.

The second factor (Fig. 4) is a shell sculpture-shape axis: taxa with sculpturing and fluting above axis one, smooth taxa below. Shell

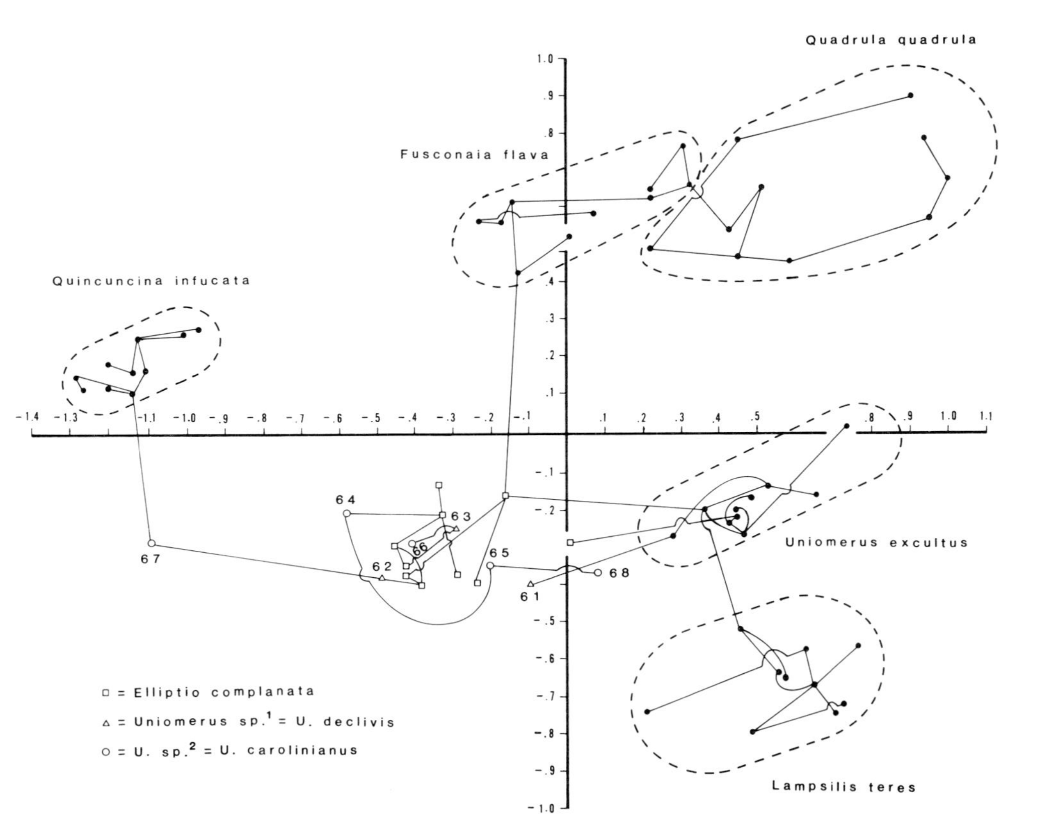

Fig. 5. As in Fig. 4, but from matrix of 68 OTUs, 25 shell characters.

Table V. The frequency of misclassification of individuals as one follows the Prim Network in Figs 4 and 5. Misclassification is with the species in ( ).

	PCA Projections		3-D scaling (not figured)
	Analysis from matrix 68 × 39	Analysis from matrix 68 × 25	matrix 68 × 25
Quadrula quadrula	0	0	0·30 (*Fusconaia flava*)
Fusconaia flava	0	0	0·10 (*Quadrula quadrula*)
Uniomerus excultus	0	0	0
Lampsilis teres	0	0	0
Elliptio complanata	0·10 (*U. declivis*)	0·10 (*U. excultus*)	0·10 (*U. declivis*)
U. carolinianus	0·20 (*U. declivis*)	0·40 (*U. declivis*)	0·40 (*U. declivis*, *E. complanata*)
U. declivis	1·00 (*U. excultus*, *E. complanata*, *U. carolinianus*)	1·00 (*U. excultus*, *E. complanata*, *U. carolinianus*)	1·00 (*E. complanata*, *U. declivis*, *U. carolinianus*)

length (and correlated characters) sorts along this axis, longest at the bottom grading to shortest at the top. Other length measurements important for shape (Nos 16–22, Table II) load heavily on this axis.

Examination of the second ordination diagram (Fig. 5) based on measurements alone, reveals a strong congruence with the first. Size, including length this time, dominates axis one, smallest to the left, largest to the right. Height loads the heaviest on axis 1 (0·909); the shortest shell to the far left grading to tallest to the far right (*Quadrula*). Shape separates taxa along the second axis, especially shape anterior to the umbo (characters 5, 7, 21, 22) and characters 16–18. The least amount of dorsal and ventral congruence coupled with greatest amount of downslope anterior to the umbo is at the top of the diagram with opposite trends towards the bottom.

2. *Electrophoresis*

Allele frequencies are given in Table VI. Nei's genetic distance (*D*) and identity (*I*) values are given in Table VII. Heterozygosity per locus, average individual heterozygosity (H), and frequency of polymorphic loci (P) are given in Table VIII. Multidimensional scaling based on Nei's *D* matrix yielded a stress of 0·001 after 50 iterations. Factor 1 accounted for 96·43% of the variance, factor 2, 2·77% (accumulative 99·2%). The matrix correlation (Nei's distances × distances in 3-dimensional space) was 0·840. The ordination diagram is shown in Fig. 6.

On genetic data, taxa group as predicted (Davis and Fuller, 1981). (1) *Lampsilis* is far removed from the set enclosing the other taxa. (2) *Fusconaia flava* and *Elliptio complanata* are closely related (as in Davis *et al.*, 1981). (3) *Quincuncina* and *Quadrula quadrula* are closely related and together diverge from the *Elliptio-Fusconaia flava-Uniomerus* grouping. (4) There are greater distances among species of *Uniomerus* than between species of different so-called generic pairs (*Elliptio-Fusconaia*; *Quincuncina-Quadrula*). (5) There is no congruence with conchological similarities except in the general relationships among species of *Elliptio* and *Uniomerus*. (6) While conchological similarities mask differences among some individuals of *Elliptio complanata* and the three species of *Uniomerus*, there is no such problem with allozyme data in the case here presented.

Table VI. Allele frequencies involving 14 loci and 70 electromorphs for the eight species.

Enzyme	Locus	Allele	Ec^{8-2}	Ff^{3}	Qq^{1}	Qi^{3}	Lt^{3}	Uc	Ud	Ue
GPI	I	18					0·02			
		17				0·02				
		15					0·98			
		14	0·20	1·0		0·98				0·17
		11	0·04						0·67	
		8	0·75					1·0	0·33	0·83
		2	0·02		0·65					
		− 4			0·35					
PGM^{I}	II	21			0·13					
		18	0·02	0·09	0·88	0·34		1·0	0·17	0·67
		15	0·66	0·91		0·66	0·98		0·83	0·33
		13					0·02			
		12	0·32							
PGM^{II}	III	34	0·07				0·02			
		32	0·11				0·08	1·0		0·11
		30	0·64	1·0			0·82		1·0	0·89
		28	0·14				0·08			
		26	0·02							
		24	0·02							
		22			0·23	0·08				
		20			0·70	0·58				
		18			0·08	0·34				
LAP	IV	40								0·36
		37						0·50	0·33	0·64
		34	0·14							
		33						0·50	0·17	
		32	0·23	0·17			0·02			
		31							0·50	
		30					0·74			
		29		0·06						
		28	0·57			0·08	0·18			
		26	0·04	0·17	0·93	0·70	0·06			
		24	0·02	0·33		0·08				
		22				0·02				
		21		0·27	0·08					
		20				0·12				

Table VI. *Continued*

MDH[I]	V	26								0·08
		23								0·31
		20						1·0		0·61
		17			0·35	1·0			1·0	
		15					1·0			
		14			0·65					
		11	0·04	1·0						
		8	0·96							
MDH[II]	VI	− 8	1·0	1·0	1·0	1·0		1·0	1·0	1·0
		− 11					1·0			
Hex	VII	36					1·0			
		34				0·94				
		31	0·32		0·03	0·06		0·90	1·0	1·0
		28	0·68	0·94	0·98			0·10		
		25		0·06						
MPI	VIII	28				0·04				
		26	1·0	1·0		0·86		0·80		0·94
		24			1·0	0·10		0·20	0·83	0·06
		23					1·0			
		22							0·17	
6PGD	IX	7		0·25						
		5	0·04	0·25	0·98	1·0		1·0	1·0	1·0
		3	0·96	0·34	0·03					
		1		0·16			1·0			
Oct	X	20			1·0	1·0	0·04			
		17	0·93	1·0			0·96	1·0	1·0	1·0
		13	0·07							
AAT	XI	12	1·0	1·0	1·0	1·0		1·0	1·0	1·0
		10					1·0			
SOD	XII	22						0·20		
		15	1·0	1·0	1·0	1·0	1·0	0·80	1·0	1·0
G3PDH	XIII	9						1·0		
		7	1·0	1·0	1·0	1·0	1·0		1·0	1·0
αGPDH	XIV	30	1·0	1·0	1·0	1·0	1·0	1·0	1·0	1·0

Table VII. Nei's *D* above the diagonal, *I* below the diagonal.

	Ec	Ff	Qq	Qi	Lt	Ue	Ud	Uc
Ec	–	0·208	0·698	0·582	0·866	0·303	0·436	0·578
Ff	0·812	–	0·642	0·437	0·773	0·341	0·428	0·725
Qq	0·498	0·526	–	0·321	1·409	0·597	0·517	0·732
Qi	0·559	0·646	0·725	–	1·215	0·476	0·458	0·742
Lt	0·421	0·461	0·244	0·297	–	0·915	0·823	1·513
Ue	0·739	0·711	0·550	0·621	0·400	–	0·216	0·209
Ud	0·647	0·652	0·596	0·633	0·439	0·806	–	0·498
Uc	0·561	0·484	0·481	0·476	0·220	0·811	0·608	–

3. Allozyme and immunological data set congruence

Comparing ordination diagrams of Davis *et al.* (1981) and Fig. 6 of this paper with that based on immunological differences (Davis and Fuller, 1981), it appears that there is good congruence among them. This is confirmed by the statistical analysis with $P = 0{\cdot}05$.

Table VIII. Heterozygosity (H) per locus, overall individual H, and P.

Enzyme	Ec^{8-2}	Ff^{3}	Qq^{1}	Qi^{3}	Lt^{3}	Uc	Ud	Ue
GPI	0·33	0	0·40	0	0·04	0	0	0·33
PGM^{I}	0·37	0·06	0·25	0·36	0·40	0	0·33	0·22
PGM^{II}	0·27	0	0·25	0·64	0·28	0	0	0·11
LAP	0·33	0·87	0·15	0	0·40	0·60	1·00	0·17
MDH^{I}	0·07	0	0·50	0	0	0	0	0·61
MDH^{II}	0	0	0	0	0	0	0	0
Hex	0·33	0·13	0	0·12	0	0·20	0	0
MPI	0	0	0	0·04	0	0·40	0·33	0·11
6PGD	0	0·50	0	0	0	0	0	0
Oct	0·13	0	0	0	0	0	0	0
AAT	0	0	0	0	0	0	0	0
SOD	0	0	0	0	0	0	0	0
G3PDH	0	0	0	0	0	0	0	0
αGPDH	0	0	0	0	0	0	0	0
Overall								
H	0·13	0·11	0·11	0·08	0·06	0·09	0·12	0·11
P	0·50	0·29	0·36	0·29	0·29	0·21	0·21	0·43

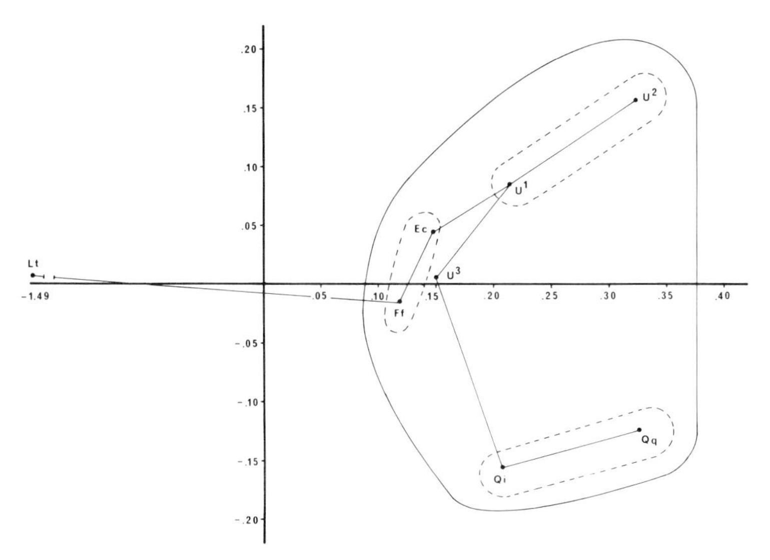

Fig. 6. Ordination diagram of OTUs on first two factors following multi-dimensional scaling. Solid and dashed lines are computer derived sets and subsets. U_1 = *Uniomerus excultus*; U_2 = *U. carolinianus*; U_3 = *U. declivis*. OTUs connected by Prim Network.

4. *Anatomy*

Distance coefficients separating the five genera of this study that belong to the Pleurobemini or Amblemini are given in Table IX. Only eight characters, two of them on the shell, serve to distinguish among the genera. Unequivocal close relationships are between *Quincuncina* and *Quadrula, Elliptio* and *Uniomerus*, *Elliptio* and *Fusconaia*. *Fusconaia*

Table IX. Distance coefficients based on percent difference between pairs of genera based on eight morphological characters (two of them shell characters) separating genera (from Davis and Fuller, 1981: 240, Table 13).

	Elliptio	Uniomerus	Fusconaia	Quincuncina	Quadrula
E	–	0·38	0·38	0·50	0·44
U		–	0·50	0·63	0·56
F			–	0·38	0·56
Qui				–	0·19
Qua					–

is more closely related to *Quincuncina* than it is to *Uniomerus*; *Quadrula* is more closely related to *Elliptio* than *Fusconaia* is to *Uniomerus*. The equivocal relationships of *Fusconaia, Quincuncina* and *Quadrula* to the other genera do not permit on the basis of morphology, clear separation of genera into the two divergent clades presented in Davis and Fuller (1981).

DISCUSSION

1. Species recognition and sample size

An objection will surely be raised by some that there were only three specimens of *Uniomerus declivis* and five of *U. carolinianus*. How can one rely on such small sample sizes to resolve issues of species status and relationships? One cannot say much about the population genetics of these species regarding H, P and allele frequencies; one can say, however, a great deal about their systematic relationships. They are sympatric congeners in an aquatic system where the potential for gene flow between them is undoubted. However, they are fixed for alternative alleles at three loci, and the overall genetic distance is very great, i.e. 0·498. It seems that there is no gene flow between these two forms.

To gain a more refined estimate of the genetic distance between two species than that based on only two individuals necessitates the use of 30 or more individuals (Sarich, 1977). Genetic distances are not much affected by sample size and thus a single individual from a wild population may be used to represent a species in such comparisons as done here (Gorman and Renzi, 1979). A sample size of two individuals will yield H within 2·5% of the H calculated for a much larger sample; there is $< 0{\cdot}1$ difference in D (Gorman and Renzi, 1979). The problem, then, is not one of sample size but the amount of D that clearly indicates specific difference.

2. Species concepts, genetic distance and species recognition

The life cycle of unionids involves a parasitic phase of larval glochidia infesting fish, and laboratory breeding is not possible. Dispersal is by fish transport. The clams are long-lived (mostly $\geqslant 7$

years) and grow slowly. It is not practical to contemplate breeding studies for genetic analyses. Reproductive isolation of populations may be assumed if populations are within completely separated drainages and distances between drainage outlets to the sea, coupled with high salinity barriers, preclude fish transport from one drainage to another. The biological species concept that places emphasis on reproductive isolation cannot be tested practically because the clams cannot be bred, in a practical way. The following species concept is used here: a species of unionid is a single lineage comprised of one or more populations that diverge from other lineages. Divergence is shown by significant morphological, cytological, reproductive biological and/or ecological differences. Differences between two lineages should be based on ten or more individuals of equivalent age per population. Equivalent age is based on counts of growth lines. The case for species status is strengthened if reproductive isolation is highly probable due to drainage system differences mentioned above.

Differences must be independent of environmental influence, i.e. must not be ecophenotypic. Shell characters are those presumably most affected by different environmental factors. Environmental influence can be assessed by correlation analysis or experimentally, procedures that may be time consuming yet are more practical than attempts at laboratory breeding. Small differences in size should not be used by themselves. While a single character may, indeed, be diagnostic in separating two species, the more characters that serve to separate lineages, the more convincing the argument for separation of lineages. Many single characters are controlled at a single genetic locus and thus involve intraspecific variability. For example, smooth versus ribbed shell sculpture in the gastropod *Oncomelania hupensis* is controlled at a single locus with multiple alleles (Davis and Ruff, 1974). Allometry and scaling are powerful tools coupling shell characters and anatomy to discern species (Davis, 1981b).

The complex of characters used here in the morphometric analysis serves to distinguish among species in many but not all instances. Differences in shape involve many characters that are not correlated. On the basis of shell morphometrics it is clear that *Uniomerus excultus* is a distinct species, morphologically separable from the Florida species of *Uniomerus*. There must be disrupted gene flow; the former is an Atlantic coast drainage species, the latter are Gulf coast drainage species. However, convergence limits the use of shell characters for discriminating among some species, as shown in the phenotypic similarities of shells of *Elliptio complanata, Uniomerus declivis* and *U. carolinianus* (Figs 4, 5). While some shells of the two Gulf coast species of *Uniomerus* are distinct, others grade into each other. Thus, a species concept based only on the shell is insufficient given our current state of knowledge.

Detailed comprehensive comparative anatomical studies for uncovering suites of characters that serve to discriminate among congeneric species have yet to be done. This field of investigation has been underused and requires development. Too often, one or two character-state differences between species have been used to erect genera (as reviewed in Davis and Fuller, 1981). The traditional naive assumption has been that there are too few anatomical characters for disciminating among unionid species. That this is not so has been demonstrated by discriminating among two species of sympatric lanceolate *Elliptio* on the basis of four anatomical characters and shell morphometrics (Coney *et al.*, 1982). The field of comparative anatomy in unionid systematics awaits full development. Given the constraints, conditions and indications discussed above, a species concept based on integrating shell morphometrics and detailed anatomy is probably justified, but problems with sibling species remain. These are being resolved using molecular genetics.

In one direction it is clear that there are unionid species as well as vertebrate species for which $I \geqslant 0{\cdot}90$ (Davis *et al.*, 1981). In the other direction it is not known when I and D values always indicate specific status for unionids. When species are sympatric, as in the case of *U. carolinianus* and *U. declivis*, one can readily discern the lack of gene flow, and thus the maintenance of genetically separated lineages by means of fixation of alternative alleles or significant differences of gene frequencies within various loci. Likewise, nonoverlapping shell phenotypes or nonoverlapping ecological parameters of sympatric taxa serve to indicate the existence of different species (e.g. unionids in Lake Waccamaw, Davis *et al.*, 1981). The problem is with allopatric populations.

It has been pointed out that allozyme analyses cannot be used to justify specific status of allopatric populations except in those cases where there are profound differences involving unique loci or complete discordance over several loci (Davis *et al.*, 1981). One can, however, with present knowledge, state the I and D values, based on

14 or more loci, that have a high probability that two taxa of Unionidae or Sphaeriidae are different species. From empirical data for these families, the probability is good if I = 0·87 or less (D = 0·139), excellent if I = 0·80 or less (D = 0·222).

Mean values of D and standard deviations are given for pairwise comparisons of conspecific populations and congeneric species of freshwater bivalves contrasted with congeneric species comparisons of two genera of fish (Table X). *A posteriori* it is clear that (1) populations of species are 90% or more similar with only a few populations only 87% similar; (2) congeneric species may be 99.9% similar yet others may differ by varying amounts to the extent that some species pairs are only 32% similar; (3) average similarities between congeneric species range from 57 to 81% with considerable variances.

In summary, in the absence of direct breeding experiments, the unionid species concept is best based on integrating shell morphometrics, detailed and comprehensive anatomy, and ecology. In many cases shell morphometrics and geography suffice. Molecular genetics has brought to light sibling species. With genetic distances of $D \geqslant$ 0·139 the probability is good that distinct species are involved. In no case is the species concept based on genetic distance alone. The ability to discriminate among species should improve considerably with increased comprehensive comparative anatomy studies.

3. Genera, clades, and methods

The congruence between the immunology and allozyme data sets confirms the clades established by Davis and Fuller (1981). Clearly *Elliptio*, *Fusconaia* and *Uniomerus* belong in one clade, *Quincuncina* and *Quadrula* in another with *Lampsilis* in yet a third. As made clear in this study, one cannot group species into genera and clades on the basis of only shell characters given current techniques. Conchologically *Uniomerus* and *Elliptio* intergrade; *Fusconaia* and *Quadrula* cluster together; *Quincuncina* seems more related to *Elliptio* than to *Fusconaia*. These results are similar to those of Hornbach *et al.* (1980) studying the relationships between species of Sphaeriidae on the basis of conchology and allozyme data. They found that there was no congruence between allozyme and shell character data sets, and that similarities among

Table X. Mean values of D and standard deviations for pairwise comparisons of conspecific populations and congeneric species. I values in ( ).

Conspecific population comparisons	No. populations	$\overline{X} \pm$ S.D.	Range	Source
Unionidae				
Elliptio complanata	7[a]	0·068 ± 0·033 (0·934)	0·023–0·144 (0·98–0·87)	Davis *et al*. (1981)
E. icterina	9	0·097 ± 0·050 (0·908)	0·031–0·184 (0·97–0·83)	Davis *et al*. (1981)
Sphaeriidae				
Sphaerium straitinum	13	0·037 ± 0·026 (0·964)	0·001–0·106 (0·999–0·90)	Hornbach *et al*. (1980)
Congeneric species comparisons	**No. of species**			
Elliptio	7	0·210 ± 0·117 (0·81)	0·010–0·446 (0·99–0·64)	Davis *et al*. (1981)
Uniomerus	3	0·308 ± 0·165 (0·73)	0·216–0·498 (0·81–0·61)	this paper
Sphaerium	4	0·568 ± 0·310 (0·57)	0·282–1·060 (0·75–0·35)	Hornbach *et al*. (1980)
Fish				
Lepomis	10	0·626 ± 0·192 (0·53)	0·15–1·05 (0·86–0·35)	Avise (1977)
Notropis	47	0·619 ± 0·245 (0·54)	0·05–1·50 (0·95–0·22)	Avise (1977)

[a] Excluding populations 7A, B; Davis *et al*. (1981).

species based on shell characters were consistent with general size and shapes of the shells.

Anatomical data provide more insight into relationships than do conchological data. However, as pointed out in the Results, one cannot make a clean separation of genera into the divergent clades Pleurobemini and Amblemini given the anatomical data of six characters. One must await intensive searches for new characters to learn if, indeed, too few anatomical characters exist to delineate clades clearly, or if additional characters can be found. What characters do exist show the close relationship of *Elliptio*, *Uniomerus* and *Fusconaia*, and the close relationship of *Quincuncina* and *Quadrula*, and thus the two shell sculptural characters and six anatomical characters available together give a much better indication of relationships than could be assessed through shell characters alone.

Unequivocally, results from analyses of protein variation far surpass those from conchological or anatomical variation in this group for uncovering sibling species, discerning species, and assessing species groups and clades. The congruence of ordination diagrams based on different protein data sets (and hence different groups of genes) using different methods, shows the genetic relationships implied by the ordination results are correct. Given these results, it is possible to state that two greatly used key characters have undergone parallel evolutions. These characters, the number of marsupial demibranchs (tetragenous vs ectobranchous conditions) and length of breeding season, had been the basis for erecting higher taxa (Heard and Guckert, 1971). By not using these characters to assess relationships, and by arguing that the direction of anatomical evolution was from simple to complex, the cladistic result was generally congruent with the relationships based on protein variation (Davis and Fuller, 1981). In using an integrated approach to systematics it is clear that the greater the congruence among data sets, the greater the probability of being correct in assessing relationships.

The use of molecular genetics has made the following possible, which have not thus far been possible using anatomy or conchology alone. (1) The Anodontinae has been shown to be a clade independent of the *Margaritifera* clade (Margaritiferinae) and the Ambleminae clade (that includes the Lampsilini, Pleurobemini, Amblemini) (Davis and Fuller, 1981). (2) Allozyme electromorphs have served to identify sympatric sibling species (Davis *et al.*, 1981; this paper). (3) Species placed in synonymy on the basis of general shell resemblance have been shown to be distinct species (Davis *et al.*, 1981; this paper). (4) Divergent clades have been demonstrated.

Molecular genetics does not distinguish between allopatric species where genetic distances are low. While allopatric populations of different drainage systems separated by an I of 0·85 are probably distinct species, proof requires additional data.

ACKNOWLEDGEMENTS

The help of Dr William Heard in obtaining specimens from Florida was essential for this study. Caryl Hesterman provided the technical laboratory help and drafted the figures. The manuscript was typed by Mrs Cindy Bogan. The manuscript was read and criticized by Drs A. J. Cain and K. E. Hoagland.

REFERENCES

Avise, J. C. (1977). Is evolution gradual or rectangular? Evidence from living fishes. *Proc. Nat. Acad. Sci., U.S.A.* **74**, 5083–5087.

Cain, A. J. (1964). The perfection of animals. *Viewpts. Biol.* **3**, 36–63.

Cain, A. J. (1982). On homology and convergence. *In* "Problems of Phylogenetic Reconstruction" (K. A. Joysey and E. A. Friday, eds), pp. 1–19. Academic Press, London.

Cain, A. J. and Harrison, G. A. (1960). Phyletic weighting. *Proc. zool. Soc. Lond.* **135**, 1–31.

Coney, C. C., Moore, R. H. and Kool, S. P. (1982). Comparative shell morphometrics and soft anatomy in disjunct populations of *Elliptio lanceolata* (Lea, 1828) complex. *Bull. Am. Malacol. Union* **1981**, 37.

Davis, G. M. (1973). Integrated approaches to molluscan systematics. *Malacol. Rev.* **6**, 42–43.

Davis, G. M. (1979). Experimental methods in Molluscan Systematics. *In* "Pulmonates Vol. 2A" (V. Fretter and J. Peake, eds), p. 100–169. Academic Press, London.

Davis, G. M. (1981a). Introduction to the second international symposium on evolution and adaptive radiation of Mollusca. *Malacologia* **21**, 1–4.

Davis, G. M. (1981b). Different modes of evolution and adaptive radiation in the Pomatiopsidae (Prosobranchia: Mesogastropoda). *Malacologia* **21**, 209–262.

Davis, G. M. and Fuller, S. L. H. (1981). Genetic relationships among Recent Unionacea (Bivalvia) of North America. *Malacologia* **20**, 217–253.

Davis, G. M. and Ruff, M. (1974). *Oncomelania hupensis* (Gastropoda: Hydrobiidae): hybridization, genetics, and transmission of *Schistosoma japonicum*. *Malacol. Rev.* **6**, 181–197.
Davis, G. M. and Silva, M. C. P. (1983). *Potamolithus*: morphology, convergence, and relationships among hydrobioid snails. *Malacologia*, in press.
Davis, G. M., Heard, W. H., Fuller, S. L. H. and Hesterman, C. (1981). Molluscan genetics and speciation in *Elliptio* and its relationships to other taxa of North American Unionidae (Bivalvia). *Biol. J. Linn. Soc.* **15**, 131–150.
Gorman, G. C. and Renzi, J. (1979). Genetic distance and heterozygosity estimates in electrophoretic studies: effects of sample size. *Copeia* **1979**, 242–249.
Heard, W. H. and Guckert, R. H. (1971). A re-evaluation of the recent Unionacea (Pelecypoda) of North America. *Malacologia* **10**, 333–355.
Hornbach, D. J., McLeod, M. J., Guttman, S. I. and Seilkop, S. K. (1980). Genetic and morphological variation in the freshwater clam, *Sphaerium* (Bivalvia: Sphaeriidae). *J. Moll. Stud.* **46**, 158–170.
Johnson, R. I. (1970). The systematics and zoogeography of the Unionidae (Mollusca: Bivalvia) of the southern Atlantic slope region. *Bull. Mus. comp. Zool. Harv.* **140**, 263–450.
Modell, H. (1942). Das natürliche System der Najaden. *Arch. Molluskenk.* **74**, 161–191.
Modell, H. (1949). Das natürliche System der Najaden. 2. *Arch. Molluskenk.* **78**, 29–48.
Modell, H. (1964). Das natürliche System der Najaden. *Arch. Molluskenk.* **93**, 71–126.
Rohlf, E. J., Kishpaugh, J. and Kirk, D. (1972). "NT-SYS; Numerical taxonomy system of multivariate statistical programs." State University of New York, Stony Brook, New York.
Sarich, V. M. (1977). Rates, sample sizes and the neutrality hypothesis for electrophoresis in evolutionary studies. *Nature, Lond.* **265**, 24–28.
Siegel, S. (1956). "Nonparametric statistics". McGraw-Hill, New York.

APPENDIX

Nomenclature and choice of species names for Uniomerus *of this study*

Uniomerus tetralasmus (Say, 1830) is the designated type species of *Uniomerus* (see review of Johnson, 1970). No types of this species have been located (Johnson, 1970) and the illustration and description (Say, 1831, pl. 23) do not fit any of the phenotypes of the three populations here studied. I selected the oldest available names for species with figured material and types agreeing with the phenotypes of our populations. These were *U. carolinianus* (Bosc., 1801): *U. declivis* (Say, 1832); *U. excultus* (Conrad, 1838).

Uniomerus carolinianus (Bosc, L.A.G., 1801 [1802]). Histoire naturelle des coquilles. Vol. III, p. 142, pl. 23, Fig. 2 [figs seen in 2nd edition, 1824, Chez Verdière, Libraire, Quai des Augustins, no. 25, Paris].

Uniomerus declivis [Say, T., 1831 (1832)]. *Transylvania J. Med.* **4**, 527.
Type locality: Bayou Teche, Louisiana.
Say, T. 1832, American Conchology No. 4 [no pagination], pl. 35.
Syntypes, ANSP 41698.

Uniomerus excultus (Conrad, T. A., 1838) Monography of the family Unionidae, or naiades of Lamarck, (fresh water bivalve shells) of North America. No. 11, 99–100, pl. 55. Type, ANSP 20427 (lost). Type locality, New Orleans, Orleans Parish, Louisiana.

13 | Protein Variation and the Origin of Parthenogenetic Forms

J. LOKKI

Department of Genetics, University of Helsinki, P. Rautatiek. 13, SF-00100 Helsinki 10, Finland

Abstract: Parthenogenetic animals have a twofold reproductive advantage over sexually reproducing forms. In spite of this advantage, parthenogenesis is an exceptional mode of reproduction in the Animal Kingdom. It represents always a condition derived from sexually reproducing ancestral forms. There are evidently very many different mechanisms that have led to parthenogenetic reproduction. Electrophoretic separation of enzymes has provided a powerful tool in elucidating the origin of parthenogenetic lizards, fishes and insects. In addition, immunological methods have been applied in vertebrates to verify results obtained by electrophoresis.

INTRODUCTION

Asexual reproduction and seed development without fertilization (agamospermy) are common in plants. In contrast to this, higher animals reproduce mainly through the union of gametes. It is commonly held that sexual reproduction is essential for the maintenance of genetic variation. The view is based on the prevalence of sexual reproduction, in particular in animals. From this view, it follows that parthenogenetic or asexual forms evolve less effectively than their bisexual competitors. A recent review (Maynard Smith, 1978) summarizes the present state of knowledge on the importance of sex versus the more efficient asexual or parthenogenetic reproduction.

Systematics Association Special Volume No. 24, "Protein Polymorphism: Adaptive and Taxonomic Significance", edited by G. S. Oxford and D. Rollinson, 1983, Academic Press, London and New York.

In general, sexual reproduction is the primary mode in higher animals. Parthenogenesis always represents a derived condition. Theoretically, a transition to parthenogenesis doubles the efficiency of reproduction in each generation. In spite of this efficiency, parthenogenetic forms rarely undergo much morphological differentiation. The doubling of chromosome number, polyploidization, is common in parthenogenetic lineages (Lokki and Saura, 1980).

Parthenogenesis differs profoundly from the normal development of a fertilized egg. The transition process from bisexuality is by no means simple and the cytological processes involved in it are quite variable in different cases. Parthenogenesis can be divided into three main types (Suomalainen, 1950): apomictic (or ameiotic), automictic (or meiotic) and generative or haploid parthenogenesis.

In automictic parthenogenesis, the early stages of meiosis proceed normally. The chromosomes pair at zygotene, crossing over occurs between them, and they form bivalents. As a result of this reduction process, the chromosome number becomes halved. The basic somatic number is restored by a fusion of two nuclei. In sexual reproduction, the nuclei originate from two different animals of separate sex. In automictic parthenogenesis, the nuclei originate from a single individual. The restoration of the normal chromosome number in the eggs of animals with automictic parthenogenesis occurs in different ways. The halves of the divided chromosomes may remain in the same nucleus. This process is typical for certain insects (Nur, 1971). It results in complete homozygosity for all genes. The two central polar nuclei may fuse and the development of the egg cell proper is suppressed. The first polar nucleus may fuse with the nucleus of the secondary oocyte (cases reviewed in Suomalainen *et al.*, 1976). A mechanism common to many parthenogenetic vertebrates involves a premeiotic doubling of the chromosome number which is followed by a normal meiosis (Cuellar, 1971; White *et al.*, 1963).

An interesting feature often associated with automixis is pseudogamy. It involves a union of a sperm with the egg cell, but the sperm contributes nothing to the egg nucleus. The sperm is needed only to activate the development of the egg. Pseudogamy characterizes certain parthenogenetic worms, insects, fishes and amphibians (Christensen, 1960; Bogart, 1980; Lokki and Saura, 1980; Schultz, 1980).

In apomictic parthenogenesis, features associated with normal meiosis are absent. In general, only one division takes place in the eggs and it is normally a typical mitosis. Apomictic parthenogenesis represents a profound deviation from normal meiosis and it quite probably represents a condition derived through automictic intermediate stages. Apomictic parthenogenesis is a simple condition. As there is no recombination, an apomictic female produces offspring genetically identical with its mother and each other. The only genetic variation in the progeny of such a stamping machine is due to new mutations. The mutant types again perpetuate their kind.

Haploid or generative parthenogenesis characterizes certain animal groups. It is the rule in Hymenoptera. Eggs may develop either through fertilization or without it. Unfertilized eggs develop into males and fertilized eggs into females.

Cyclic parthenogenesis combines the features of both sexual and parthenogenetic reproduction. Typically, several parthenogenetically reproducing generations are followed by a bisexual generation. Such forms, e.g. cladocerans and aphids, combine the evolutionary advantages of sexual reproduction with the efficient reproduction associated with parthenogenesis.

PARTHENOGENETIC VERTEBRATES

Parthenogenetic vertebrates are believed to have arisen through hybridization of two related species. The parthenogenetic hybrid is in general triploid, with two genomes derived from one and one genome from the other ancestral species. The family Poeciliidae comprises the best known cases among fishes. Parthenogenesis occurs in two genera: *Poecilopsis* and *Poecilia*, both from Mexico. *Poecilopsis monacha* has hybridized with several species, e.g. *P. lucida* and *P. occidentalis*, to form pseudogamic derivatives (Schultz, 1977). They are in general designated by the mechanism of their origin, e.g. a *P. monacha-lucida* (diploid) or *P. 2 monacha-lucida* (triploid). Vrijenhoek (1972) established the genetic makeup through electrophoretic studies: i.e. that the sperm do not contribute anything to the offspring of pseudogamic females. Triploids always have two doses of an allele of one parent and one from the other (Vrijenhoek, 1975). The hypotheses concerning the origin of these forms have been verified through a laboratory hybrid synthesis of parthenogenetic forms (Schultz, 1973); furthermore, tissue grafting studies have been used to supplement data derived by other means (Angus and Schultz, 1979).

In the genus *Poecilia* the situation is superficially similar with diploid and triploid parthenogenetic forms. Here the parthenogenetic species has been given a separate name, *P. formosa* (Abramoff *et al.*, 1968). Triploidy was established with electrophoretic techniques (Balsano *et al.*, 1972). There is evidence that *P. formosa* may not be a simple hybrid between *P. latipenna* and *P. mexicana*, as it contains unique alleles not present in either of these two presumed parental species (Turner *et al.*, 1980). Phenotypically similar parthenogenetic fish forms have evidently originated several times. They are geographically more widespread and ecologically less restricted than the bisexual parental species (Schultz, 1977).

The pseudogamic salamander species (*Ambystoma*) have originated in a fashion similar to that in Poeciliidae (Uzzell and Goldblatt, 1967; Bogart, 1980). The situation in amphibians is treated theoretically by Asher and Nace (1971).

Parthenogenetic lizards (*Cnemidophorus* and *Lacerta*) have also originated through hybridization of the bisexual species *C. tigris* and *C. septemvittatus*. The diploid parthenogenetic forms are polyphyletic, i.e. they have originated several times, while the triploids have originated only once (Parker and Selander, 1976). The parthenogenetic *Lacerta* species live in Caucasia. The situation is similar to that in the American *Cnemidophorus* (Uzzell and Darevsky, 1975). Triploid parthenogenetic females of *Cnemidophorus* and *Lacerta* may be fertilized in matings with diploid males. Diploid parthenogenetic females produce triploid offspring upon fertilization; again triploids give rise to tetraploids (Darevsky and Danielyan, 1968; Cole, 1979).

PARTHENOGENETIC INSECTS

There are large insect groups that are characterized by parthenogenetic reproduction, namely aphids and the Hymenoptera. Again, parthenogenetic forms are exceptional in certain other orders. In the Lepidoptera many flightless forms are characterized by parthenogenesis, again in the Coleoptera, such forms have been concentrated in certain subfamilies of weevils (Curculionidae). The majority of insects are bisexual (see Suomalainen *et al.*, 1976 for a compilation of established cases).

1. *Automictic parthenogenesis*

Representative cases of automictic parthenogenesis are the bagworm moths *Solenobia triquetrella* (Lokki *et al.*, 1975) and *S. lichenella* (Suomalainen *et al.*, 1981). The mechanism of automixis differs in the two species. In *S. triquetrella* the two central polar nuclei fuse, while the egg cell degenerates. This process should restore the entire parental genotype in each generation (Seiler, 1963). In *S. lichenella* the first polar nucleus fuses with the nucleus of the secondary oocyte (Narbel-Hofstetter, 1950). The process should have genetic consequences similar to those in *S. triquetrella*.

S. triquetrella comprises diploid bisexual, diploid parthenogenetic and tetraploid parthenogenetic races. The distribution of these races is typical for parthenogenetic insects in Europe. The diploid bisexual form lives in the Alps and adjacent areas (Seiler, 1961), the diploid parthenogenetic forms have spread somewhat further, while the tetraploid parthenogenetic forms have spread over central and northern Europe to about lat. 63° N (the situation is shown in Fig. 1). Electrophoresis proved to be a suitable method to study the relationships of these forms (Lokki *et al.*, 1975). Both diploid and tetraploid parthenogenetic populations proved to be distinct from each other. There

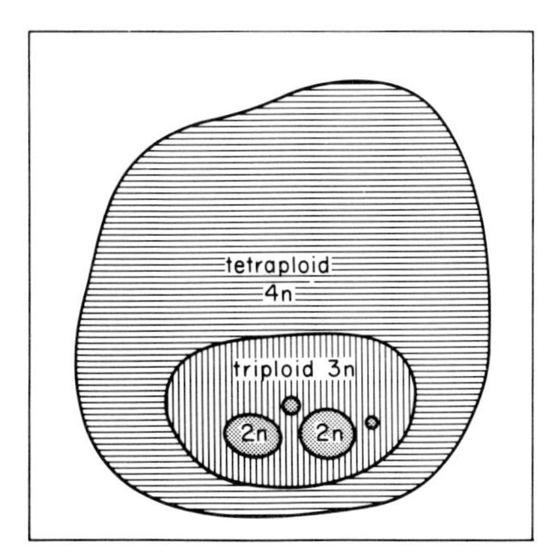

Fig. 1. A diagram of the distribution of different degrees of ploidy in a parthenogenetic insect species.

were, in addition, types that could be accounted for by mutations that had occurred in a parthenogenetic lineage following the parthenogenetic mode of reproduction. Evidently, parthenogenesis and polyploidy had originated several times in *S. triquetrella*. The allozyme composition of the bisexual populations was in accord with the ones in parthenogenetic populations. This agreed completely with Seiler's view on the multiple origin of parthenogenesis within this species (Seiler, 1964).

S. lichenella has no obvious bisexual ancestor. The enzyme phenotypes of populations from central and northern Europe are rather similar. In fact, they are so similar that we may be dealing with a monophyletic parthenogenetic lineage that has only diverged through mutations after the origin of parthenogenesis (Suomalainen *et al.*, 1981).

Certain *Drosophila* species have proved to be suitable material for the study of automictic parthenogenesis. *D. mangabeirai* is a species with obligatory parthenogenesis. The females of several species, e.g. *D. mercatorum*, lay numbers of eggs which give rise to parthenogenetic reproducing individuals (Templeton *et al.*, 1976). Templeton (1979) has collected evidence of the genetic changes that must accompany the transition to parthenogenesis. He considers it within the framework of his concept of "genetic revolution". It is, however, evident that vestiges of the "revolution" are not visible on the electrophoretically detectable level.

2. *Apomictic parthenogenesis*

Apomictic parthenogenesis represents a radical departure from the normal course of meiosis. There is evidence suggesting that apomixis has evolved through stages of automictic parthenogenesis (e.g. Seiler, 1947). Accordingly, it is the simplest and most advanced stage of parthenogenesis (see, however, Bernstein *et al.*, 1981 for a contradictory view).

Otiorrhynchus scaber is a representative case of apomictic parthenogenesis. The diploid bisexual race has a relict distribution in the eastern Alps. Bisexual populations occur only in localities in southeastern Austria that were not covered by ice during the last Ice Age – typically rugged and narrow canyons. Apparently the bisexual populations have been unable to spread much following the retreat of the Alpine ice sheet.

In contrast to this restricted distribution, the triploid parthenogenetic race has spread over the mountain ranges of central Europe. The tetraploid parthenogenetic race is distributed over the mountains of central Europe and has a continuous distribution over the conifer forests of northern Europe (cf. Fig. 1). Morphologically, the three degrees of polyploidy are different to the extent that they have been given separate names.

All three stages of ploidy (diploid bisexual, triploid and tetraploid) may occur together in a single Austrian population (unpublished). A single, apparently successful, tetraploid lineage has diverged through mutations into an array of separate biotypes that have colonized central and northern Europe (Saura *et al.*, 1976a). There are certain tetraploid types that have a very limited distribution.

The triploid populations also consist of some distinct lineages. Virtually all enzyme variation found in these lineages is found also in the diploids. Suomalainen and Saura (1973) argued, therefore, that the widespread parthenogenetic types were monophyletic. This is undoubtedly true. The local Austrian parthenogenetic types may be younger than the widespread type. They might, of course, also represent less adapted types.

The argument of monophyly in parthenogenetic organisms rests in the occurrence of several permanent heterozygotes. Several identical allele combinations cannot arise polyphyletically – monomorphism for common alleles would be less convincing. Similar monophyletic situations are found in numerous other species of beetles (Lokki *et al.*, 1976a, b; Saura *et al.*, 1976b).

The hypothesis for the origin of parthenogenesis in the above apomictic beetles rests on electrophoretic evidence and on earlier cytological observations (Suomalainen, 1940). Parthenogenesis would first proceed through a relatively brief diploid automictic stage, followed by apomixis and polyploidy. Alternatively, parthenogenesis would originate in an allopolyploid hybrid individual that would escape sterility through automictic parthenogenesis. The early stages of this automixis might be associated with pseudogamy. This course of events was probably followed in parthenogenetic fishes, lizards and very many plants. Now all available evidence favours the first hypothesis in beetles.

There are three diploid, 39 triploid, 17 tetraploid, 5 pentaploid and two hexaploid parthenogenetic races or species of weevils (Lokki and

Saura, 1980; Takenouchi, 1980a). The rarity of diploidy is striking in parthenogenetic weevils, which all seem to be apomictic.

Takenouchi (1980b) and Takenouchi *et al.* (1981) have subjected eggs of parthenogenetic weevil species to low temperature and observed that embryos with both higher and lower degrees of polyploidy result from this treatment. We have analysed, by electrophoresis, specimens from Japanese natural populations and the results agree with Takenouchi's observations: identical enzyme phenotypes are found in weevils with various degrees of polyploidy (unpublished).

Takenouchi (1976, 1981) has also attempted to cross diploid and polyploid parthenogenetic females with conspecific males or males of related species. Apparently the sperm contribute nothing to the developing eggs.

The above results explain certain phenomena that have been observed in our earlier studies. We have noted (Lokki *et al.*, 1976b) that an identical enzyme phenotype was found in diploid and triploid *Polydrusus mollis*. The distribution of triploid parthenogenetic forms is confined in central Europe, while diploid parthenogenetic forms have spread to northern Europe. The geographic distribution of these forms suggests that the diploid form would be the derived one. The distributions of the two other diploid parthenogenetic weevils do not contribute to the geographic distribution problem, as they are both confined to northern Japan.

There are, accordingly, at least two hypotheses to account for the origin of parthenogenesis and polyploidy in weevils, which is the insect group with by far the highest number of well-established parthenogenetic forms. The first one is a transition from bisexuality to diploid automictic parthenogenesis, followed by apomixis and polyploidization. The moth *Solenobia* would represent this course of events (even though it remains at an automictic stage). The second hypothesis envisages a situation, in which an unreduced diploid gamete is fertilized by a haploid one. This results in the formation of a triploid. The only escape from sterility for the triploid is parthenogenesis, the initial mechanism of which remains unknown. Templeton (Templeton *et al.*, 1976; Templeton, 1979) has argued that certain genotypes of *Drosophila mercatorum* are preadapted for parthenogenesis. This preadaptation is not, however, visible in the electrophoretically detectable phenotype. Now parthenogenesis is detected in weevils at the cytological and electrophoretical levels and potential preadapted complexes escape notice.

One may obtain a theoretical estimate for the increase of heterozygosity following the origin of apomictic parthenogenesis through a chance fertilization within a species (Lokki, 1976a, b). Assuming that the average degree of heterozygosity per locus per individual is 0·2, then the corresponding heterozygosity in a triploid is 0·36 and in a tetraploid 0·49. An inspection of Table I shows that the heterozygosity has, indeed, increased in polyploid parthenogenetic beetles

Table I. The observed degrees of heterozygosity per locus per individual in different diploid and polyploid populations of parthenogenetic insects.

Species	Bisexual race	Parthenogenetic races 2n	3n	4n	References
Bromius obscurus	0·18		0·34		Lokki *et al.* (1976a)
Polydrusus mollis	0·14	0·36	0·37		Lokki *et al.* (1976b)
Otiorrhynchus scaber	0·31		0·25	0·38	Suomalainen and Saura (1973) Saura *et al.* (1976a)
Otiorrhynchus salicis	0·12		0·24		Saura *et al.* (1976b)
Otiorrhynchus singularis			0·37		Suomalainen and Saura (1973)
Strophosomus melanogrammus			0·30		Suomalainen and Saura (1973)
Solenobia triquetrella	0·23	0·23		0·23	Lokki *et al.* (1975)

with the exception of *Otiorrhynchus scaber*. The degree of heterozygosity of diploid bisexual *O. scaber* is, however, too high. As for *Solenobia*, heterozygosity values for each of the three types and two grades of polyploidy are equal. This, of course, fits well with the postulated origin of parthenogenesis in *Solenobia* (Seiler, 1961). As for weevils with two different degrees of polyploidy, the results for *Polydrusus mollis* support the contention that triploidy has been the first parthenogenetic stage. In addition to the circumstance that the degrees of heterozygosity are virtually identical in diploid and triploid parthenogenetic races, their enzyme phenotypes are also very similar (cf. Lokki *et al.*, 1976b).

The distribution and ecology of parthenogenetic animals indicates that they often represent apparently successful forms that are capable of spreading further than their bisexual progenitors. This seems to hold true in such groups as vertebrates, insects and annelids. The evolutionary potential of parthenogenetic animals has probably been dismissed too lightly by stating that they represent evolutionary dead ends. The dismissal is based on the relative rarity of parthenogenetic forms among animals. The converse, the evolutionary advantage of sexual reproduction is not easy to prove (as exemplified by Maynard Smith, 1978).

REFERENCES

Abramoff, P., Darnell, R. M. and Balsano, J. S. (1968). Electrophoretic demonstration of the hybrid origin of the gynogenetic teleost *Poecilia formosa*. *Am. Nat.* **102**, 555–558.

Angus, R. A. and Schultz, R. J. (1979). Clonal diversity in the unisexual fish *Poeciliopsis monacha-lucida*: a tissue graft analysis. *Evolution, Lancaster, Pa.* **33**, 27–40.

Asher, J. H. and Nace, G. W. (1971). The genetic structure and evolutionary fate of parthenogenetic amphibian populations as determined by Markovian analysis. *Am. Zool.* **11**, 381–398.

Balsano, J. S., Darnell, R. F. and Abramoff, P. (1972). Electrophoretic evidence of triploidy associated with populations of the gynogenetic teleost *Poecilia formosa*. *Copeia* **1972**, 292–297.

Bernstein, H., Byers, G. S. and Michod, R. E. (1981). Evolution of sexual reproduction: importance of DNA repair, complementation, and variation. *Am. Nat.* **117**, 537–549.

Bogart, J. P. (1980). Evolutionary implications of polyploidy in amphibians and reptiles. *In* "Polyploidy, Biological Relevance" (W. H. Lewis, ed.), pp. 341–378. Plenum Press, New York.

Christensen, B. (1960). A comparative cytological investigation of the reproductive cycle of an amphimictic diploid and a parthenogenetic triploid form of *Lumbricillus lineatus* (O.F.M.) (Oligochaeta, Enchytraeidae). *Chromosoma* **11**, 365–379.

Cole, C. J. (1979). Chromosome inheritance in parthenogenetic lizards and evolution of allopolyploidy in reptiles. *J. Hered.* **70**, 95–102.

Cuellar, O. (1971). Reproduction and the mechanism of meiotic restitution in the parthenogenetic lizard *Cnemidophorus uniparens*. *J. Morph.* **133**, 139–165.

Darevsky, I. S. and Danielyan, F. D. (1968). Diploid and triploid progeny arising from natural mating of parthenogenetic *Lacerta armeniaca* and *L. unisexualis* with bisexual *L. saxicola valentini*. *J. Herpetol.* **2**, 65–69.

Lokki, J. (1976a). Genetic polymorphism and evolution in parthenogenetic animals. VII. The amount of heterozygosity in diploid populations. *Hereditas* **83**, 57–64.

Lokki, J. (1976b)., Genetic polymorphism and evolution in parthenogenetic animals. VIII. Heterozygosity in relation to polyploidy. *Hereditas* **83**, 65–72.

Lokki, J. and Saura, A. (1980). Polyploidy in insect evolution. *In* "Polyploidy, Biological Relevance" (W. H. Lewis, ed.), pp. 277–312. Plenum Press, New York.

Lokki, J., Suomalainen, E., Saura, A. and Lankinen, P. (1975). Genetic polymorphism and evolution in parthenogenetic animals. II. Diploid and polyploid *Solenobia triquetrella* (Lepidoptera: Psychidae). *Genetics* **79**, 513–525.

Lokki, J., Saura, A., Lankinen, P. and Suomalainen, E. (1976a). Genetic polymorphism and evolution in parthenogenetic animals. V. Triploid *Adoxus obscurus* (Coleoptera: Chrysomelidae). *Genet. Res., Camb.* **28**, 27–36.

Lokki, J., Saura, A., Lankinen, P. and Suomalainen, E. (1976b). Genetic polymorphism and evolution in parthenogenetic animals. VI. Diploid and triploid *Polydrusus mollis* (Coleoptera: Curculionidae). *Hereditas* **82**, 209–216.

Maynard Smith, J. (1978). "The Evolution of Sex". Cambridge University Press, Cambridge.

Narbel-Hofstetter, M. (1950). La cytologie de la parthénogénèse chez *Solenobia sp.* (*lichenella* L.?) (Lépidoptères, Psychides). *Chromosoma* **4**, 56–90.

Nur, U. (1971). Parthenogenesis in Coccoids (Homoptera). *Am. Zool.* **11**, 301–308.

Parker, E. D. and Selander, R. K. (1976). The organization of genetic diversity in the parthenogenetic lizard *Cnemidophorus tesselatus*. *Genetics* **84**, 791–805.

Saura, A., Lokki, J., Lankinen, P. and Suomalainen, E. (1976a). Genetic polymorphism and evolution in parthenogenetic animals. III. Tetraploid *Otiorrhynchus scaber* (Coleoptera: Curculionidae). *Hereditas* **82**, 79–100.

Saura, A., Lokki, J., Lankinen, P. and Suomalainen, E. (1976b). Genetic polymorphism and evolution in parthenogenetic animals. IV. Triploid *Otiorrhynchus salicis* (Coleoptera: Curculionidae). *Entomol. Scand.* **7**, 1–6.

Schultz, R. J. (1973). Origin and synthesis of a unisexual fish. *In* "Genetics and Mutagenesis of Fish" (J. H. Schröder, ed.), pp. 207–211. Springer-Verlag, Berlin.

Schultz, R. J. (1977). Evolution and ecology of unisexual fishes. *Evol. Biol.* **10**, 277–331.

Schultz, R. J. (1980). Role of polyploidy in evolution of fishes. *In* "Polyploidy, Biological Relevance" (W. H. Lewis, ed.), pp. 313–340. Plenum Press, New York.

Seiler, J. (1947). Die Zytologie eines parthenogenetischen Rüsselkäfers, *Otiorrhynchus sulcatus* F. *Chromosoma* **3**, 88–109.

Seiler, J. (1961). Untersuchungen über die Entstehung der Parthenogenese bei *Solenobia triquetrella* F. R. (Lepidoptera, Psychidae). III. *Z. VererbLehre* **92**, 261–316.

Seiler, J. (1963). Untersuchungen über die Entstehung der Parthenogenese bei *Solenobia triquetrella* F. R. (Lepidoptera, Psychidae). IV. *Z. VererbLehre* **94**, 29–66.

Seiler, J. (1964). Untersuchungen über die Entstehung der Parthenogenese bei *Solenobia triquetrella* F. R. (Lepidoptera, Psychidae). V. *Chromosoma* **15**, 503–539.

Suomalainen, E. (1940). Beiträge zur Zytologie der parthenogenetischen Insecten. I. Coleoptera. *Ann. Acad. Sci. fenn.*, A **54**, 1–144.

Suomalainen, E. (1950). Parthenogenesis in animals. *Adv. Genet.* **3**, 193–253.

Suomalainen, E. and Saura, A. (1973). Genetic polymorphism and evolution in parthenogenetic animals. I. Polyploid Curculionidae. *Genetics* **74**, 489–508.

Suomalainen, E., Saura, A. and Lokki, J. (1976). Evolution of parthenogenetic insects. *Evol. Biol.* **9**, 209–257.

Suomalainen, E., Lokki, J. and Saura, A. (1981). Genetic polymorphism and evolution in parthenogenetic animals. X. *Solenobia* species (Lepidoptera: Psychidae). *Hereditas* **95**, 31–35.

Takenouchi, Y. (1976). A chromosome study on the eggs obtained from the trial cross between a tetraploid parthenogenetic *Catapionus gracilicornis* and a bisexual *Catapionus obscurus* (Coleoptera: Curculionidae). *J. Hokkaido Univ. Education,* II B **27**, 1–3.

Takenouchi, Y. (1980a). A diploid parthenogenetic race of the weevil species *Catapionus gracilicornis* from Japan (Coleoptera: Curculionidae). *Entomol. Generalis* **6**, 367–369.

Takenouchi, Y. (1980b). Experimental study on the evolution of parthenogenetic weevils (Coleoptera: Curculionidae). *J. Hokkaido Univ. Education,* II B **31**, 1–12.

Takenouchi, Y. (1981). A chromsome study of eggs produced by trial crosses between diploid parthenogenetic females and diploid bisexual males of *Catapionus gracilicornis. Zool. Mag. Tokyo* **90**, 231–233.

Takenouchi, Y., Okamoto, H. and Sugawara, H. (1981). A study on the influence of low temperatures on the eggs of the tetraploid parthenogenetic *Catapionus gracilicornis* Roelofs (Coleoptera: Curculionidae). *J. Hokkaido Univ. Education,* II B **32**, 1–15.

Templeton, A. R. (1979). The unit of selection in *Drosophila mercatorum*. II. Genetic revolution and the origin of coadapted genomes in parthenogenetic strains. *Genetics* **92**, 1265–1282.

Templeton, A. R., Carson, H. L. and Sing, C. F. (1976). The population genetics of parthenogenetic strains of *Drosophila mercatorum*. II. The capacity for parthenogenesis in a natural bisexual population. *Genetics* **82**, 527–542.

Turner, B. J., Brett, B.-L. H., Rasch, E. M. and Balsano, J. S. (1980). Evolutionary genetics of a gynogenetic fish, *Poecilia formosa*, the Amazon molly. *Evolution, Lancaster, Pa.* **34** 246–258.

Uzzell, T. and Darevsky, I. S. (1975). Biochemical evidence for the hybrid origin of the parthenogenetic species of the *Lacerta saxicola* complex (Sauria: Lacertidae), with a discussion of some ecological and evolutionary implications. *Copeia* **1975**, 204–222.

Uzzell, T. M. and Goldblatt, S. M. (1967). Serum proteins of salamanders of the *Ambystoma jeffersonianum* complex, and the origin of the triploid species of this group. *Evolution, Lancaster, Pa.* **21**, 345–354.

Vrijenhoek, R. C. (1972). Genetic relationships of unisexual-hybrid fishes to their progenitors using lactate dehydrogenase isozymes as gene markers (*Poeciliopsis,* Poeciliidae). *Am. Nat.* **106**, 754–766.

Vrijenhoek, R. C. (1975). Gene dosage in diploid and triploid fishes. *In* "Isozymes. Vol. IV, Genetics and Evolution" (C. L. Markert, ed.), pp. 463–475. Academic Press, New York.

White, M. J. D., Cheney, J. and Key, K. H. L. (1963). A parthenogenetic species of grasshopper with complex structural heterozygosity (Orthoptera: Acridoidea). *Aust. J. Zool.* **11**, 1–19.

Part III | Adaptive Significance of Protein Variation

14 | Adaptive Significance of Protein Variation

E. NEVO

Institute of Evolution, University of Haifa, Haifa, Israel

THE PROBLEM

1. The analysis of nature

The structure of nature is often analysed in dichotomous alternatives, as rightly pointed out by Levins and Lewontin (1980). Is nature at equilibrium, or in constant change? Is the world causal or random? Is protein variation adaptive or neutral? Yet, nature is often complex, and although simplistic models and tests are indispensable in order to understand it conceptually, pluralistic rather than dichotomous alternative models best explain nature's complexity at both the physical and biological levels, particularly at the latter. While reductionist programmes are indispensable in our attempt to understand holistic structures, it should be recalled that parts and wholes interact consistently at all levels of organization, dictating a systems approach to the analysis of nature.

The history of evolutionary theory is notoriously abundant with often sterile but fierce conflicts, based on human tendency to view the world in sharp alternatives, between single principle explanatory models of organic nature (Wright, 1955). They involved the following, not always mutually exclusive, alternatives: mutationism vs

Systematics Association Special Volume No. 24, "Protein Polymorphism: Adaptive and Taxonomic Significance", edited by G. S. Oxford and D. Rollinson, 1983, Academic Press, London and New York.

selectionism, gradualism vs saltationism or micro- vs macromutationism, Lamarckism vs Darwinism or Mendelism, deterministic vs random genetic drift phenomena, allopatric vs sympatric speciation, essentialism vs populationism, molecular vs organismic or reductionist vs holistic biology; and now, phenetics vs evolutionary taxonomy, gradualism vs punctuationism, and neutrality vs selection at the molecular level.

Eventually, synthetic or multidisciplinary approaches often unify the dichotomous alternatives into richer synthetic theoretical structures as dramatically demonstrated by the synthetic theory of evolution (reviewed by Mayr and Provine, 1980), where variation and selection, stochastic and deterministic processes interact and drive evolutionary change (Wright, 1931, 1932). Ultimately, science, including evolutionary biology, strives at discovering unity in diversity, and human analytical and reductionist endeavours are aimed at creating a synthetic and unified world view.

2. *The evolutionary forces: their roles and importance*

The major forces causing evolutionary change in organic nature at both the phenotypic and molecular levels, are certainly natural selection, mutation, migration, and random genetic drift. Yet their roles and relative importance in the overall evolutionary process are either unknown or controversial even now, particularly at the molecular level. To what extent are the patterns and dynamics of allele frequencies determined by factors of selection, migration, recombination, as against those of initial conditions, founder effects, sample and random drift fluctuations?

The discovery of vast amounts of protein polymorphism in nature (reviewed by Powell, 1975; Selander, 1976; Nevo, 1978), primarily due to the spectacular developments of molecular biology and the application of electrophoresis in population genetics, did not resolve the problem. On the contrary, the debate was only transferred from the phenotypic to the molecular level, where evolution was explained contrastingly in Darwinian and non-Darwinian terms. Currently, two major theories compete for explaining genetic diversity in nature (Lewontin, 1974; Nei, 1975): the neutral and selectionist theories, each claiming priority in explaining polymorphisms and molecular evolution. A third theory, the shifting balance theory of evolution (Wright,

1970) attempts at combining the stochastic and deterministic forces into a synthetic evolutionary model.

3. The neutral theory

The neutral theory of molecular evolution (Kimura, 1968; Kimura and Ohta, 1971; Nei, 1975; and modifications in Kimura, 1979a, b) explains intraspecific variability and evolutionary change at the molecular level not by selection but by random drift of mutant genes that are selectively equivalent in functional efficiency. Neutrality theory allows for negative or purifying selection of deleterious mutations and also for the selection of a small proportion of advantageous mutations which cause slow adaptive phenomena, and which are primarily noticeable over geological time (Nei, 1975). However, it asserts that most molecular diversity in nature is nonselective and is maintained in populations through mutational input and random fixation. Hence, polymorphism is simply an incidental, transient and unimportant phase of molecular evolution. The level of observed genic diversity in nature is believed to be in rough agreement with the expected value of neutral mutations, when mutation rate is inferred from the rate of amino acid substitutions (Kimura, 1968, 1969, 1979a, b; Kimura and Crow, 1964; Ohta and Kimura, 1973; King and Ohta, 1975; Nei, 1975, 1980; Yamazaki and Maruyama, 1975).

The neutral theory holds that organic nature is sharply dichotomized in terms of the operating forces into two almost isolated levels. At the phenotypic level, largely deterministic positive Darwinian selection acts on phenotypes primarily through environmental effects on polygenes. In contrast, at the genotypic level, molecular polymorphisms are almost invisible to natural selection and are governed primarily by random genetic drift (Kimura, 1979b). The neutral theory, original or modified, continues the heritage of the classical view (pre-Mendelian Darwinism, and Muller, 1950) for which variation is constantly being removed by purifying selection and genetic drift, rather than being preserved in the population as asserted by the balanced view. In other words, while for neutrality theory protein variation in nature reflects essentially evolutionary noise, and has no significance in adaptation or speciation, for the selectionist theory it is the essence of organic nature and the basis of adaptive evolution.

4. The Selectionist Theory

The selectionist theory proposes a variety of selective mechanisms that can maintain genetic polymorphism similarly at both the phenotypic and molecular levels (see Hartl, 1980: 225, Table X, for 13 mechanisms that can maintain genetic polymorphism). Most of these are varieties of balancing selection mechanisms in which one homozygote is favoured under some conditions (i.e. niche, habitat, season, density, frequency, life cycle, fitness component, gamete, sex, group, etc.), but disfavoured under others. The most likely general selective mechanisms that can maintain genetic polymorphism are spatially and/or temporally heterogeneous environments, and epistasis (Hartl, 1980 and references therein; but see opposing view in Nei, 1980). Selectionists believe that a large amount of protein polymorphism is maintained in nature by some form or a combination of balancing selection mechanisms (i.e. Richmond, 1970; Clarke, 1970, 1975, 1979; Antonovics, 1971; Ayala, 1972, 1974, 1977; Ayala and Gilpin, 1974; Milkman, 1976, 1978; Wills, 1973, 1981; Gillespie, 1977a, b, 1978, 1979).

5. The shifting balance theory

The shifting balance theory, the synthetic theory of evolution developed by Wright (1931, 1932, 1970, 1977, 1978) is based upon the interaction of stochastic and deterministic forces operating at the genotypic level. The theory assumes three conditions: (i) populations are subdivided, (ii) most loci are polymorphic, and (iii) most loci interact epistatically. Evolution is considered to occur when a population moves from one adaptive peak on a multidimensional surface of fitness determined by many loci, to another higher adaptive peak, by gene frequency shift rather than by complete gene substitutions, in three major phases. First, random genetic drift predominates, and local populations explore the adaptive landscape producing by chance a new balanced genotype that is attracted to a higher fitness peak. Second, a mass-selection follows in favour of the new balanced genotype in which a peak shift has occured. Third, interdeme selection follows and the population attaining the higher fitness peak increases in size and through dispersion it shifts the allele frequencies in nearby local populations into the attraction domain of the higher fitness peak. Eventually, the favourable genotype spreads throughout the species until it is outcompeted by superior adaptive peaks, and so on throughout the entire adaptive topography. The shifting balance theory is certainly pluralistic. It attempts to synthesize the major forces of evolution and ascribe to each force a specific evolutionary role and importance. However, it remains unknown whether the theory indeed describes reality not only at the phenotypic but also at the molecular level. Recently, Wright (1978) stated that some data on protein polymorphism support this theory.

6. *Objective and outline of the review*

Now that the theoretical framework has been described, the specific problem to be analysed in this review can be outlined. Genetic diversity is doubtlessly central to the evolutionary process. However, while its origin is relatively clear, its maintenance and evolutionary significance remain enigmatic and controversial even now. It has been argued that the validity of the neutral theory should be tested by examining many loci and involve simultaneously the complementary aspects of protein polymorphism and molecular evolution (Nei, 1980; Kimura, 1976). I accept the challenge with two reservations. First, while I agree that ideally one should examine as many loci as possible, a complementary approach may be also favoured, namely, the analysis of few loci in as many unrelated species as possible. If certain loci tested in depth demonstrate an effect on fitness across higher taxa, this demonstration cannot be dismissed as trivial. By contrast, it must be accepted as a serious challenge to neutrality theory (Nevo, 1983a) as should also be the demonstration of parallel genetic patterns in sympatric species (Harrison, 1977; and other cases discussed later). Secondly, I trust that the complementary aspects of protein polymorphism and molecular evolution are best analysed separately due to their different time scales, structures, nature of evidence, and the different methodologies of their testabilities, such as directness of observation and experimentation. Notably, however, molecular evidence indicates nonuniform evolutionary rates in the same protein in different taxa negating the original concept of the neutral theory (Fitch and Langley, 1976). Furthermore, even if one insists on the linkage of polymorphism and molecular evolution, both can be analysed successfully under natural selection models (Van Valen, 1974; Gillespie, 1978, 1979, 1980).

The objective of the present review will, therefore, focus primarily on the evolutionary significance of protein polymorphism. In order to draw some general conclusions, I will review the following aspects: (i) tests of neutrality; (ii) evidence on protein frequency patterns in natural populations derived from field studies; (iii) evidence on biochemical and physiological properties of enzyme variants derived from laboratory studies; (iv) matching of theory and evidence.

THE THEORETICAL TESTING OF NEUTRAL MODELS

To assess the validity of the neutral theory, estimates derived from the theory must be compared with those from natural populations. A critical analysis of such tests appears in Ewens (1979) and other studies (i.e. Ewens, 1972; Ewens and Feldman, 1976; Weir *et al.*, 1976; Kingman, 1977; Watterson, 1977; Schaeffer *et al.*, 1977). The implicit assumptions of all tests involve panmixia and stationarity, without consideration of geographical subdivision. The analysis is based on three different mathematical models: (i) the "charge state" model of Ohta and Kimura (1973); (ii) the "infinite allele" model of Kimura and Crow (1964) and (iii) the "infinite site" model (Kimura, 1969). Caution should be exercised in matching the mathematical model and data set. Ewens (1979) concludes that a variety of practical circumstances are testable under the more complete testing theory existing for the infinite allele model, and that so far in all tests, neutrality is still rejected. A milder assessment is that of Hartl (1980) that, after many tests have been conducted, it is ambiguous whether actual data support or refute the neutral theory.

A contrary conclusion is drawn by Nei and colleagues (Nei, 1975, 1980; Nei *et al.*, 1976; Fuerst *et al.*, 1977) using in their analyses the relationship between the theoretical mean and variance of various quantities such as average heterozygosity for neutral genes as the null hypothesis to be tested against real data. Their general conclusion is that most data on protein polymorphism can be accounted for by the neutral theory. They admit, however (Nei and Graur, pers. comm.) that these tests were conducted without knowing the effective population size (N_e) and mutation rate (v) which are both estimated and involved in the testing. To circumvent this problem, they tested recently the relationship between mean gene diversity , *He* (Nei, 1975) and the product of crude estimates of population size (*N*) and generation time (*g*), and again found highly significant correlation between *He* and *Ng*. However, this conclusion ignores the causal relationship between genotype, environment and population numbers (Ayala, 1968). They criticize all other tests as not being sufficiently quantitative and as involving *ad hoc* assumptions. They also assert that since all other tests are based on some type of balancing selection, their allele frequency distribution is theoretically expected to be similar to overdominant selection (Maruyama and Nei, 1981; Nei, pers. comm.), which is almost never fulfilled. Their modified conclusion is that the available data on protein polymorphism are most easily explained by a modified form of neutral mutation theory in which the effects of bottlenecking and fluctuating selection are taken into account.

The neutrality test of Nei *et al.* (1976) has been applied to an electrophoretic data set of 38 species analysed in Israel (Nevo, 1983a) (Fig. 1). Obviously, most points appear below the lowest theoretical expected curve which is based on the stepwise mutation model. Twenty-seven out of 38 species showed lower variance in gene diversity (*He*) or higher *He* values than the theoretical expectation. The probability of obtaining 27 or more points below the curve is $P = 0{\cdot}007$. Assuming $P = 0{\cdot}50$ below the curve, a sign test indicates a significant deviation from the expected ($P < 0{\cdot}02$). Thus, the null hypothesis of neutral mutations is rejected in this case, whereas ecological and populational parameters (discussed later) explain a substantial amount of gene diversity.

The problem of discriminating between selective and non-selective models is difficult, although some outlets are available (Moran, 1976). Yet ambiguity in the results of tests derives from similar expectations of the neutral and selective models, particularly selection models involving spatially heterogeneous or temporally varying environments. Hartl (1980) concludes that, to date, no overall statistical test has provided convincing evidence either for or against the neutral theory. The recent flood of data of DNA polymorphisms may possibly endow future tests greater resolving power and efficiency (Ewens, 1979).

Finally, diverse, less formal tests were suggested to test the neutral theory. These include comparison of observed and expected fre-

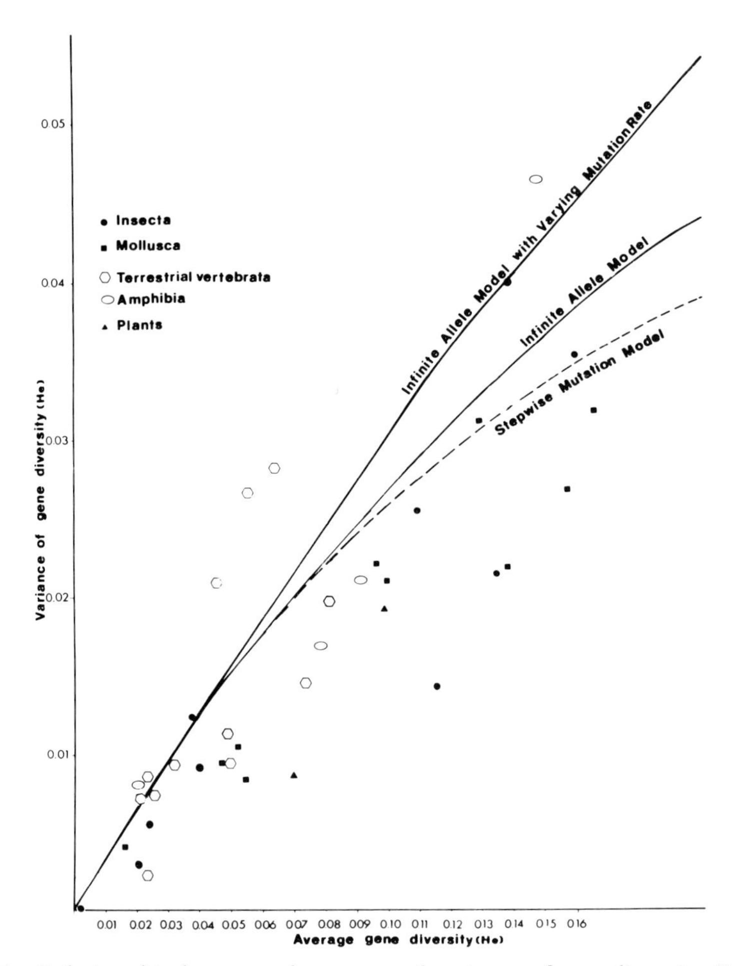

Fig. 1. Relationship between the mean and variance of gene diversity (*He*) of the 38 species described by Nevo (1983a).

quency patterns (Ewens, 1969; Maruyama and Yamazaki, 1974), tests based on estimated and observed values of θ (Ayala *et al.*, 1974), on genetic distance between taxa (Ayala, 1977) and tests correlating various genetic observations with environmental phenomena (Mitton and Koehn, 1973; Schaeffer and Johnson, 1974; Christiansen and Frydenberg, 1974; Nevo *et al.*, 1979 etc.). As indicated by Ewens (1979), such informal tests are often better than the formal procedures, especially in ecological circumstances where a useful mathematical theory is unlikely.

If current tests appear to have insufficient power to resolve the neutrality question, there are other promising and desirable alternatives. A possible approach is the development of a, not necessarily genetic, theory of phenotype (Lewontin, 1974), which allows direct correlation with the environment. This, as pointed out by Ewens and Feldman (1976), "clearly involves more ecology than population geneticists have been willing to use". Complementarily, the indirect correlative approach can provide promising guidelines for later indispensable direct biochemical, physiological and biological tests that may unveil the potential contribution of allozyme variants to fitness, to be discussed later.

THE EVIDENCE FROM PROTEIN POLYMORPHISM IN NATURE

Considerable evidence has now accumulated indicating that a substantial amount of protein polymorphism in nature is nonrandom and displays ecogeographical patterns. Description and discussion of allele frequency and polymorphism patterns of proteins in populations of plants, animals and humans appear in books (i.e. Cavalli-Sforza and Bodmer, 1971; Lewontin, 1974; Nei, 1975; Ayala, 1976; Wright, 1978; Ewens, 1979; Hartl, 1980; Wills, 1981) and in reviews (i.e. Powell, 1975; Selander, 1976; Nevo, 1978, 1983a, b; Hamrick *et al.*, 1979; Brown, 1979).

1. Macrogeographical patterns

The techniques demonstrating genic-environmental patterns include geographical mapping of allele frequencies (i.e. Bodmer and Cavalli-Sforza, 1976) and gene-environment associations (i.e. Bryant, 1974; Nevo, 1978). The correlative techniques involve two basic categories. First, linear regression and correlation analyses are used in many studies to test for geographic covariance between allele frequencies and environmental variables. Secondly, multivariate analyses, both metric and nonmetric, are used primarily on electrophoretic data involving multiple loci and many populations and species (i.e. Nevo, 1983a). They reduce or minimize dimensionality of a massive amount of complicated sets of data that are otherwise uninterpretable. In addition, multiple regresssion techniques are used to generate the best ecological variable-combination that may explain or predict the pattern of genetic variation.

The methodology of genetic-environmental correlation was extensively criticized on many grounds, including, among others, nonexistence, spuriousness, indirect relationship, nongenetic mechanism, noncausality, etc. (Schnell and Selander, 1981; Nei and Graur, pers. comm.). Caution in conducting and interpreting correlation analysis is certainly warranted (Clarke, 1978; Levins and Lewontin, 1980). However, it is a far cry between the necessary caution and operational safeguards, and the wholesale pessimism and negation of the technique. Genuine associations between allelic frequencies and quantitative environmental factors may indeed require the following considerations, among others: a sufficiently large number of independent replicates, choice and diversity of environmental factors and their proper quantification, randomization tests (i.e. comparisons of the correlations obtained with the correlations generated by randomizing the data set) etc. Furthermore, the associated factors discovered in the analysis should be retested in new settings in nature, including microgeographic analysis, but most importantly under controlled laboratory conditions. It should be recalled that genic-environmental associations are *post hoc* findings of past complex forces and processes, the final disentanglement of which may necessitate *pre hoc* experimentation to unveil unequivocally the cause-effect relationships (Nevo *et al.*, 1983b). In other words, although correlations are not causations (Wright, 1921), if genuine, they may be excellent guides to infer promising testable causations by direct experimentation to be discussed later. It should be obvious by now that the causes of genetic variation must be first explored in the ecology of organisms (Clarke, 1978) as promising guidelines to critical biochemical, physiological and biological experimentation to unveil fitness differentials among protein variants.

(a) Enzyme polymorphisms in plant populations The genetic structure, variation and correlates of life history characteristics of enzyme polymorphisms in more than 100 species of plants is now reviewed, with evolutionary and ecogenetic perspectives, exploiting the diversity of breeding systems characterizing plants (Allard *et al.*, 1978; Brown, 1978, 1979; Brown and Moran, 1979; Hamrick *et al.*, 1979) and permitting us to compare and contrast their pattern with those of animal

species. In general, a high level of genetic polymorphism seems to prevail in most species, while monomorphic natural populations occur particularly in inbreeding colonizers, or at extreme margins of distribution (Brown, 1978; Hamrick *et al.*, 1979). Domestication of cultivated species has generally reduced variation within populations. Hence, genetic resources might be essential in breeding programmes targeted at increasing the genetic basis of greater ecological amplitude and tolerance, disease resistance and other agronomically important characteristics (Brown, 1978, 1979; Nevo *et al.*, 1979, 1982b).

The maintenance of relatively high levels of genetic variation in plants is explicable on ecological and life history grounds (Hamrick *et al.*, 1979). First, plants are sedentary, hence, they are expected to respond to their environments as coarse grained. They may respond to spatiotemporal heterogeneity either through individual developmental plasticity (i.e. Bradshaw *et al.*, 1965; Bradshaw, 1972; Harper, 1977), or by the evolution of locally adapted phenotypes (i.e. Antonovics, 1971; Jain, 1976). Notably, inbreeding plant species show more intense geographic and microgeographic differentiation and multilocus association than outbreeders (Brown, 1979; Brown *et al.*, 1980). Secondly, plant population size, particularly that of perennials nearing climax, is often large and relatively stable, hence, less likely to be affected by genetic drift and more likely to maintain higher levels of genetic variation.

Genetic variation in plant species varies nonrandomly, being associated with ecological and life history characteristics (Hamrick *et al.*, 1979). Species characterized by large ranges, high fecundities, outbreeding, wind pollination, a long generation time, and nearing climax have more genetic variation than do species with contrasting characteristic profiles. Presumably, a diversity of genotypes permits range extension to a wider variety of microhabitats. Geographic differences among natural populations are less evident in trees than in herbaceous plants (Brown and Moran, 1979), but long living perennials contain the highest levels of allozyme variation, possibly explicable by differences in their life history characteristics (Hamrick *et al.*, 1979). Many tree species are characterized by long generation times, outcrossing, wind pollination, high fecundity and winged seed dispersal, in contrast to many herbaceous perennials characterized by higher turnover rates, a mixture of asexual and sexual reproduction, a range of pollination mechanisms and mating systems, and lower fecundities.

The former syndrome is associated with higher, the latter with lower amounts of genetic variation.

Geographic variation. Spatial allozyme variation in plants has been summarized and discussed by Brown (1979). Outbreeders are about twice as diverse per locus on average as compared to inbreeders. However, inbreeders exhibit far greater diversity between populations than outbreeders: about fivefold greater relative to their intrapopulation variation. Both inbreeder and outbreeder plants, however, are much more highly differentiated than are highly vagile animals. Spatial allozymic differentiation has been described in *Avena barbata* (Clegg and Allard, 1972; Kahler *et al.*, 1980), in *Hordeum spontaneum* and in *Triticum dicoccoides* (Nevo *et al.*, 1979, 1982c), and in *Oenothera* (Levy *et al.*, 1975; Levy and Levin, 1975) associated with climatic factors. Clinal patterns in allozymes have been described in *Lycopersicon pimpinellifolium* (Rick *et al.*, 1977) and in *Oenothera* (Levy *et al.*, 1975). By contrast, endemism of allozyme variants was described in *Lycoperiscon cheesmanii* (Rick and Fobes, 1975) in the Galapagos islands. Genetic differentiation in space is, therefore, distinct in plants involving local, regional and global differentiations. These patterns are often nonrandomly associated with environmental variables, primarily with climate and soil, but also with biotic factors. Finally, the niche width variation hypothesis which predicts a positive correlation between genetic and environmental variation (Van Valen, 1965) has been exemplified generally in plants (Hamrick *et al.*, 1979) as well as in specific cases such as in *Lupinus* (Babbel and Selander, 1974), but not in *Silene* (Baker *et al.*, 1975). In sum, the relationships between degree of environmental heterogeneity and genetic polymorphism are prevalent in plants, but they are often complex and their unravelling sometimes requires information of the genetic system on a multilocus basis coupled with the environmental response of populations (Allard *et al.*, 1978).

Maintenance of genetic polymorphism. The evidence derived from plants was critically discussed by Brown (1979). Various forms of balancing selection involving differential viability and fertility among homozygous genotypes based on unilocus and multilocus estimates were either found, suggested or hypothesized. The following biological parameters have been discussed among others: life cycle stages (Clegg and Allard, 1973; Schaal and Levin, 1976; Clegg *et al.*, 1978); environmental heterogeneity (Brown *et al.*, 1976; Steiner and Levin, 1977; Cavener and Clegg, 1978; Nevo *et al.*, 1981c); temporal heterogeneity (Kerster and Levin, 1970; Templeton and Levin, 1979); cross compatibility (Levin, 1973); heterostyly (Levin, 1975); zygotes (Workman and Jain, 1966); sexes (Kidwell *et al.*, 1977); mating structure (Clegg and Allard, 1973; Clegg *et al.*, 1976; reviews and references in Allard *et al.*, 1977 and in Brown, 1979); frequency (Jain, 1967; Jain and Jain, 1970), and disruptive selection (Stewart *et al.*, 1979), etc.

The evolutionary (Bradshaw, 1972), genetic (Levin, 1978) and demographic (Clegg *et al.*, 1978) consequences of being a plant are varied and complex. These involve, among others, strong and diverse selection, restricted migration, sedentariness in space and time, flexibility of the breeding system, and genetic organization. Demographic structures such as fecundity and reproductive schedules and modes, seed pool properties and diverse patterns of population subdivision affect the genetic structure. Large viability and fertility components of selection are revealed in annual plant populations, both determining total fitness. Since differential fitness is associated with different stages of life cycles, the measure of net selection over the entire life cycle underestimates differential fitness. Fitness must be recognized and studied as a functional and ecological problem, hence, its complexity depends on both environmental heterogeneity and developmental stages. Finally, the climax of selection for a multiplicity of alleles in plants is known in the pollen incompatibility loci discouraging self-fertilization (Wright, 1939).

(b) Enzyme polymorphism in animal populations The genetic structure, variation and correlates of ecological and life history characteristics of enzyme polymorphisms in which more than 14 loci have been studied are now available for more than 1100 animal species (Nevo *et al.*, 1983), a lower species sample was reviewed by Powell (1975) Selander (1976) and Nevo (1978). The levels of polymorphism (P) and heterozygosity (H) of 243 largely animal species in which 14 or more loci were tested are reviewed in Nevo (1978). The amounts of P and H vary nonrandomly between loci, populations, species, habitats and life zones, and are correlated with ecological heterogeneity. On

average, based on 242 species, $P = 0{\cdot}26 \pm$ s.d. $0{\cdot}15$, and $H = 0{\cdot}07 \pm$ s.d. $0{\cdot}05$. Loci coding for nonspecific enzymes utilizing external metabolites are genetically more diverse than their counterparts utilizing more specific internal metabolites (see also Johnson, 1979). Likewise, habitat generalist species are significantly more polymorphic than habitat specialist species.

Cosmopolitan and widespread species are in general drastically more variable than narrow ranging and endemic species. Thus, in animals, as in plants, genic diversity is nonrandom in space and in time and is associated with ecological and life history characteristics as exemplified in the following two studies of 38 species in Israel (Nevo, 1983a) and of about 130 mammalian species (Nevo, 1983b).

Allozymic variation encoded by 15 shared loci out of a mean of 26 nonshared ones was studied in 38 mostly unrelated plant, invertebrate and vertebrate species involving 162 populations sampled along transects of increasing aridity in Israel (Nevo, 1983a). The results indicate that the level and pattern of genic diversity either of the 15 individual loci or of the overall genetic indices of polymorphism and heterozygosity vary nonrandomly between loci, populations, species and habitats. Genic diversity is significantly explained by ecological (i.e. generalists vs specialists, mesic vs xeric, overground vs underground species), populational (i.e. patchy vs continuous population structure, population size, gene flow, fecundity) and taxonomic (rodents, amphibians, insects, landsnails, plants) variables. About a third of the genetic variance is explained by each of the ecologic, population and taxonomic subdivisions and about half of the variance by all three biotic categories combined in the analysis of 38 species. Randomization tests suggest that the correlations found are real and biologically meaningful.

When all 162 populations are subdivided into taxonomic, ecologic and populational categories climatic factors, primarily related to the availability of water and to the temperature regime, explain about 30% of the variance in the genetic indices and in 14 out of 15 loci. Since some of the variance explained by the biotic and climatic factors is complementary rather than overlapping, together they explain well above half of the genetic variance of the 38 species and 162 populations. The repetitive patterns of allozyme polymorphism or heterozygosity and their intimate link with climatic variation over many unrelated species and populations strongly implicates selection in the genetic differentiation of populations. Natural selection in several forms, but most likely through the mechanisms of spatiotemporally varying environments and epistasis at the various life cycle stages of organisms appears to be a major force causing evolutionary change at the molecular level. Other factors, including mutation, migration and genetic drift, may certainly interact with natural selection either directly or indirectly and thereby contribute differentially according to circumstance to population genetic differentiation at the molecular level.

In a second review study (Nevo, 1983b), the patterns and evolutionary significance of allozymic variation in more than 130 species of mammals were studied. As we found in the previous analysis (Nevo, 1983a; Hamrick *et al.*, 1979), the levels and patterns of genic diversity are nonrandom. In the mammalian species analysed (Nevo, 1983b), genic diversity varies largely and significantly among loci, populations, species, life zones (cosmopolitans and tropical species > temperate or arctic); geographical ranges (widespread and regional > narrow or endemic); habitat type (overground > underground); habitat width (generalist > specialist); aridity index (mesic and arid > mesic > arid); population structure (patchy > isolate); population size (large > small); gene flow (medium > low). These patterns are largely in line with those reported by Nevo (1978, 1983a), and suggest that the rough categorization of ecological, populational and life history variables explains about a quarter of the gene diversity, although another part remains as yet unexplained and may involve both stochastic and deterministic components that the analysis cannot yet resolve. To generalize about the role and relative importance of each evolutionary force, and establish a direct cause–effect relationship between ecologic and genetic factors, we need many more carefully designed field observations on many populations and species, coupled with critical laboratory testing of many loci at the biochemical, physiological and biological levels to assess the contribution of allozyme polymorphisms to fitness.

The genic-environmental niche width variation hypothesis was validated for allozymic variation in animals by laboratory experimentation (i.e. Powell, 1971; McDonald and Ayala, 1974; Powell and Wistrand, 1978) and field observations (i.e. Nevo, 1976, 1978, 1979, 1983a; Steiner, 1977), though refuted in some cases (i.e. Somero and Soule, 1974; Lester and Selander, 1979; Jones *et al.*, 1980; McCracken and Selander, 1980; Varvio-Aho, 1981; Varvio-Aho and Pamilo, 1981).

Jaenike *et al.* (1980) argue that in a parthenogenetic earthworm living under fluctuating environments, selection favours "general purpose" genotypes which are relatively insensitive to environmental changes, and that only broad-niched clones can become widespread. Finally, in a review of environmental and morphological correlates of genetic variation in mammals, Schnell and Selander (1981) claim that the intensive research on protein polymorphisms yielded great insights into the genetic structure of populations but little understanding of the mechanisms maintaining the polymorphisms in nature. As was mentioned earlier (Nevo, 1983b), this evaluation seems overpessimistic.

Levels of gene diversity were correlated with the following characteristics of proteins: (1) quarternary structure (Zouros, 1976; Harris *et al.*, 1977; Ward, 1977; Koehn and Eanes, 1978); (2) subunit molecular weight (Brown and Langley, 1979; Koehn and Eanes, 1978; Nei *et al.*, 1978; Ward, 1978; Turner *et al.*, 1979); (3) relative mutation rates (Zouros, 1979) and (4) relative effective population size (Nei, 1975; Soulè, 1976; Zouros, 1979; Nei and Graur, pers. comm.). Points 1–3 suggest molecular constraints for mutability. These correlations can be accommodated by both the selectionist (Ward, 1977; Harris *et al.*, 1977) and neutralist (Kimura and Ohta, 1974) models and cannot distinguish between them. Furthermore, all the above correlations and any potential future ones do not exclude environmental molecular correlations involving each of the aforementioned categories. Therefore, neither the quarternary structure, subunit molecular weight, mutation rate, nor effective population size affect the finding of substantial environmental correlates of allozymes with both physical and biotic factors as demonstrated in several reviews (Nevo, 1978, 1983a, b; Hamrick *et al.*, 1979). Remarkably, the level of enzyme variation is not only dependent on the external, but also on the internal subcellular environment, as shown for *PGI* in plants (Gottlieb and Weeden, 1981), and for multigene families (Hood *et al.*, 1975).

Geographic variation and genetic-environmental correlations. Clinal and other geographic structuring of allele frequencies of polymorphic genes are abundant and are often significantly associated with spatial or temporal variation of environmental factors (Koehn and Rasmussen, 1967; Koehn, 1969; Johnson *et al.*, 1969; Merritt, 1972; Johnson and

Schaffer, 1973; Bryant, 1974; McKechnie *et al.*, 1975; Nevo and Bar, 1976; Johnson, 1976; Hedrick *et al.*, 1976; Carson and Kaneshiro, 1976; Avise *et al.*, 1979; Hartl, 1980; Koehn *et al.*, 1980a; Carson *et al.*, 1981; Nevo *et al.*, 1981b, 1982a; Oakeshott *et al.*, 1982, among others). In some cases, the clinal and geographic patterning of allozymes are explicable by environmental, physical and biotic factors (i.e. Nevo, 1983a and references therein; Carson *et al.*, 1981).

Genetic-environmental correlations are reinforced if conducted and evaluated properly. This is particularly true if parallel genetic–environmental correlations are sought between allele frequencies at homologous loci in closely related sympatric species (i.e. Clarke, 1975; Harrison, 1977; Borowsky, 1977; Newkirk and Doyle, 1979; but see reservations in Varvio-Aho and Pamilo, 1982), between environmental heterogeneity and multilocus organization (Allard *et al.*, 1978), or between allele frequencies and metabolically homologous or at least analogous shared loci in many unrelated species (Nevo, 1983a). The latter study demonstrates not only inferrential adaptive correlations of enzyme polymorphisms. Since it was conducted on many loci, populations and species sharing the same ecological background, it may roughly quantify the proportion of gene diversity involved in adaptive evolution, at least of structural genes coding for soluble enzymes. The remarkable result is that most loci analysed are similarly correlated with the environment in many unrelated species (Nevo, 1983a).

The maintenance of allozyme polymorphism in animals. Natural selection, in some form, may often be a major determinant of genetic population structure and differentiation in animals (Hedrick *et al.*, 1976; Nevo, 1978; Clarke, 1979; Hartl, 1980; Wills, 1981) and a substantial factor in molecular evolution and regulation (Whitt, 1981). Balancing selection in animals has been either found or inferred for a variety of biological factors, but the relative importance of each is yet unknown (see Table X in Hartl, 1980: 225). The balancing factors involve, among others, environmental heterogeneity (i.e. Hedrick *et al.*, 1976 and references therein; Felsenstein, 1976; Lewontin *et al.*, 1978; Koehn, 1978; Powell and Taylor, 1979 for multiple niche and habitat selection; Koehn *et al.*, 1980c), all cases of stress environments to be discussed later, e.g. at the alcohol dehydrogenase locus, *ADH* (Gibson, 1970; Bijlsma-Meeles and Van Delden, 1974; Van Delden *et al.*, 1975; Clarke, 1975; Cavener and Clegg, 1978; Clarke *et al.*, 1979), overdominance (Mitton and Koehn, 1975; Clegg *et al.*, 1976; Tosic and Ayala, 1980), frequency and density dependent selection including ecological factors such as predation, parasitism and competition, among others (Ayala and Tracey, 1974; Clarke, 1975, 1979 and references therein; Clarke and Allendorf, 1979; Snyder and Ayala, 1979), sexual and gametic selection (Marinkovic and Ayala, 1977; Gilbert *et al.*, 1981), sexes (Eanes and Koehn, 1978), epistasis (Mitton and Koehn, 1973), regulation (McDonald and Ayala, 1978) and rank-order or truncation selection (Wills, 1978, 1981).

Milkman (1978) presented evidence that the cost of selection does not limit genic polymorphism; that truncation selection is not necessary for high cost-efficiency; and that the opposing directions of selection in a heterogeneous environment do not reduce cost-efficiency critically. Obviously, the deterministic mechanisms outlined above interact with stochastic processes including founder effects and random drift of populations. We are still largely ignorant as to the relative proportions of deterministic and stochastic factors in the genetic differentiation of populations.

(c) Protein polymorphisms in humans. This vast and dramatically expanding field will be discussed here only very briefly (reviewed in Cavalli-Sforza and Bodmer, 1971; Bodmer and Cavalli-Sforza, 1976; Hood *et al.*, 1975; Harris and Hopkinson, 1976; Wills, 1981). A variety of selective mechanisms has been found or inferred for the maintenance of red blood cell protein polymorphisms associated with malarial infection. Thus, conditional and geographically localized heterosis accounts for the high frequency of the sickle-cell anaemia allele (*Hb-S*), and for some of the heterogeneous thalassemias, whereas possibly weak heterosis maintains locally the *Hb-E* allele, and selection for the homozygote *Hb-C* leads to its high local frequency. *G-6PD* deficiencies, also associated with malarias, may possibly be maintained by heterosis in the common alleles, and by selection for the hemizygotes in the rare alleles.

The major human histocompatibility, or *HLA* system, the surface leucocyte antigens of man (or *H-2* in mice), consists of a family of closely-linked, highly polymorphic multiple allele loci controlling cell surface specificities, components of the complement system and immune response. The proportion of gamete types varies geographically and ethnically predisposing to resistance or susceptibility to a variety of neoplastic, autoimmune and infectious diseases. The presumed function of the system is cell to cell recognition and defense against pathogens (Bodmer, 1972). Heterozygotes may have a greater range of defense against host antigens and they are approximately equally fit and superior to all homozygotes suggesting that heterosis may be involved in this remarkably stable polymorphism (Wills, 1981). Additional mechanisms of selection including frequency dependent selection associated with disease resistance may be involved. Population data suggest the action of "bottleneck" natural selection, perhaps via linkage disequilibrium involving linked immune response genes (Bodmer, 1975; Bodmer and Bodmer, 1978).

Immunologically detected polymorphisms, particulary the *GM* system, represent a fascinating family of multiallelic protein polymorphisms determining differences in the constant regions of immunoglobulin heavy chains. The *GM* system is highly polymorphic, and it distinctly varies geographically, possibly selected in accord with immune response differences, i.e. differential resistance or susceptibility to infectious disease. Other immunoglobulin polymorphisms, such as *Inv*, control differences in the light chains.

New molecular genetics at the DNA level (Gilbert, 1978) revealed recently the diverse structures and mechanisms generating the enormous amount of immunoglobulin antibodies (Leder, 1982). The potential of the immune system to generate enormous diversity by shuffling a few hundred genes coupled with somatic DNA recombination and selective RNA transcription results in an astronomic amount of antigenic binding specificity. Antibody diversity is further amplified by clonal selection of the best-fitting antigen combining immunoglobulin site, and by heavy chain class switching. Natural selection has been shown to be quite effective in increasing genetic variation in a gene family such as that of the immunoglobulins (Hood *et al.*, 1975; Ohta, 1978).

Enzyme polymorphism in man. Polymorphism data are now available for more than 100 enzymatic loci (Harris *et al.*, 1977) and for 121 enzymatic and nonenzymatic proteins (Nei and Roychoudhury, 1982). The estimates for the proportion of loci polymorphic within the three major races are 45–52%, and the average heterozygosity per locus is 13–16% for protein loci, twice the estimate (7·3%) obtainable from enzyme loci (Harris *et al.*, 1977). The estimations of diversity for blood groups are also high: for 67, 35 and 25 blood group loci in Caucasoid, Negroid and Mongoloid, respectively, the proportion of loci polymorphic within the three major races is 34–56%, and the average heterozygosity is 11–20%. Natural selection appears to be involved also in the genetic differentiation of enzyme loci in man, as suggested by the relationship between polymorphism and subunit structure of enzymes (Harris *et al.*, 1977).

The genetic diversity in man is, therefore, very high for blood groups, protein, histocompatibility and immunoglobulin loci. This high diversity involves in a variety of gene families, a substantial effect of natural selection coupled with admixture due to massive historical migrations of human populations.

2. Microgeographic patterns

Microgeographic allozymic differentiation can take place over remarkably short distances, despite considerable gene flow, and is therefore adaptive at the morphological, physiological (Bradshaw, 1972) and allozymic (Nevo *et al.*, 1977, 1981c, 1982a) levels. The relatively few studies of microgeographic differentiation at the level of enzyme polymorphism largely suggest that the differences in allele frequencies between subpopulations are statistically significant, and partially related to environmental factors (reviewed for plants in Brown, 1979; for animals in Selander and Kaufman, 1975; and for humans in Nei and Roychoudhury, 1982; Smouse *et al.*, 1981, 1982). In some cases, such as in *Helix aspersa* (Selander and Kaufman, 1975), an examination of the gross features of the habitat types of the colonies yield no evidence of association with allele frequencies, and intercolony variation may by interpreted by either selection or drift, i.e. the evidence cannot distinguish between the two explanatory models. But in *Drosophila*, significant microspatial differentiation in chromosome inversion and allozyme frequencies are the rule rather than the exception (Richmond, 1978). High genetic differentiation between populations was also reported for *Thomomys bottae* pocket gophers (Patton and Feder, 1981). Behavioural habitat selection may be important in maintaining microspatial differences in gene frequencies in some species (Powell and Taylor, 1979; Taylor and Powell, 1977).

The degree of genetic differentiation within and between populations of 13 vagile species (insects and vertebrates) as compared to ten sedentary species (plants and landsnails) was analysed by the G_{ST} method of Nei (1973) which extends the Wright fixation index analysis F_{ST} (Wright, 1965) to multiallelic structures (Nevo, 1983a). The results indicate distinct differential patterns between the two groups. Of the total genetic variation found in vagile organisms, 20% existed between populations as against 39% in sedentary ones, i.e. genic divergence is about twicefold in the latter (Wilcoxon test, $P < 0 \cdot 01$). This pattern is in line with that found for inbreeding plants (Brown, 1979), sedentary animals and snails (Selander and Kaufman, 1975; McCracken and Brussard, 1980; Patton and Feder, 1981), and vagile organisms, including man (Nei, 1975). Thus, genetic differentiation appears to be at least partly correlated with environmental grain (Levins, 1968; Gillespie, 1974b); it is significantly higher in coarse grained as compared to fine grained environments. This pattern implicates ecological factors in population differentiation. However, besides some broad patterns of climatic factors very little is known about the factors of local environmental heterogeneity either physically or biotically (Jain, 1976). In the few cases in which specific factors such as temperature alone can be singled out in nature, local genetic differentiation may be substantive and causally interpreted. In barnacles near power plants, 8 out of 11 allozyme loci changed significantly and repetitively in an adaptive fashion due to a 10°C warmer environment (Nevo *et al.*, 1977).

BIOCHEMICAL AND PHYSIOLOGICAL STUDIES OF ALLOZYME POLYMORPHISM

A substantial amount of circumstantial evidence has accumulated indicating that genetic polymorphisms are associated with environmental heterogeneity. Clearly, however, correlations are descriptive rather than deterministic. They do not establish directly causal relationships between genetic variation and environment. Yet, if consistent over many loci and species, they score indirectly useful, presumably genuine associations between allozymes and environment which may guide further critical experiments to substantiate the causal mechanisms involved in the polymorphism. The unequivocal demonstration of the adaptive significance of an enzyme polymorphism depends on interrelated multidisciplinary evidence. The latter must show that the enzyme phenotypic diversity varies with the environment and involves *in vitro* biochemical and physiological differences among allozyme variants that contribute to fitness. A beautiful demonstration was presented for leucine aminopeptidase (*LAP*) colleagues (Koehn, 1978; Koehn *et al.*, 1980a, b, c; Koehn and Immerman, 1981; Koehn and Siebenaller, 1981.)

1. In vitro *and* in vivo *enzyme studies*

Numerous studies have demonstrated *in vitro* kinetic differences between isozymes (many articles in Markert, 1975) and allozyme variants (i.e. Koehn, 1969; Gibson, 1970, 1972; Merritt, 1972; Harper and Armstrong, 1973; Day *et al.*, 1974a, b; Koehn, 1978; Terri and Peet, 1978; Place and Powers, 1979; Bijlsma and Meulen-Bruijns, 1979; Narise, 1979; Danford and Beardmore, 1979; Grossman, 1980; Charles and Lee, 1980; McDonald *et al.*, 1980; Hoffman, 1981; Graves and Somero, 1982, among others). Although *in vitro* kinetic differences between allozymes could indicate that selection is operating, only the demonstration of *in vivo* physiological differences that ultimately determine fitness may substantiate the selection hypothesis (Lewontin, 1974). Several studies have indeed described such *in vivo* differences suggesting that at least some of the *in vitro* kinetic differences may give rise to *in vivo* differences affecting fitness (i.e. Koehn, 1978; Papel *et al.*, 1979; Anderson and McDonald, 1981; Cavener and Clegg, 1981a, b; DiMichele and Powers, 1982). Finally, environmental and physiological stresses may unveil fitness differences among allozyme variants of several enzymes such as amylase (DeJong and Scharloo, 1976; Scharloo *et al.*, 1977; Hickey, 1977; Hoorn, 1978, but see reservations in Yardley, 1978), alcohol dehydrogenase (Van Delden *et al.*, 1978), esterase (Ghandi, 1978), glucose-6-phosphate dehydrogenase and 6-phosphogluconate dehydrogenase (Bijlsma and Meulen-Bruijns, 1979), lactate dehydrogenase (Powers *et al.*, 1979), aminopeptidase-I (Koehn, 1978; Koehn *et al.*, 1980a; Koehn and Immerman, 1981; Koehn and Siebenaller, 1981). Following is a short summary of our observations in nature and critical laboratory experiments at the Institute of Evolution, University of Haifa, utilizing pollution biology of marine organisms as a stress environment to test for potential fitness differentials of allozyme variants (Nevo *et al.*, 1983b).

2. Pollution selection of allozyme polymorphisms in marine organisms

The evolutionary significance of allozyme polymorphisms in several Mediterranean marine organisms was tested first by *post hoc* gene frequency analysis at 11–15 gene loci in natural populations of barnacles, *Balanus amphitrite*, under thermal (Nevo *et al.*, 1977) and chemical (Nevo *et al.*, 1978) pollutions. Secondly, we tested by *pre hoc* controlled laboratory experiments the effects of heavy metal pollution (Hg, Zn, Cd) on genotypic frequencies of 15 phosphoglucomutase (*PGM*) genotypes in thousands of individuals of the shrimp *Palaemon elegans* (Nevo *et al.*, 1980, 1981a, 1983b). Similarly, we tested the effects of Hg, Zn, Cd, Pb, Cu pollutions on the genotypic and allelic frequencies of five phosphoglucose isomerase (*PGI*) genotypes in the two close species of marine gastropods, *Monodonta turbinata* and *M. turbiformis* (Lavie and Nevo, 1982; Nevo *et al.*, 1983b).

In both the thermal and chemical pollution studies, we established in repetitive experiments statistically significant differences in allele frequencies at 8 of 11 (73%) and 10 of 15 (67%) gene loci, respectively, between the contrasting environments in each. While no specific function could be singled out in the *post hoc* chemical study due to the complex nature of polluted marine water, *temperature* could be specified as the single selective agent in the thermal study. The strongest, direct and specific evidence for significant differential survivorship among allozyme genotypes was obtained in the *pre hoc* studies in *Palaemon* and *Monodonta* (Nevo *et al.*, 1983b). Their differential viability was probably associated with the different degree of heavy metal inhibition, *uniquely* related to each specific pollutant. Furthermore, we demonstrated in the two *Monodonta* species *parallel genetic differentiation* as a response to pollution.

Our results are inconsistent with the neutral theory of allozyme polymorphisms and appear to reflect the adaptive nature of the allozyme polymorphisms studied. Allozyme genotypes may be sensitive to and vary with the quality and quantity of specific pollutants. Therefore, they can provide precise genetic indicators of pollution for the *short and long term* genetic changes populations undergo due to pollution. Ideally, in different marine species specific genetic loci either singly or in combination, may prove sensitive markers to different pollutants and can easily be assayed by quick electrophoretic tests and be used as genetic monitors. An extensive search for the appropriate enzymatic systems in various primarily sedentary marine species exposed to single and multiple pollutants is therefore urgent.

THE THEORY AND ITS ELUCIDATION OF PROTEIN POLYMORPHISM IN NATURE

Population genetic theory incorporates now a variety of models that can explain the maintenance of the vast amount of genetic polymorphism in nature and generate predictions that can be verified or falsified by observation and critical experiments (Hedrick *et al.*, 1976; Felsenstein, 1976). Single locus theory indicates that selection acting differentially in space coupled with limited migration (Karlin and McGregor, 1972a, b, 1973; Gillespie, 1975) and/or habitat selection (Taylor and Powell, 1978; Powell and Taylor, 1979; Templeton and Rothman, 1981) will maintain a substantial amount of polymorphism. Polymorphism is much more likely in coarse as compared with fine-grained environments (Gillespie, 1974b). Present mathematical models predict a stable polymorphism due to environmental heterogeneity in space in various situations (Levene, 1953; Christiansen and Feldman, 1975; Hedrick *et al.*, 1976; Karlin and Feldman, 1981), or in time (Bryant, 1974, 1976), but Hedrick (1978) concluded that spatial heterogeneity is more efficient. In other words, different selective pressures vary with the environment, including frequency dependent selection (Kojima and Yarbrough, 1967; Hedrick, 1972) primarily by biotic factors (Clarke, 1979). A classification of some of the mechanisms that can maintain polymorphism by selection is given in Hartl (1980: 225, Table X). Obviously, a variety of factors, either singly or in combination, operating at various times in the life cycle of organisms and unveiled by selection component analysis (Ostergaard and Christiansen, 1981) can contribute to a stable polymorphism. Finally, heterozygosity *per se* may be advantageous due to the higher flexibility of the heterozygotes which are better buffered against environmental perturbations in space and time (Haldane, 1954; Lerner, 1954; Gillespie and Langley, 1974; Bryant, 1974, 1976; Sved, 1976; Gillespie, 1977a; but see reservations in Lewontin *et al.*, 1978 and Clarke, 1979).

Gene clusters, multilocus structures, and epistasis in general, have received recently an increasing theoretical consideration (i.e. Wright,

1964; Lewontin, 1974; Feldman *et al.*, 1974; Hood *et al.*, 1975; Hedrick *et al.*, 1978; Li, 1978; Karlin, 1979, 1981; Bodmer, 1979). The total fitness of the individual relates not only to individual fitness at single loci, but to the epistatic interaction of the entire genetic system, hence, the great potential importance of multilocus structures to the maintenance of genetic polymorphism. Theoretically, there is an increased likelihood for a globally stable (unique) equilibrium for the structured fitness models of multilocus forms (Karlin, 1979, 1982; Karlin and Lieberman, 1979; Karlin and Avni, 1981). However, the analysis of multilocus variation in natural populations is just beginning (Hedrick *et al.*, 1978), and its potential contribution to polymorphism studies remains a future challenge. Nevertheless, the paramount evolutionary significance of multigene families is already evident (Hood *et al.*, 1975; Bodmer, 1979; Leder, 1982).

The Levene model (1953) has been extended recently to multilocus and multiniche structures (reviews in Hedrick *et al.*, 1976; Felsenstein, 1976; Hedrick *et al.*, 1978; Karlin, 1979, 1982). Gillespie and colleagues (Gillespie and Langley, 1974; Gillespie, 1974a, b, 1975, 1976a, b, 1977a, b, c, 1979, 1980; Gillespie and Guess, 1978) concluded on theoretical grounds that allozymic polymorphism is primarily due to selection acting on environmental variation in gene function. Genetic variation will be more likely in spatiotemporally more variable environments than in constant ones. Furthermore, heterozygote intermediacy plus random environmental fluctuations are sufficient elements to explain genetic variability, and the conditions for polymorphism in heterogeneous environments are less stringent than those of overdominant selection in multiple allelic systems, as also concluded by Lewontin *et al.* (1978). Theoretically, nonoverdominant stable equilibria may be generated even in a single niche if selection coefficients vary from generation to generation in a specific manner in infinitely large populations (Dempster, 1955; Haldane and Jayakar, 1963; Karlin and Levikson, 1974; Gillespie, 1978).

The most likely general mechanisms maintaining genetic polymorphisms by selection are spatially and temporally varying environments and epistasis (see classification in Hartl 1980: 225). However, the type of selection operating in individual cases is often unclear and the analysis of the proportion and interaction of deterministic and stochastic processes in natural populations remains a future challenge of evolutionary biology. Finally, the neutral theory provides a model that accounts simultaneously for enzyme polymorphism and molecular evolution. Under this model, enzyme polymorphism is viewed as a "phase of molecular evolution" (Kimura and Ohta, 1971). The first attempts to account for both polymorphism and molecular evolution by natural selection models were provided by Gillespie (1978, 1979, 1980). His SAS-CFF model, based on multiallelic selection in a random environment, appears to predict roughly the actual allele frequency configurations observed in nature, and the transient properties of the model behave similarly to the neutral model.

Where are we in terms of the selection, neutrality and shifting balance evolutionary theories? The neutralist-selectionist controversy is as alive today (Nei, 1980) as it was since the proposal of the neutral theory (Kimura, 1968), although some feel that it is largely over (Clarke, 1979; Berry, 1980). It will certainly not be settled by any single study, no matter how exhaustive, whether correlative or biochemical-physiological. In fact, it may never be resolved completely since the sharp dichotomy of either school may be unrealistic. The real problem is how much adaptiveness is implicated at the molecular level. Even this target may be unrealistic since as Wills (1981) noted "it seems unlikely that we will ever be able to determine what proportion of all alleles in a population is neutral and what proportion is subject to selection". Moreover, as pointed out by Hartl (1981) the issue was oversimplified and the sharp dichotomy between neutral and selected alleles is artificial, primarily since alleles may change their sign differentially not only with the external environment but also with the internal genetic background (Dykhuizen and Hartl, 1980; Hartl and Dykhuizen, 1981) and hierarchy of selective units (Bargiello and Grossfield, 1979) analogous to Wright's shifting balance theory of evolution (Wright, 1969, 1970, 1978). However, the aforementioned considerations may be overly pessimistic. It seems likely that a substantial amount of the general organization of the genetic material of higher organisms is into gene clusters of duplicated genes and that multigene families, rather than the number of individual genes, reflect the general complexity and dynamics of higher organisms (Hood *et al.*, 1975; Bodmer, 1979), and this genetic structure is more tractable.

The importance of the shifting balance theory (Wright, 1970, 1978) is not in its truth, since this is yet unknown, but in its synthetic attempt to assign evolutionary role and importance to the major interacting deterministic and stochastic forces causing evolutionary change. Real progress in our understanding of nature cannot depend on single principal explanatory models and dichotomous alternatives since the latter ridicule the complexity of nature, rather than resolve it. The analysis of diversity in nature must depend on multidisciplinary diverse factors and methodologies. Understanding of the relative roles and importance of stochastic and deterministic factors at the molecular level of protein and DNA depends on intensive and extensive field and critical laboratory studies at single and multilocus systems in many species with a full consideration of the ecological, biochemical, physiological and historical interacting determinants. The evidence reviewed here of protein variation in nature demonstrates clearly that either/or extreme views are unrealistic, and the problem is essentially a quantitative one: how much of each. Much of protein diversity in nature is adaptive and propelled by natural selection but we can hardly estimate at the present time how much. The elucidation of the roles, importance and interaction of the evolutionary forces at the molecular level must proceed through unbiased observations and critical testing of nature's complexity.

SUMMARY

Protein variation in natural populations of plants, animals and humans is briefly reviewed. Explanatory models attempting to explain genetic diversity and evolutionary change in nature, including the selectionist, neutral and shifting balance theories are compared and contrasted in view of the evidence based on both macro- and microgeographic correlative analyses of field studies and critical biochemical and physiological laboratory tests affecting fitness.

The analysis suggests that the levels and pattern of protein diversity vary nonrandomly within and between species and are often associated with ecological heterogeneity and are, therefore, at least partly adaptive. Natural selection, in its various forms and acting at various times in the life cycle of organisms, may often be a major determinant of genetic differentiation of populations and species at enzymatic and nonenzymatic (i.e. histocompatibility, immunoglobulin and other multigene families) protein levels. However, it is suggested that a synthetic view of the major forces maintaining genetic diversity and evolutionary change, i.e. natural selection, mutation, migration and random drift, may be more realistic in explaining the structure of organic nature.

The future challenge of evolutionary biology at the molecular level is to study the role, importance and interaction of the major evolutionary forces in an attempt to explain genetic diversity and evolutionary change of proteins and DNA in nature. This challenge calls for extensive field observations and critical laboratory experimentation at both single and multilocus systems in many species involving the full consideration of ecological, physiological, biochemical, genetical and historical interacting determinants affecting fitness and thereby causing diversity and evolutionary change.

ACKNOWLEDGEMENTS

My deep gratitude is extended to A. Beiles for statistical assistance and many productive discussions on evolutionary problems. I thank A. Beiles, D. Adler and R. Ben-Shlomo for reading, commenting and assisting in the final shape of the manuscript. This research was partly supported by a grant from the United States Israel Binational Science Foundation (BSF), Jerusalem, Israel.

REFERENCES

Allard, R. W., Kahler, A. L. and Clegg, M. T. (1977). Estimation of mating cycle components of selection in plants. *In* "Measuring Selection in Natural Populations" (F. B. Christiansen and T. Fenchel, eds) pp. 1–19, Lecture notes in Biomathematics. Springer-Verlag, Heidelberg.

Allard, R. W., Miller, R. D. and Kahler, A. L. (1978). The relationship between degree of environmental heterogeneity and genetic polymorphism. *Verh. K. ned. Akad. Wet.,* II **70**, 49–73.

Anderson, S. M. and McDonald, J. F. (1981). Effect of environmental alcohol on *in vivo* properties of *Drosophila* alcohol dehydrogenase. *Biochem. Genet.* **19**, 421–430.

Antonovics, J. (1971). The effects of a heterogeneous environment on the genetics of natural populations. *Am. Scient.* **59**, 593–599.

Avise, J. C., Smith, M. H. and Selander, R. K. (1979). Biochemical polymorphism and systematics in the genus *Peromyscus* VII. Geographic differentiation in members of the *truei* and *maniculatus* species groups. *J. Mammal.* **60**, 177–192.

Ayala, F. J. (1968). Genotype, environment, and population numbers. *Science, N.Y.* **162**, 1453–1459.

Ayala, F. J. (1972). Darwinian versus non-Darwinian evolution in natural populations of *Drosophila*. Proc. 6th Berkeley Symp. Mathematical Statistics and Probability **5**, 211–236.

Ayala, F. J. (1974). Biological evolution: Natural selection or random walk? *Am. Scient.* **62**, 692–701.

Ayala, F. J. (1976). Protein evolution in related species: Adaptive "Foci". *Johns Hopkins J.* **138**, 262–278.

Ayala, F. J. (1977). Protein evolution in different species: is it a random process? *In* "Molecular Evolution and Polymorphism" (M. Kimura, ed.), pp. 73–102. Proceedings of the Second Taniguchi International Symposium on Biophysics, National Institute of Genetics, Mishima, Japan.

Ayala, F. J. and Giplin, M. E. (1974). Gene frequency comparisons between taxa: Support for the natural selection of protein polymorphisms. *Proc. Nat. Acad. Sci., U.S.A.* **71**, 4847–4849.

Ayala, F. J. and Tracey, M. L. (1974). Genetic differentiation within and between species of the *Drosophila willistoni* group. *Proc. Nat. Acad. Sci., U.S.A.* **71**, 999–1003.

Ayala, F. J., Tracey, M. L., Barr, L. G., McDonald, J. F. and Perez-Sales, S. (1974). Genetic variation in natural populations of five *Drosophila* species and the hypothesis of the selective neutrality of protein polymorphisms. *Genetics* **77**, 343–384.

Babbel, G. R. and Selander, R. K. (1974). Genetic variability in edaphically restricted and widespread plant species. *Evolution, Lancaster, Pa.* **28**, 619–630.

Baker, J., Maynard Smith, J. and Strobeck, C. (1975). Genetic polymorphism in the bladder campion, *Silene maritima*. *Biochem. Genet.* **13**, 393–410.

Bargiello, T. and Grossfield, J. (1979). Biochemical polymorphisms: The unit of selection and the hypothesis of conditional neutrality. *BioSystems* **11**, 183–192.

Berry, R. J. (1980). Genes, pollution and monitoring. *Rapp. P.-v. Réun. Cons. perm. int. Explor. Mer* **179**, 253–257.

Bijlsma, R. S. and Meulen-Bruijns, C. (1979). Polymorphism at the *G6PD* and *6PGD* loci in *Drosophila melanogaster*. III. Developmental and biochemical aspects. *Biochem. Genet.* **17**, 1131–1144.

Bijlsma-Meeles, E. and Van Delden, W. (1974). Intra- and interpopulation selection concerning the alcohol dehydrogenase locus in *Drosophila melanogaster*. *Nature, Lond.* **247**, 369–371.

Bodmer, W. F. (1972). Evolutionary significance of the HL-A system. *Nature, Lond.* **237**, 139–145.

Bodmer, W. F. (1975). Evolution of HL-A and other major histocompatibility systems. *Genetics* **79**, 293–304.

Bodmer, W. F. (1979). Gene clusters and the HLA system in human genetics: Possibilities and realities. *CIBA Foundation Series* **66**, 205–229.

Bodmer, W. F. and Cavalli-Sforza, L. L. (1976). "Genetics, Evolution, and Man". W. H. Freeman, San Francisco.

Bodmer, W. F. and Bodmer, J. G. (1978). Evolution and function of the HLA system. *Br. med. Bull.* **34**, 309–316.

Borowsky, R. (1977). Detection of the effects of selection on protein polymorphisms in natural populations by means of a distance analysis. *Evolution, Lancaster, Pa.* **31**, 341–346.

Bradshaw, A. D. (1972). Some evolutionary consequences of being a plant. *Evol. Biol.* **5**, 25–47.

Bradshaw, A. D., McNeilly, T. S. and Gregory, R. P. G. (1965). Industrialization, evolution and the development of heavy metal tolerance in plants. *Symp. Br. Ecol. Soc.* **5**, 327–343.

Brown, A. H. D. (1978). Isozymes, plant population genetic structure and genetic conservation. *Theor. Appl. Genet.* **52**, 145–157.

Brown, A. H. D. (1979). Enzyme polymorphism in plant populations. *Theor. Pop. Biol.* **15**, 1–42.

Brown, A. H. D. and Moran, G. F. (1979). Isozymes and the genetic resources of forest trees. Presented at the Symposium on Isozymes of North American Forest Trees and Forest Insects. July 27, 1979, Berkeley, Calif.

Brown, A. H. D., Feldman, M. W. and Nevo, E. (1980). Multilocus structure of natural populations of *Hordeum spontaneum*. *Genetics* **96**, 523–536.

Brown, A. H. D., Marshall, D. R. and Munday, J. (1976). Adaptedness of variants at an alcohol dehydrogenase locus in *Bromus mollis*. *Aust. J. Biol. Sci.* **29**, 389–396.

Brown, A. J. L. and Langley, C. H. (1979). Correlation between heterozygosity and subunit molecular weight. *Nature, Lond.* **277**, 649–651.

Bryant, E. H. (1974) On the adaptive significance of enzyme polymorphisms in relation to environmental variability. *Am. Nat.* **108**, 1–19.

Bryant, E. H. (1976). A comment on the role of environment variation in maintaining polymorphisms in natural populations. *Evolution, Lancaster, Pa.* **30**, 188–190.

Carson, H. L. and Kaneshiro, K. Y. (1976). *Drosophila* of Hawaii: systematics and ecological genetics. *A. Rev. Ecol. Syst.* **7**, 311–345.

Carson, H. L., Craddock, E. M., Johnson, W. E., Newman, L. J., Paik, K., Steiner, W. W. M. and Sung, K. C. (1981). Genetic studies of natural populations. *In* "Island Ecosystems" (D. Mueller-Dombois, K. W. Bridges and H. L. Carson, eds), pp. 438–470. Hutchinson, London.

Cavalli-Sforza L. L. and Bodmer, W. F. (1971). "The Genetics of Human Populations". W. H. Freeman, San Francisco.

Cavener, D. R., and Clegg, M. T. (1978). Dynamics of correlated genetic systems. IV. Multilocus effects of ethanol stress environments. *Genetics* **90**, 629–644.

Cavener, D. R. and Clegg, M. T. (1981a). Evidence for biochemical and physiological differences between enzyme genotypes in *Drosophila melanogaster*. *Proc. Nat. Acad. Sci., U.S.A.* **78**, 4444–4447.

Cavener, D. R. and Clegg, M. T. (1981b). Multigenic response to ethanol in *Drosophila melanogaster*. *Evolution, Lancaster, Pa.* **35**, 1–10.
Charles, D. J. and Lee, C. Y. (1980). Biochemical and immunological characterization of genetic variants of phosphoglucose isomerase from mouse. *Biochem. Genet.* **18**, 153–169.
Christiansen, F. B. and Feldman, M. W. (1975). Subdivided populations: A review of the one and two locus deterministic theory. *Theor. Pop. Biol.* **7**, 13–38.
Christiansen, F. B. and Frydenberg, O. S. (1974). Geographical patterns of four polymorphisms in *Zoarces viviparus* as evidence of selection. *Genetics* **77**, 765–770.
Clarke, B. (1970). Darwinian evolution of proteins. *Science, N.Y.* **168**, 1009–1011.
Clarke, B. (1975). The contribution of ecological genetics to evolutionary theory: detecting the direct effects of natural selection on particular polymorphic loci. *Genetics, Suppl.* **79**, 101–113.
Clarke, B. (1978). Some contributions of snails to the development of ecological genetics. *In* "Ecological Genetics: The Interface" (P. F. Brussard, ed.), pp. 159–170. Springer-Verlag, New York.
Clarke, B. (1979). The evolution of genetic diversity. *Proc. R. Soc. Lond.* B **205**, 453–474.
Clarke, B. and Allendorf, F. W. (1979). Frequency-dependent selection due to kinetic differences between allozymes. *Nature, Lond.* **279**, 732–734.
Clarke, B., Camfield, R. G., Galvin, A. M. and Pitts, C. R. (1979). Environmental factors affecting the quantity of alcohol dehydrogenase in *Drosophila melanogaster*. *Nature, Lond.* **280**, 516–518.
Clegg, M. T. and Allard, R. W. (1972). Patterns of genetic differentiation in the slender wild oat species *Avena barbata*. *Proc. Nat. Acad. Sci., U.S.A.* **69**, 1820–1824.
Clegg, M. T. and Allard, R. W. (1973). Viability versus fecundity selection in the slender wild oat, *Avena barbata L. Science, N.Y.* **181**, 667–668.
Clegg, M. T., Kidwell, J. F., Kidwell, M. G. and Daniel, N. J. (1976). Dynamics of correlated genetic systems. I. Selection in the region of the glued locus of *Drosophila melanogaster*. *Genetics* **83**, 793–810.
Clegg, M. T., Kahler, A. L. and Allard, R. W. (1978). Genetic demography of plant populations. *In* "Ecological Genetics: The Interface" (P. F. Brussard, ed.), pp. 173–188. Springer-Verlag, New York.
Danford, N. D. and Beardmore, J. A. (1979). Biochemical properties of Esterase-6 in *Drosophila melanogaster*. *Biochem. Genet.* **17**, 1–22.
Day, T. H., Hillier, P. C. and Clarke, B. (1974a). Properties of genetically polymorphic isozymes of alcohol dehydrogenase in *Drosophila melanogaster*. *Biochem. Genet.* **11**, 141–153.
Day, T. H., Hillier, P. C. and Clarke, B. (1974b). The relative quantities and catalytic activities of enzymes produced by alleles at the alcohol dehydrogenase locus in *Drosophila melanogaster*. *Biochem. Genet.* **11**, 155–165.

DeJong, G. and Scharloo, W. (1976). Environmental determination of selective significance or neutrality of amylase variants in *Drosophila melanogaster*. *Genetics* **84**, 77–94.

Dempster, E. (1955). Maintenance of genetic heterogeneity. *Cold Spring Harb. Symp. quant. Biology* **20**, 25–32.

DiMichele, L. and Powers, D. A. (1982). LDH-B genotype-specific hatching times of *Fundulus heteroclitus*. *Nature, Lond.* **296**, 563–564.

Dykhuizen, D. and Hartl, D. L. (1980). Selective neutrality of 6PGD allozymes in *E. coli* and the effects of genetic background. *Genetics* **96**, 801–817.

Eanes, W. F. and Koehn, R. K. (1978). An analysis of genetic structure in the monarch butterfly, *Danaus plexippus* L. *Evolution, Lancaster, Pa.* **32**, 784–797.

Ewens, W. J. (1969). The transient behaviour of stochastic processes, with application in the natural sciences. *Bull. 36th session I.S.I.*, 603–622.

Ewens, W. J. (1972). The sampling theory of selectively neutral alleles. *Theor. Pop. Biol.* **3**, 87–112.

Ewens, W. J. (1979). "Mathematical Population Genetics". Springer-Verlag. Berlin.

Ewens, W. J. and Feldman, M. W. (1976). The theoretical assessment of selective neutrality. *In* "Population Genetics and Ecology" (S. Karlin and E. Nevo, eds), pp. 303–337. Academic Press, New York.

Feldman, M. W., Franklin, I. and Thomson, G. J. (1974). Selection in complex genetic systems. I. The symmetric equilibria of the three-locus symmetric viability model. *Genetics* **76**, 135–162.

Felsenstein, J. (1976). The theoretical population genetics of variable selection and migration. *Ann. Rev. Genet.* **10**, 253–280.

Fitch, W. M. and Langley, C. H. (1976). Evolutionary rates in proteins: Neutral mutations and the molecular clock. *In* "Molecular Anthropology" (M. Goodman, R. E. Tashian and J. H. Tashian, eds), pp. 197–219. Plenum, New York.

Fuerst, P. A., Chakraborty, R. and Nei, M. (1977). Statistical studies on protein polymorphism in natural populations. I. Distribution of single locus heterozygosity. *Genetics* **86**, 455–483.

Ghandi, Y. (1978). "A study of the Esterase-6 enzyme polymorphism in *Drosophila melanogaster*." Ph.D. thesis, University of Nottingham.

Gibson, J. (1970). Enzyme flexibility in *Drosophila melanogaster*. *Nature, Lond.* **227**, 959–960.

Gibson, J. (1972). Differences in the number of molecules produced by two allelic electrophoretic variants in *D. melanogaster*. *Experientia* **28**, 975–976.

Gilbert, B. G., Richmond, R. C. and Sheehan, K. B. (1981). Studies of Esterase-6 in *Drosophila melanogaster*. V. Progeny production and sperm use in females inseminated by males having active or null alleles. *Evolution, Lancaster, Pa.* **35**, 21–37.

Gilbert, W. (1978). Why genes in pieces? *Nature, Lond.* **271**, 501.

Gillespie, J. H. (1974a). Polymorphism in patchy environments. *Am. Nat.* **108**, 145–151.
Gillespie, J. H. (1974b). The role of environmental grain in the maintenance of genetic variation. *Am. Nat.* **108**, 831–836.
Gillespie, J. H. (1975). The role of migration in the genetic structure of populations in temporarily and spatially varying environments. I. Conditions for polymorphism. *Am. Nat.* **109**, 127–135.
Gillespie, J. H. (1976a). The role of migration in the genetic structure of populations in temporally and spatially varying environments. II. Island models. *Theor. Pop. Biol.* **10**, 227–238.
Gillespie, J. H. (1976b). A general model to account for enzyme variation in natural populations. II. Characterization of the fitness function. *Am. Nat.* **110**, 809–821.
Gillespie, J. H. (1977a). A general model to account for enzyme variation in natural populations. III. Multiple alleles. *Evolution, Lancaster, Pa.* **31**, 85–90.
Gillespie, J. H. (1977b). A general model to account for enzyme variation in natural populations. IV. The quantitative genetics of fitness traits. *In* "Lecture Notes in Biomathematics" (F. B. Christiansen and T. M. Fenchel, eds), pp. 301–314. Springer-Verlag, Berlin.
Gillespie, J. H. (1977c). Sampling theory for alleles in a random environment. *Nature, Lond.* **266**, 443–445.
Gillespie, J. H. (1978). A general model to account for enzyme variation in natural populations. V. The SAS-CFF model. *Theor. Pop. Biol.* **14**, 1–45.
Gillespie, J. H. (1979). Molecular evolution and polymorphism in a random environment. *Genetics* **93**, 737–754.
Gillespie, J. H. (1980). The stationary distribution of an asymmetrical model of selection in a random environment. *Theor. Pop. Biol.* **17**, 129–140.
Gillespie, J. H. and Guess, H. A. (1978). The effect of environmental autocorrelation on the progress of selection in a random environment. *Am. Nat.* **112**, 897–909.
Gillespie, J. H. and Langley, C. H. (1974). A general model to account for enzyme variation in natural populations. *Genetics* **76**, 837–848.
Gottlieb, L. D. and Weeden, N. F. (1981). Correlation between subcellular location and phosphoglucose isomerase variability. *Evolution, Lancaster, Pa.* **35**, 1019–1022.
Graves, J. E. and Somero, G. N. (1982). Electrophoretic and functional enzymic evolution in four species of eastern Pacific barracudas from different thermal environments. *Evolution, Lancaster, Pa.* **36**, 97–106.
Grossman, A. (1980). Analysis of genetic variation affecting the relative activities of Fast and Slow ADH dimers in *Drosophila melanogaster* heterozygotes. *Biochem. Genet.* **18**, 765–780.
Haldane, J. B. S. (1954). "The Biochemistry of Genetics". Allen and Unwin, London.
Haldane, J. B. S. and Jayakar, S. D. (1963). Polymorphism due to selection of varying direction. *J. Genet.* **58**, 237–242.

Hamrick, J. L., Linhart, Y. B. and Mitton, J. B. (1979). Relationships between life history characteristics and electrophoretically detectable genetic variation in plants. *A. Rev. Ecol. Syst.* **10**, 173–200.

Harper, J. L. (1977). "Population Biology of Plants". Academic Press, London.

Harper, R. A. and Armstrong, F. B. (1973). Alkaline phosphatase of *Drosophila melanogaster*. II. Biochemical comparison among four allelic forms. *Biochem. Genet.* **10**, 29–38.

Harris, H. and Hopkinson, D. A. (1976). "Handbook of Enzyme Electrophoresis in Human Genetics". North-Holland, Amsterdam.

Harris, H., Hopkinson, D. A. and Edwards, Y. H. (1977). Polymorphism and the subunit structure of enzymes: A contribution to the neutralist-selectionist controversy. *Proc. Nat. Acad. Sci., U.S.A.* **74**, 698–701.

Harrison, R. G. (1977). Parallel variation at an enzyme locus in sibling species of field crickets. *Nature, Lond.* **266**, 168–170.

Hartl, D. L. (1980). "Principles of Population Genetics". Sinauer Press Associates, Sunderland, Mass.

Hartl, D. L. (1981). Polymorphism and selection. *Science, N.Y.* **213**, 1369–1370.

Hartl, D. L. and Dykhuizen, D. E. (1981). Potential for selection among nearly neutral allozymes of 6-phosphogluconate dehydrogenase in *Escherichia coli*. *Proc. Nat. Acad. Sci., U.S.A.* **78**, 6344–6348.

Hedrick, P. W. (1972). Maintenance of genetic variation with a frequency dependent selection model as compared to the overdominant model. *Genetics* **72**, 771–775.

Hedrick, P. W. (1978). Genetic variation in a heterogeneous environment. V. Spatial heterogeneity in finite populations. *Genetics* **89**, 389–401.

Hedrick, P. W., Ginevan, M. E. and Ewing, E. P. (1976). Genetic polymorphism in heterogeneous environments. *A. Rev. Ecol. Syst.* **7**, 1–32.

Hedrick, P. W., Jain, S. and Holden, L. (1978). Multilocus systems in evolution. *Evol. Biol.* **11**, 101–185.

Hickey, D. A. (1977). Selection for amylase allozymes in *Drosophila melanogaster*. *Evolution, Lancaster, Pa.* **31**, 800–804.

Hoffmann, R. (1981). Evolutionary genetics of *Metridium senile*. I. Kinetic differences in phosphoglucose isomerase allozymes. *Biochem. Genet.* **19**, 129–144.

Hood, L., Campbell, J. H. and Elgin, S. C. R. (1975). The organization, expression and evolution of antibody genes and other multigene families. *A. Rev. Genet.* **9**, 305–353.

Hoorn, A. J. W. (1978). "The functional significance of amylase polymorphism in *Drosophila melanogaster*." Ph.D. thesis, University of Utrecht.

Jaenike, J., Parker, E. D. Jr. and Selander, R. K. (1980). Clonal niche structure in the parthenogenetic earthworm *Octolasion turtaeum*. *Am. Nat.* **116**, 196–205.

Jain, S. K. (1967). Population dynamics of a gametophytic factor controlling selective fertilization. *Genetica* **38**, 485–503.

Jain, S. K. (1976). Patterns of survival and microevolution in plant populations. *In* "Population Genetics and Ecology" (S. Karlin and E. Nevo, eds), pp. 49–89. Academic Press, New York.

Jain, K. B. L. and Jain, S. K. (1970). Frequency-dependent selection under mixed selfing and random mating. *Heredity* **25**, 217–221.

Johnson, G. B. (1976). Polymorphism and predictability at the α-glycerophosphate dehydrogenase locus in *Colias* butterflies: gradients in allele frequency within single populations. *Biochem. Genet.* **14**, 403–426.

Johnson, G. B. (1979). Genetic polymorphism among enzyme loci. *In* "Physiological Genetics" (J. G. Scandalios, ed.), pp. 239–273. Academic Press, New York.

Johnson, F. M. and Schaffer, H. E. (1973). Isozyme variability in species of the genus *Drosophila*. VII. Genotype environment relationships in populations of *D. melanogaster* from the eastern United States. *Biochem. Genet.* **10**, 149–163.

Johnson, F. M., Schaffer, H. E., Gillaspy, J. E. and Rockwood, E. S. (1969). Isozyme genotype-environment relationships in natural populations of the harvester ant, *Pogonomyrmex barbatus*, from Texas. *Biochem. Genet.* **3**, 429–450.

Jones, J. S., Selander, R. K. and Schnell, G. D. (1980). Patterns of morphological and molecular polymorphism in the land snail *Cepaea nemoralis*. *Biol. J. Linn. Soc.* **14**, 359–387.

Kahler, A. L., Allard, R. W., Krzakowa, M., Wehrhahn, C. F. and Nevo, E. (1980). Associations between isozyme phenotypes and environment in the slender wild oat (*Avena barbata*) in Israel. *Theor. Appl. Genet.* **56**, 31–47.

Karlin, S. (1979). Principles of polymorphism and epistasis for multilocus systems. *Proc. Nat. Acad. Sci., U.S.A.* **75**, 541–546.

Karlin, S. (1981). Some natural viability systems for a multiallelic locus: a theoretical study. *Genetics* **97**, 457–473.

Karlin, S. (1982). Classifications of selection-migration structures and conditions for a protected polymorphism. *Evol. Biol.* **14**, 61–204.

Karlin, S. and Avni, H. (1981). Analysis of central equilibria in multilocus systems: A generalized symmetric viability regime. *Theor. Pop. Biol.* **20**, 241–280.

Karlin, S. and Feldman, M. W. (1981). A theoretical and numerical assessment of genetic variability. *Genetics* **97**, 475–493.

Karlin, S. and Lieberman, U. (1979). Central equilibria in multilocus systems. I. Nonepistatic selection. *Genetics* **91**, 777–798.

Karlin, S. and Levikson, B. (1974). Temporal fluctuations in selection intensities: Case of small population size. *Theor. Pop. Biol.* **6**, 383–412.

Karlin, S. and McGregor, J. L. (1972a). Application of method of small parameters to multiniche population genetic models. *Theor. Pop. Biol.* **3**, 186–209.

Karlin, S. and McGregor, J. L. (1972b). Polymorphisms for genetics and ecological systems with weak coupling. *Theor. Pop. Biol.* **3**, 210–238.

Karlin, S. and McGregor, J. L. (1973). Equilibria for genetic systems with weak interactions. *In* "Proc. Sixth Berkeley Symp. Math. Stat. and Prob." (L. M. Le Cam, J. Neyman, and E. L. Scott, eds), pp. 79–87. University of California Press, Berkeley.

Kerster, H. W. and Levin, D. A. (1970). Temporal phenotypic heterogeneity as a substrate for selection. *Proc. Nat. Acad. Sci., U.S.A.* **66**, 370–376.

Kidwell, J. F., Clegg, M. T., Stewart, F. M. and Prout, T. (1977). Regions of stable equilibria for models of differential selection in the two sexes under random mating. *Genetics* **85**, 171–183.

Kimura, M. (1968). Evolutionary rate at the molecular level. *Nature, Lond.* **217**, 624–626.

Kimura, M. (1969). The number of heterozygous nucleotide sites maintained in a finite population due to steady flux of mutations. *Genetics* **61**, 893–903.

Kimura, M. (1976). Population genetics and molecular evolution. *Johns Hopkins Med. J.* **138**, 253–261.

Kimura, M. (1979a). Model of effectively neutral mutations in which selective constraint is incorporated. *Proc. Nat. Acad. Sci., U.S.A.* **76**, 3440–3444.

Kimura, M. (1979b). The neutral theory of molecular evolution. *Scient. Am.* **241**, 98–126.

Kimura, M. and Crow, J. F. (1964). The number of alleles that can be maintained in a finite population. *Genetics* **49**, 725–738.

Kimura, M. and Ohta, T. (1971). "Theoretical Aspects of Population Genetics". Princeton University Press, New Jersey.

Kimura, M. and Ohta, T. (1974). On some principles governing molecular evolution. *Proc. Nat. Acad. Sci., U.S.A.* **71**, 2848–2852.

King, J. L. and Ohta, T. H. (1975). Polyallelic mutational equilibria. *Genetics* **79**, 681–691.

Kingman, J. F. C. (1977). A note on multi-dimensional models of neutral mutation. *Theor. Pop. Biol.* **11**, 285–290.

Koehn, R. K. (1969). Esterase heterogeneity: Dynamics of a polymorphism. *Science, N.Y.* **163**, 943–944.

Koehn, R. K. (1978). Physiology and biochemistry of enzyme variation: The interface of ecology and population genetics. *In* "Ecological Genetics: The Interface" (P. F. Brussard, ed.), pp. 51–72. Springer-Verlag, New York.

Koehn, R. K. and Eanes, W. F. (1978). Molecular structure and protein variation within and among populations. *Evol. Biol.* **11**, 39–100.

Koehn, R. K. and Immerman, F. W. (1981). Biochemical studies of aminopeptidase polymorphism in *Mytilus edulis*. I. Dependence of enzyme activity on season, tissue, and genotype. *Biochem. Genet.* **19**, 1115–1142.

Koehn, R. K. and Rasmussen, D. I. (1967). Polymorphic and monomorphic serum esterase heterogeneity in Catostomid fish populations. *Biochem. Genet.* **1**, 131–144.

Koehn, R. K. and Siebenaller, J. F. (1981). Biochemical studies of aminopeptidase polymorphism in *Mytilus edulis*. II. Dependence of reaction rate on physical factors and enzyme concentration. *Biochem. Genet.* **19**, 1143–1162.

Koehn, R. K. Bayne, B. L., Moore, M. N. and Siebenaller, J. F. (1980a). Salinity related physiological and genetic differences between populations of *Mytilus edulis*. *Biol. J. Linn. Soc.* **14**, 319–334.

Koehn, R. K., Hall, J. G. and Zera, A. J. (1980b). Parallel variation of genotype-dependent amino peptidase-I activity between *Mytilus edulis* and *Mercenaria mercenaria*. *Marine Biology Letters* **1**, 245–253.

Koehn, R. K., Newall, R. I. E. and Immerman. F. (1980c). Maintenance of an aminopeptidase allele frequency cline by natural selection. *Proc. Nat. Acad. Sci., U.S.A.* **77**, 5385–5389.

Kojima, K. and Yarbrough, K. M. (1967). Frequency-dependent selection at the esterase-6 locus in *Drosophila melanogaster*. *Proc. Nat. Acad. Sci., U.S.A.* **57**, 645–649.

Lavie, B. and Nevo, E. (1982). Heavy-metal selection of phosphoglucose isomerase allozymes in marine gastropods. *Marine Biology*, **71**, 17–22.

Leder, P. (1982). The genetics of antibody diversity. *Scient. Am.* **246**, 72–83.

Lerner, J. M. (1954). "Genetic Homeostasis". Wiley, New York.

Lester, L. J. and Selander, R. K. (1979). Population genetics of haplodiploid insects. *Genetics* **92**, 1329–1345.

Levene, H. (1953). Genetic equilibrium when more than one ecological niche is available. *Am. Nat.* **87**, 331–333.

Levin, D. A. (1973). Polymorphism for interspecific cross-compatibility in *Phlox*. *Proc. Nat. Acad. Sci., U.S.A.* **70**, 1149–1150.

Levin, D. A. (1975). Spatial segregation of pins and thrums in populations of *Hedyotis nigricans*. *Evolution, Lancaster, Pa.* **28**, 648–655.

Levin, D. A. (1978). Some genetic consequences of being a plant. *In* "Ecological Genetics: The Interface" (P. F. Brussard, ed.), pp. 189–212. Springer-Verlag, New York.

Levins, R. (1968). "Evolution in Changing Environments". Princeton University Press, New Jersey.

Levins, R. and Lewontin, R. (1980). Dialectics and reductionism in ecology. *Synthese* **43**, 47–78.

Levy, M. and Levin, D. A. (1975). Genic heterozygosity and variation in permanent translocation heterozygotes of the *Oenothera biennis* complex. *Genetics* **79**, 493–512.

Levy, M., Steiner, E. and Levin, D. A. (1975). Allozyme genetics in the permanent translocation heterozygotes of the *Oenothera biennis* complex. *Biochem. Genet.* **13**, 487–500.

Lewontin, R. C. (1974). "The Genetic Basis of Evolutionary Change". Columbia University Press, New York.

Lewontin, R. C., Ginzburg, L. R. and Tuljapurkar, S. D. (1978). Heterosis as an explanation for large amounts of genic polymorphism. *Genetics* **88**, 149–169.

Li, W. H. (1978). Maintenance of genetic variability under the joint effect of mutation, selection and random drift. *Genetics* **90**, 349–382.

Marinković, D. and Ayala, F. J. (1977). Fitness of allozyme variants in *Drosophila pseudoobscura* III. Possible factors contributing to the maintenance of polymorphisms in nature. *Genetica* **47**, 65–70.

Markert, C. L. (ed.) (1975). "Isozymes" (4 vols). Academic Press, New York.

Maruyama, T. and Nei, M. (1981). Genetic variability maintained by mutation and overdominant selection in finite populations. *Genetics* **98**, 441–459.

Maruyama, T. and Yamazaki, T. (1974). Analysis of heterozygosity in regard to the neutrality theory of protein polymorphisms. *J. Mol. Evol.* **4**, 195–199.

Mayr, E. and Provine, W. B. (eds) (1980). "The Evolutionary Synthesis". Harvard University Press, Cambridge, Mass.

McCracken, G. F. and Brussard, P. F. (1980). The population biology of the white-lipped land snail, *Triodopsis albolabris*: genetic variability. *Evolution, Lancaster, Pa.* **34**, 92–104.

McCracken, G. F. and Selander, R. K. (1980). Self-fertilization and monogenic strains in natural populations of terrestrial slugs. *Proc. Nat. Acad. Sci., U.S.A.* **77**, 684–688.

McDonald, J. F. and Ayala, F. J. (1974). Genetic response to environmental heterogeneity. *Nature, Lond.* **250**, 572–574.

McDonald, J. F. and Ayala, F. J. (1978). Gene regulation in adaptive evolution. *Can. J. Genet. Cytol.* **20**, 159–175.

McDonald, J., Anderson, S. and Santos, M. (1980). Biochemical differences between products of the *Adh*-locus in *Drosophila*. *Genetics* **95**, 1013–1022.

McKechnie, S. W., Ehrlich, P. R. and White, R. R. (1975). Population genetics of *Euphydryas* butterflies. I. Genetic variation and the neutrality hypothesis. *Genetics* **81**, 571–594.

Merritt, R. B. (1972). Geographic distribution and enzymatic properties of lactate dehydrogenase allozymes in the fathead minnow, *Pimephales proleus*. *Am. Nat.* **196**, 173–184.

Milkman, R. D. (1976). Selection is the major determinant. *Trends in Biochemical Science* **1**, N152–N154.

Milkman, R. D. (1978). The maintenance of polymorphisms by natural selection. *In* "Marine Organisms, Genetics, Ecology and Evolution" (B. Battaglia and J. A. Beardmore, eds), pp. 3–22. Plenum Press, New York and London.

Mitton, J. B. and Koehn, R. K. (1973). Population genetics of marine pelecypods. III. Epistasis between functionally related isoenzymes of *Mytilus edulis*. *Genetics* **73**, 487–496.

Mitton, J. B. and Koehn, R. K. (1975). Genetic organization and adaptive response of allozymes to ecological variables in *Fundulus heteroclitus*. *Genetics* **79**, 97–111.

Moran, P. A. P. (1976). A selective model for electrophoretic profiles in protein polymorphisms. *Genet. Res., Camb.* **28**, 47–53.
Muller, H. J. (1950). Our load of mutations. *Am. J. Hum. Genet.* **2**, 111–176.
Narise, S. (1979). Biochemical differences between cytoplasmic malate dehydrogenase allozymes of *Drosophila virilis*. *Biochem. Genet.* **17**, 433–444.
Nei, M. (1973). Analysis of gene diversity in subdivided populations. *Proc. Nat. Acad. Sci., U.S.A.* **70**, 3321–3323.
Nei, M. (1975). "Molecular Population Genetics and Evolution". North Holland, Amsterdam.
Nei, M. (1980). Stochastic theory of population genetics and evolution. *In* "Vito Volterra Symposium on Mathematical Models in Biology" (C. Barigozzi ed.), pp. 17–47, Springer-Verlag, Berlin.
Nei, M. and Roychoudhury, A. K. (1982). Genetic relationship and evolution of human races. *Evol. Biol.* **14**, 1–59.
Nei. M., Fuerst, P. A. and Chakraborty, R. (1976). Testing the neutral mutation hypothesis by distribution of single locus heterozygosity. *Nature, Lond.* **262**, 491–493.
Nei, M., Fuerst, P. A. and Chakraborty, R. (1978). Subunit molecular weight and genetic variability of proteins in natural populations. *Proc. Nat. Acad. Sci., U.S.A.* **75**, 3359–3362.
Nevo, E. (1976). Adaptive strategies of genetic systems in constant and varying environments. *In* "Population Genetics and Ecology" (S. Karlin and E. Nevo, eds), pp. 141–158. Academic Press, New York.
Nevo, E. (1978). Genetic variation in natural populations: patterns and theory. *Theor. Pop. Biol.* **13**, 121–177.
Nevo, E. (1979). Adaptive convergence and divergence of subterranean mammals. *A. Rev. Ecol. Syst.* **10**, 269–308.
Nevo, E. (1982a). Population genetics and ecology. The interface. *In* "Evolution of Molecules and Men" (D. S. Bendall, ed.), pp. 287–321. Cambridge University Press.
Nevo, E. (1982b). Adaptive allozyme polymorphisms in mammals. *Proc. 3rd Intern. Theriological Congr., Helsinki, Acta Zool. Finn.* in press.
Nevo, E. and Bar, Z. (1976). Natural selection of genetic polymorphisms along climatic gradients. *In* "Population Genetics and Ecology" (S. Karlin and E. Nevo, eds), pp. 159–184, Academic Press, New York.
Nevo, E., Shimony, T. and Libni, M. (1977). Thermal selection of allozyme polymorphisms in barnacles. *Nature, Lond.* **267**, 699–701.
Nevo, E., Shimony, T. and Libni, M. (1978). Pollution selection of allozyme polymorphisms in barnacles. *Experientia* **34**, 1562–1564.
Nevo, E., Zohary, D., Brown, A. H. D. and Haber, M. (1979). Genetic diversity and environmental associations of wild barley, *Hordeum spontaneum*, in Israel. *Evolution, Lancaster, Pa.* **33**, 815–833.
Nevo, E., Pearl, T., Beiles, A. and Wool, D. (1981a). Mercury selection of allozyme genotypes in shrimps. *Experientia* **37**, 1152–1154.
Nevo, E., Bar-El, Ch., Bar, Z. and Beiles, A. (1981b). Genetic structure and climatic correlates of desert landsnails. *Oecologia* **48**, 199–208.

Nevo, E., Bar-El, Ch., Beiles, A. and Yom-Tov, Y. (1982a). Adaptive microgeographic differentiation of allozyme polymorphism in landsnails. *Genetica*, **59**, 61–57.

Nevo, E., Beilles, A. and Ben-Shlomo, R. (1983a). The evolutionary significance of genetic diversity: Ecological demographic and life history correlates. Lecture notes in Biomathematics, in press.

Nevo, E., Lavie, B. and Ben-Shlomo, R. (1983b). Selection of allozyme *In* "Isozymes: Current Topics in Biological and Medial Research" (M. C. Ratayyi, J. G. Scandalios and G. S. Whitt, eds). Liss Pub, New York.

Nevo, E., Brown, A. H. D., Zohary, D. Storch, N. and Beiles, A. (1981c). Microgeographic edaphic differentiation in allozyme polymorphisms of wild barley (*Hordeum spontaneum*, Poaceae). *Pl. Syst. Evol.* **138**, 287–292.

Nevo, E., Golenberg, E., Beiles, A., Brown, A. H. D. and Zohary, D. (1982b). Genetic diversity and environmental associations of wild wheat. *Triticum dicoccoides*, in Israel. *Theor. Appl. Genet.*, **62**, 241–254.

Nevo, E., Pearl, T., Beiles, A., Wool, D., and Zoller, U. (1980). Genetic structure as a potential monitor of marine pollution. *V^es Journées Étude Pollutions*, 61–68. Cagliari, C.I.E.S.M.

Newkirk, G. F. and Doyle, R. W. (1979). Clinal variation at an esterase locus. *Littorina saxatilis* and *L. obtusata. Canad. J. Genet. Cytol.* **21**, 505–513.

Oakeshott, J. G., Gibson, J. B., Anderson, P. R., Knibb, W. R., Anderson, D. G. and Chambers, G. K. (1982). Alcohol dehydrogenase and glycerol-3-phosphate dehydrogenase clines in *Drosophila melanogaster* on different continents. *Evolution, Lancaster, Pa.* **36**, 86–96.

Ohta, T. (1978). Theoretical study on genetic variation in multigene families. *Genet. Res., Camb.* **31**, 13–28.

Ohta, T. and Kimura, M. (1973). A model of mutation appropriate to estimate the number of electrophoretically detectable alleles in a finite population. *Genet. Res., Camb.* **22**, 201–204.

Ostergaard, H. and Christiansen, F. B. (1981). Selection component analysis of natural polymorphisms using population samples including mother-offspring combinations, II. *Theor. Pop. Biol.* **19**, 378–419.

Papel, I., Henderson, M., Herrewege, J., David, J. and Sofer, W. (1979). *Drosophila* alcohol dehydrogenase activity *in vitro* and *in vivo*: effects of acetone feeding. *Biochem. Genet.* **17**, 553–563.

Patton, J. L. and Feder, J. H. (1981). Microspatial genetic heterogeneity in pocket gophers: non-random breeding, and drift. *Evolution, Lancaster, Pa.* **35**, 912–920.

Place, A. R. and Powers, D. (1979). Genetic variation and relative catalytic efficiencies: Lactate dehydrogenase B allozymes of *Fundulus heteroclitus. Proc. Nat. Acad. Sci., U.S.A.* **76**, 2534–2538.

Powell, J. R. (1971). Genetic polymorphism in varied environments. *Science, N.Y.* **174**, 1035–1036.

Powell, J. R. (1975). Protein variation in natural populations of animals. *Evol. Biol.* **8**, 79–119.
Powell, J. R. and Taylor, C. E. (1979). Genetic variation in ecologically diverse environments. *Am. Scient.* **67**, 590–596.
Powell, J. R. and Wistrand, H. (1978). The effect of heterogeneous environments and a competitor on genetic variation in *Drosophila*. *Am. Nat.* **112**, 935–947.
Powers, D. A., Greaney, G. S. and Place A. R. (1979). Physiological correlation between lactate dehydrogenase genotype and haemoglobin function in Killi fish. *Nature, Lond.* **277**, 240–241.
Richmond, R. C. (1970). Non-Darwinian evolution: a critique. *Nature, Lond.* **225**, 1025–1028.
Richmond, R. C. (1978). Microspatial genetic differentiation in natural populations of *Drosophila*. *In* "Ecological Genetics: The Interface" (P. F. Brussard, ed.), pp. 127–142. Springer-Verlag, New York.
Rick, C. M. and Fobes, J. F. (1975). Allozymes of Galapagos tomatoes: polymorphism, geographic distribution, and affinities. *Evolution, Lancaster, Pa.* **29**, 443–457.
Rick, C. M., Fobes, J. F. and Holle, M. (1977). Genetic variation in *Lycopersicon pimpinellifolium*: evidence of evolutionary change in mating systems. *Plant Syst. Evol.* **127**, 139–170.
Schaal, B. A. and Levin, D. A. (1976). The demographic genetics of *Liatris cylindracea* Michx. (Compositae). *Am. Nat.* **110**, 191–206.
Schaeffer, H. E and Johnson, F. M. (1974). Isozyme allelic frequencies related to selection and gene-flow hypotheses. *Genetics* **77**, 163–168.
Schaeffer, H. E., Yardley, D. G. and Anderson, W. W. (1977). Drift or selection: A statistical test of gene frequency over generations. *Genetics* **87**, 371–379.
Schnell, G. D. and Selander, R. K. (1981). Environmental and morphological correlates of genetic variation in mammals. *In* "Mammalian Population Genetics" (M. H. Smith and J. Joule, eds), pp. 60–99. University of Georgia Press, Athens.
Scharloo, W., van Dijken, F. R., Hoorn, A. J. W., de Jong, G. and Thorig, G. F. W. (1977). Functional aspects of genetic variation. *In* "Measuring Selection in Natural Populations" (F. B. Christiansen and T. M. Fenchel, eds), pp. 132–147. Springer-Verlag, Berlin.
Selander, R. K. (1976). Genic variation in natural populations. *In* "Molecular Evolution" (F. J. Ayala, ed.), pp. 21–45. Sinauer Press Associates, Sunderland, Mass.
Selander, R. K. and Kaufman, D. W. (1975). Genetic structure of populations of the brown snail (*Helix aspersa*). I. Microgeographic variation. *Evolution, Lancaster, Pa.* **29**, 385–401.
Smouse, P. E., Vitzthum, V. J. and Neel, J. V. (1981). The impact of random and lineal fission on the genetic divergence of small human groups. A case study among the Yanomama. *Genetics* **98**, 179–197.

Smouse, P. E., Spielman, R. S. and Park, M. H. (1982). Multiple-locus allocation of individuals to groups as a function of the genetic variation within and differences among human populations. *Am. Nat.* **119**, 445–463.

Snyder, T. P. and Ayala, F. J. (1979). Frequency-dependent selection at the *PGM-1* locus of *Drosophila pseudoobscura*. *Genetics* **92**, 995–1003.

Somero, G. N. and Soulè, M. (1974). Genetic variation in marine fishes as a test of the niche-variation hypothesis. *Nature, Lond.* **249**, 670–673.

Soulè, M. (1976). Allozyme variation: its determinants in space and time. *In* "Molecular Evolution" (F. J. Ayala, ed.), pp. 60–70. Sinauer Press Associates, Sunderland, Mass.

Steiner, E. and Levin, D. A. (1977). Allozyme, *Si* gene, cytological, and morphological polymorphisms in a population of *Oenothera biennis*. *Evolution, Lancaster, Pa.* **31**, 127–133.

Steiner, W. W. M. (1977). Niche width and genetic variation in Hawaiian *Drosophila*. *Am. Nat.* **111**, 1037–1045.

Stewart, F. M., Clegg, M. T. and Kidwell, J. F. (1979). Two locus models of selection and mutation within and among full-sib lines. *Theor. Appl. Genet.* **54**, 133–139.

Sved, J. A. (1976). The relationship between genotype and fitness for heterotic models. *In* "Population Genetics and Ecology" (S. Karlin and E. Nevo, eds), pp. 441–464. Academic Press, New York.

Taylor, C. E. and Powell, J. R. (1977). Microgeographic differentiation of chromosomal and enzyme polymorphisms in *Drosophila persimilis*. *Genetics* **85**, 681–695.

Taylor, C. E. and Powell, J. R. (1978). Habitat choice in natural populations of *Drosophila*. *Oecologia* **37**, 69–75.

Templeton, A. R. and Levin, D. A. (1979). Evolutionary consequences of seed pools. *Am. Nat.* **114**, 232–249.

Templeton, A. R. and Rothman, E. D. (1981). Evolution in fine-grained environments. II. Habitat selection as a homeostatic mechanism. *Theor. Pop. Biol.* **19**, 326–340.

Terri, J. A. and Peet, M. M. (1978). Adaptation of malate dehydrogenase to environmental temperature variability in two populations of *Potentilla glandulosa* Lintl. *Oecologia* **34**, 133–142.

Tosic, M. and Ayala, F. J. (1980). "Overcompensation" at an enzyme locus in *Drosophila pseudoobscura*. *Genet. Res., Camb.* **36**, 57–67.

Turner, J. R. G., Johnson, M. S. and Eanes, W. F. (1979). Contrasted modes of evolution in the same genome: allozymes and adaptive change in *Heliconius*. *Proc. Nat. Acad. Sci., U.S.A.* **76**, 1924–1928.

Van Delden, W., Kamping, A. and Van Dijk, H. (1975). Selection at the alcohol dehydrogenase locus in *Drosophila melanogaster*. *Experientia* **31**, 418–419.

Van Delden, W., Boerema, A. C. and Kamping, A. (1978). The alcohol dehydrogenase polymorphism in populations of *Drosophila melanogaster*. I. Selection in different environments. *Genetics* **90**, 161–191.

Van Valen, L. (1965). Morphological variation and width of ecological niche. *Am. Nat.* **99**, 377–390.
Van Valen, L. (1974). Molecular evolution as predicted by natural selection. *J. Mol. Evol.* **3**, 89–101.
Varvio-Aho, S. L. (1981). The effects of ecological differences on the amount of enzyme gene variation in Finnish water-strider (*Gerris*) species. *Hereditas* **94**, 35–39.
Varvio-Aho, S. L. and Pamilo, P. (1981). Spatio-temporal microdifferentiation of water-strider (*Gerris*) populations. *Genet. Res., Camb.* **37**, 253–263.
Varvio-Aho, S. L. and Pamilo, P. (1982). Searching for parallel enzyme gene variation among sympatric congeners. *Evolution, Lancaster, Pa.* **36**, 200–203.
Ward, R. D. (1977). Relationship between enzyme heterozygosity and quarternary structure. *Biochem. Genet.* **15**, 123–135.
Ward, R. D. (1978). Subunit size of enzymes and genetic heterozygosity in vertebrates. *Biochem. Genet.* **16**, 799–810.
Watterson, G. A. (1977). Heterosis or neutrality? *Genetics* **85**, 789–814.
Weir, B. S., Brown, A. H. D. and Marshall, D. R. (1976). Testing for selective neutrality of electrophoretically detectable protein polymorphisms. *Genetics* **84**, 639–659.
Whitt, G. S. (1981). Evolution of isozyme loci and their differential regulation. *In* "Evolution Today", Proceedings of the Second International Congress of Systematic and Evolutionary Biology (G. G. E. Scudder and J. L. Reveal, eds), pp. 271–289.
Wills, C. (1973). In defense of naive pan-selectionism. *Am. Nat.* **107**, 23–34.
Wills, C. (1978). Rank-order selection is capable of maintaining all genetic polymorphisms. *Genetics* **89**, 403–417.
Wills, C. (1981). "Genetic Variability". Clarendon Press, Oxford.
Workman, P. L. and Jain, S. K. (1966) Zygotic selection under mixed random mating and self-fertilization: Theory and problems of estimation. *Genetics* **54**, 159–171.
Wright, S. (1921). Correlation and causation. *J. Agric. Res.* **20**, 557–565.
Wright, S. (1931). Evolution in Mendelian populations. *Genetics* **16**, 97–159.
Wright, S. (1932). The roles of mutation, inbreeding, crossbreeding and selection in evolution. *Proc. 6th Int. Congress Genetics* **1**, 356–365.
Wright, S. (1939). The distribution of self sterility alleles in populations. *Genetics* **24**, 538–552.
Wright, S. (1955). Classification of the factors of evolution. *Cold Spring Harb. Symp. quant. Biol.* **20**, pp. 16–24.
Wright, S. (1964). The distribution of self-incompatibility alleles in populations. *Evolution, Lancaster, Pa.* **18**, 609–619.
Wright, S. (1965). The interpretation of population structure by F-statistics with special regard to systems of mating. *Evolution, Lancaster, Pa.* **19**, 355–420.
Wright, S. (1969). "Evolution and the genetics of populations. Vol. 2. The Theory of Gene Frequencies". University of Chicago Press, Chicago.

Wright, S. (1970). Random drift and the shifting balance theory of evolution. *In* "Mathematical Topics in Population Genetics" (K. Kojima, ed.), pp. 1–31. Springer-Verlag, New York.
Wright, S. (1977). "Evolution and the Genetics of Populations. Vol. 3. Experimental results and evolutionary deductions". University of Chicago Press, Chicago.
Wright, S. (1978). "Evolution and the Genetics of Populations. Vol. 4. Variability Within and Among Natural Populations". University of Chicago Press, Chicago.
Yamazaki, T. and Maruyama, T. (1975). Isozyme polymorphism maintenance mechanisms viewed from the standpoint of population genetics. *In* "Isozymes IV Genetics and Evolution" (C. L. Markert, ed.), pp. 103–114. Academic Press, New York.
Yardley, D. G. (1978). Selection at the amylase locus of *D. melanogaster*. A word of caution. *Evolution, Lancaster, Pa.* **32**, 920–921.
Zouros, E. (1976). The distribution of enzyme and inversion polymorphisms over the genome of *Drosophila*: Evidence against balancing selection. *Genetics* **83**, 169–179.
Zouros, E. (1979). Mutation rates, population sizes and amount of electrophoretic variation of enzyme loci in natural populations. *Genetics* **92**, 623–646.

15 Natural Selection in Hybrid Mussel Populations

D. O. F. SKIBINSKI

Department of Genetics, University College of Swansea, Singleton Park, Swansea SA2 8PP, U.K.

Abstract: A study has been made of variation at three diagnostic allozyme loci and variation in shell characters at two localities in south-west England which occur within a zone of intergradation between the mussels *Mytilus edulis* and *M. galloprovincialis*. At both localities there is marked size dependent variation in allele frequencies with larger mussels having much higher frequencies of alleles characteristically at high frequency in *M. galloprovincialis*. Three hypotheses to explain these results are considered: (1) differential growth; (2) an historical change in the genetic composition of the mussel populations; (3) differential viability. Evidence appears to favour the third alternative.

Factor analysis clearly demonstrates a statistical association between genotypes at the diagnostic loci and variation in shell characters. Differential viability is thus unlikely to be acting only or primarily through the allozyme variants.

INTRODUCTION

The common mussel *Mytilus* has been widely used in population genetic studies and numerous examples exist of geographic or clinal variation in the frequencies of allozyme variants (e.g. Koehn *et al.*, 1976; Thiesen, 1978; Skibinski and Beardmore, 1979; Gosling and

Systematics Association Special Volume No. 24, "Protein Polymorphism: Adaptive and Taxonomic Significance", edited by G. S. Oxford and D. Rollinson, 1983, Academic Press, London and New York.

Wilkins, 1981). The variation has, however, been interpreted in different ways. In North America and the Baltic, geographic variation has been viewed as occurring within a single species, *Mytilus edulis* (Koehn *et al.*, 1976; Thiesen, 1978) and has been interpreted in terms of natural selection acting at individual allozyme loci. Temperature and salinity have been identified as possible selective factors (Levinton and Suchanek, 1978; Koehn *et al.*, 1980) and evidence for epistatic interactions in fitness between different allozyme loci has also been reported (Mitton and Koehn, 1973). By contrast, geographic variation in Great Britain and Ireland has been interpreted as resulting from the mixture of *Mytilus edulis* with a second species or form of mussel which resembles the Mediterranean mussel *Mytilus galloprovincialis* (Ahmad and Beardmore, 1976; Skibinski and Beardmore, 1979; Gosling and Wilkins, 1981), and which, in fact, has a high genetic identity with *Mytilus galloprovincialis* (Skibinski *et al.*, 1980). Geographic variation in allozyme frequencies is thought to reflect differences in the distribution of the two forms of mussel which have characteristic but different allele frequencies at several loci. In many respects the electrophoretic work confirms the results of earlier studies based on morphological and physiological characters, where it was also suggested that *Mytilus galloprovincialis* occurs sympatrically with *edulis* in parts of Britain, Ireland and France (Lewis and Seed, 1969; Seed, 1971, 1972, 1978).

Where the two types of mussel do occur sympatrically, random samples of individuals often show large deviations from Hardy-Weinberg equilibrium and inter-locus correlations at diagnostic loci, and associations between these loci and diagnostic morphological characters (Skibinski *et al.*, 1978a; Skibinski *et al.*, 1978b; Skibinski and Beardmore, 1979) – observations symptomatic of mixtures of different species. At some localities, however, intra- and inter- locus allelic correlations are weak or absent (Skibinski and Beardmore, 1979) which suggests local extensive hybridization between the two forms. Again, similar results were obtained through the study of morphological characters alone (Seed, 1978).

While clines or geographic variation in *Mytilus* may have different explanations in different places, it is important in studying the adaptive significance of allozyme variation to show that physiological or ecological differences between genotypes at a locus under investigation cannot be attributed to other loci for which two, perhaps cryptic, species also show a difference. Even if hybridization has reduced or eroded allelic correlations between allozyme loci, strong linkage disequilibrium may still occur between allozyme loci and other closely linked loci. A cline resulting from selection at one or more loci may be an effective barrier to gene flow past the cline at a linked locus even if variation at this locus is neutral, the effectiveness depending on the closeness of the linkage (Barton, 1979).

Whatever the role of selection at individual loci, the importance and nature of the ecological and selective forces controlling the distributions of *edulis* and *galloprovincialis* and the extent of hybridization are not yet well understood. Gosling and Wilkins (1981) provided evidence that *galloprovincialis* is more prevalent in exposed than sheltered localities in Ireland. This was confirmed for the British mainland by Skibinski *et al.* (1982), where evidence was found that *galloprovincialis* is less prevalent than *edulis* in estuarine environments.

Most reported studies of allozyme variation in British *Mytilus* have been concerned with macrogeographic variation. This paper considers a more detailed study of size-dependent allozyme variation at two localities within a zone of intergradation of *edulis* and *galloprovincialis*. Allozyme loci have proved to be of value as diagnostic markers in the analysis of hybrid zones and introgression between related species (e.g. Hunt and Selander, 1973; Avise and Smith, 1974; Blackwell and Bull, 1978) and it is in this role that the allozyme variation studied here is primarily viewed. Evidence is present allowing the tentative conclusion that strong selection acting through differential viability is operating at these localities.

METHODS

The two localities which provided mussels for investigation in this study are at Mevagissey and Whitsand's Bay on the southern coast of south-west England (Fig. 1). To the west, the *galloprovincialis* form predominates; to the east, the *edulis* form predominates. Both localities are on open coastline with the mussels attached in patches to rock faces. Numerous small sub-samples (clumps of mussels around $15cm^2$) scattered over areas of several thousand m^2 were removed for investigation from Mevagissey in August 1979 and August and September 1981 and from Whitsand's Bay in November and July 1980 and February and May 1981. Unless stated otherwise, the results for a

Fig. 1. The location of the mussel sample sites at Mevagissey and Whitsand's Bay in south-west England.

given locality are obtained by pooling data over sub-samples and collection dates.

Mevagissey was represented in the survey reported by Skibinski and Beardmore (1979). Samples of mussels showed strong inter-locus correlations at diagnostic loci but quite small deviations from Hardy-Weinberg equilibrium. This was interpreted (Skibinski and Beardmore, 1979; Skibinski *et al.*, 1982) as evidence of considerable hybridization but not complete mixing of gene pools. The situation at Whitsand's Bay is very similar. At both localities, there is no evidence of any bimodality in morphology and mussels showing varying degrees of resemblance to *edulis* and *galloprovincialis* are found.

Three allozyme loci *Esterase-D* (*Est-D*), *Leucine aminopeptidase-1* (*Lap-1*) and *Octopine dehydrogenase* (*Odh*) were used. *Lap-1* was used only in studying the earlier samples of mussels and thereafter was replaced by *Odh*. The methods of electrophoretic assay for *Est-D* and *Lap-1* were essentially those used by Ahmad *et al.* (1977). *Odh* was stained using a method similar to that of Beaumont *et al.* (1980). Each of the three loci has six or more alleles and, to facilitate analysis, the alleles at each locus are pooled to form two synthetic alleles, one consisting of alleles typically at high frequency in *edulis* (*Est-D*^{E} = 0·97, *Lap-1*^{E} = 0·94, *Odh*^{E} = 0·90, approximately), the other consisting of alleles typically at high frequency in *galloprovincialis* (*Est-D*^{G} = 0·97, *Lap-1*^{G} = 0·93, *Odh*^{G} = 0·86, approximately). For convenience, these are often referred to simply as *E* and *G*. It should be noted from the frequency values given in parentheses that the alleles are not perfectly diagnostic; for example *G*/*G* *edulis* individuals and *E*/*E* *galloprovincialis* individuals are occasionally found.

A morphometric analysis was also performed using some of the shell characters identified by Seed and co-workers which are useful in distinguishing between typical *edulis* and typical *galloprovincialis* forms.

(1) Shell height, which is greater in *galloprovincialis*. (2) The distance of the maximum width of the shell from the ventral side of the shell, which is shorter in *galloprovincialis*. (3) The amount of curvature of the ventral edge of the shell, which tends to be greater in *galloprovincialis*. (4 and 5) The length and width of the anterior adductor muscle scar, which is longer and more oval in shape in *edulis*. (6) The proportion of the area of the anterior adductor scar, which is coloured blue. In *edulis*, this proportion is high, caused by the prismatic layer of the shell. In *galloprovincialis*, the blue colour is often absent or covered by the white nacrous layer of the shell. (7) Shell length was also measured. The measurements were made in mm on one valve of the two shells of each individual mussel examined.

The morphometric analysis was made on a pooled sample of mussels from Whitsand's Bay collected in July 1980 and February and May 1981. A single valve from each individual mussel was taken and the valves sorted into three groups according to the di-locus genotypes at the *Est-D* and *Odh* loci. The three groups were (1) *Est-D*^{E}/*Est-D*^{E} *Odh*^{E}/*Odh*^{E}. (2) *Est-D*^{E}/*Est-D*^{G} *Odh*^{E}/*Odh*^{G}. (3) *Est-D*^{G}/*Est-D*^{G} *Odh*^{G}/*Odh*^{G}. The shells in each group were then ranked according to shell length in mm and triads of shells, one from each group, were taken such that in each triad, shells did not differ in length by more than one mm. 25 triads, (75 shells) were obtained in this way. The characters described above were then measured on each shell, the values for characters 1–5 being standardized by dividing by shell length, character 7. A factor analysis using the method of principal components was then made on the characters 1–7 for the 75 shells.

RESULTS AND DISCUSSION

Variation in genotype frequency with shell length for the two localities is shown in Fig. 2. In this and subsequent analyses, the mussels were grouped into 5 mm size classes according to shell length. A consistent pattern can be seen for all loci and both localities – the frequency of *G/G* increases with shell length at the expense of a decrease in the frequency of *E/E*. *E/G* also shows a less marked increase in frequency with shell length. A similar pattern of change was found if the size distribution was divided into percentiles. Three possible explanations of these changes are:

(1) *G/G* and *E/G* grow faster and have a greater maximum growth than *E/E*.
(2) There is an historical change occurring in the composition of the mussel population with *E/E* gradually replacing other genotypes.
(3) *G/G* and *E/G* have higher viability than *E/E*, assuming that length is roughly proportional to age.

Differential viability is the most appealing hypothesis and, perhaps, the most plausible. The other two hypotheses will, however, be considered first. A test for differential growth was made in two ways using mussels from Whitsand's Bay. Individuals in a sample of mussels collected in August 1980 were lightly scored with a saw, using the method of Thiesen (1975), so that new growth could be distinguished, and then transplanted in submerged net cages to Swansea Docks. Food is abundant in the Docks and potential for growth easily measured. After two months the mussels were removed; at this time, little mortality was recorded. Individuals were typed electrophoretically at all three loci and new growth measured. Length before and after growth is plotted in Fig. 3 for mussels homozygous *E/E* at all loci and mussels homozygous *G/G* at all loci. The growth rate (e^{-k}) and maximum growth (L_{∞}) of the von Bertalanffy growth equation were computed and no significant differences found between the two groups of mussels. The analysis was also repeated using the same shells but by measuring the natural annual growth on those of the shells where growth rings could be clearly identified. Although non-annual check rings on the shell may be laid down as a result of environmental stress, and rings may be obliterated by erosion, it is reasonable to

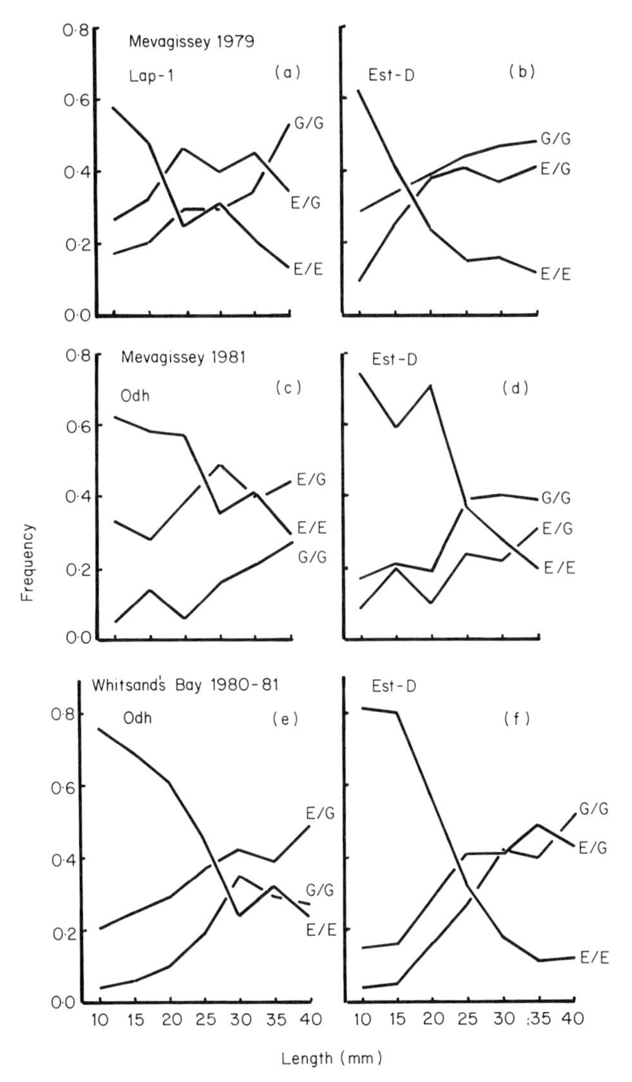

Fig. 2. Genotype frequencies at three loci *Lap-1, Est-D* and *Odh* plotted against size class for Mevagissey and Whitsand's Bay. (a) Mevagissey 1979 *Lap-1* (sample size, n = 767 individuals), (b) Mevagissey 1979 *Est-D* (n = 637), (c) Mevagissey 1980–81 *Odh* (n = 597), (d) Mevagissey 1980–81 *Est-D* (n = 583), (e) Whitsand's Bay 1980–81 *Odh* (n = 1042), (f) Whitsand's Bay 1980–81 *Est-D* (n = 1211). The numbers on the abscissa indicate size classes as follows 10 (10–14 mm), 15 (15–19 mm), 20 (20–24 mm), 25 (25–29 mm), 30 (30–34 mm), 35 (35–39 mm), 40 ($\geq$ 40 mm).

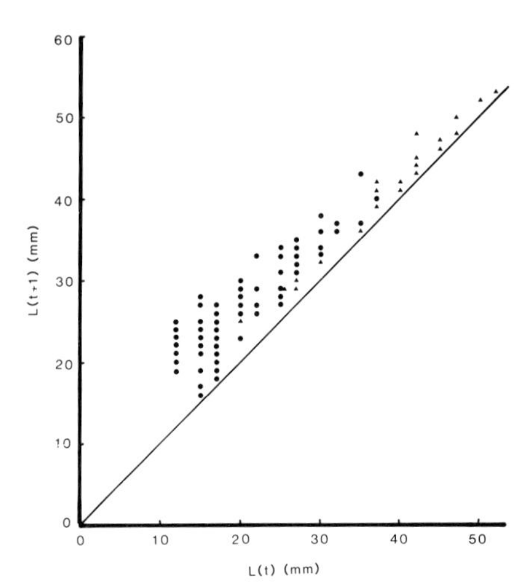

Fig. 3. Growth of Whitsand's Bay mussels in Swansea Docks. L_t, initial shell length in mm, L_{t+1}, shell length two months later. ●, mussels homozygous *E/E* at all three loci, ▲, mussels homozygous G/G at all three loci. The von Bertalanffy equation is $L_{t+1} = L_t e^{-k} + L_\infty (1 - e^{-k})$, where L_∞ is the maximum length attainable and k is the growth constant. For *E/E E/E E/E*, $e^{-k} = 0{\cdot}81$ and $L_\infty = 54$, for *G/G G/G G/G*, $e^{-k} = 0{\cdot}94$ and $L_\infty = 83$. The differences between the groups are not significant. For natural growth of mussels (see text), for *E/E E/E E/E*, $e^{-k} = 0{\cdot}80$ and $L_\infty = 60$, for *G/G G/G G/G*, $e^{-k} = 0{\cdot}76$ and $L_\infty = 61$. Differences between groups are not significant.

assume no systematic bias between the two groups when comparing mussels from the same locality. The values of the growth parameters are given in the legend of Fig. 3; once again, no significant differences between groups were observed. A large difference in growth would be required to explain the results of Fig. 2, and it is quite clear from inspection of Fig. 3 that no sufficiently large difference in growth could exist within the bounds of statistical error associated with the data presented.

The second hypothesis, that of historical change, could occur in a number of ways. A takeover by the *edulis* form by immigration of zygotes from the east where *edulis* is prevalent could be occurring. Another possibility is that the *edulis* form now has an adaptive advantage with respect to viability among the smaller size classes which it lacked in the past, or, the fecundity of *edulis* may now be greater than in the past. The only sure way of testing the hypothesis is to compare samples taken over a number of years. Some relevant data are given in Fig. 4 where the frequency of *Est-D*^{G} is plotted against size class for samples taken from approximately the same place high on the shore at Mevagissey in 1978, 1979 and 1981. Comparison of allele frequency between years in the smaller size classes should be most informative as the annual amount of growth at this stage will be greater. Comparing the 1979 and 1981 data in the 10–30 mm size range, there is no indication of a large decrease in the frequency of *Est-D*^{G}. There is also very little difference between the results for 1978 and 1981. Thus, these data do not provide strong support for the hypothesis that *edulis* is gradually replacing *galloprovincialis*. Future sampling may clarify the situation.

Providing good direct evidence for the third hypothesis, that of differential viability, is difficult and ideally requires following a sample of tagged mussels over time in the field. This is not easy without holding the mussels captive within a net cage, and because they are naturally attached to exposed rocks the experimental procedure would result in alteration in environment. In the absence of good evidence for differential growth or historical change, differential viability does, however, receive indirect support. Evidence for some selection acting through differential viability is presented in Fig. 5. This shows the frequency of the G alleles for samples of mussels taken

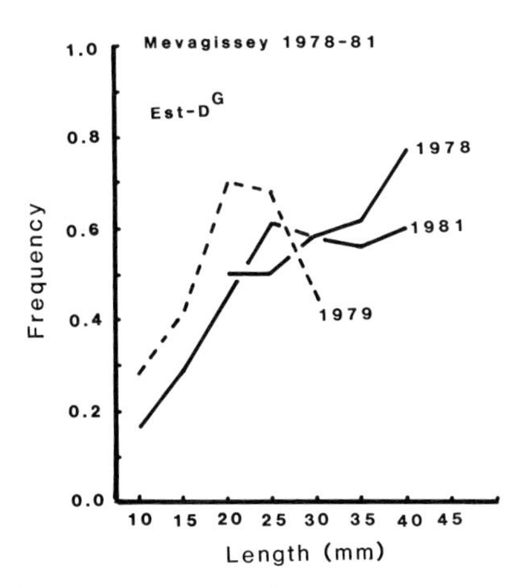

Fig. 4. Size-dependent variation in frequency of *Est-D*^{G} in mussels taken high on the shore at Mevagissey in 1978 ($n = 120$), 1979 ($n = 241$) and 1981 ($n = 120$). Labelling of abscissa as in Fig. 2.

from high up and low down in the tidal zone. Two samples, one high and one low, were taken along a number of transects at each locality and Fig. 5 gives pooled values over transects for high and low. Most of the transects were taken on vertical or near vertical rock faces. The difference between high and low is marked with the frequency of the G allele being generally greater in the high than low samples. It is difficult to see how this could be the result of an historical change

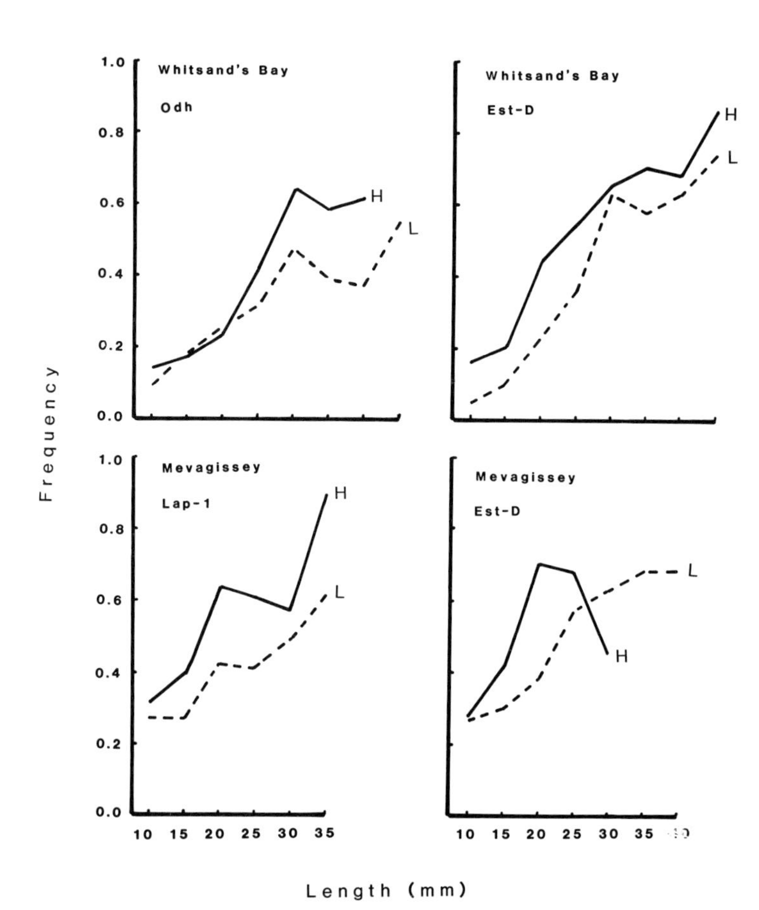

Fig. 5. Size-dependent variation in frequency of allele *G* at three loci *Est-D*, *Lap-1* and *Odh* in high shore (H) and low shore (L) samples for Mevagissey and Whitsand's Bay. Sample sizes, Mevagissey – *Lap-1*, (High, n = 274), (Low, n = 238), *Est-D*, (High, n = 241), (Low, n = 184); Whitsand's Bay – *Est-D*, (High, n = 814), (Low, n = 647), *Odh*, (High, n = 814), (Low, n = 647). Labelling of abscissa as in Fig. 2.

alone, and although growth is generally more rapid lower down the shore, the difference persists in the largest size classes. The most plausible explanation at present is that differential viability is acting in the two micro-environments. It has been suggested that the nature of the morphological differences between *edulis* and *galloprovincialis* indicates that the latter is better adapted for powerful anchorage to the substrate (Seed, 1978) in line with trends observed in other bivalve lineages (Yonge and Campbell, 1968). This suggestion would be consistent with the observed allele frequency differences if conditions higher up the shore do require more powerful anchorage. Mussels that become detached high up the shore could either die or become re-attached lower down. In both cases, the observed allele frequency differences could be explained.

Assuming that age is proportional to size, a quantitative estimate of relative viability in the transition from a small to a large size class can be made. If the frequencies of *E/E*, *E/G* and *G/G* before selection are *a, b* and *c*, respectively, and the frequencies after selection are *A, B* and *C*, respectively, and the relative viabilities are w_1, w_2 and 1, respectively, it can be shown that $w_2 = Bc/bC$ and $w_1 = Ac/aC$. Applying these equations in a comparison of the Whitsand's Bay 10–19 mm and ⩾ 40 mm size classes gives $w_2 = 0{\cdot}182$ and $w_1 = 0{\cdot}009$ for *Est-D* and $w_2 = 0{\cdot}374$ and $w_1 = 0{\cdot}058$ for *Odh*. The variation in relative viability is high, and there is an indication of allelic interaction as, for both loci, the square of the viability of E/G is greater than the viability of *E/E*. The sampling error will be high, however, and the method of estimation must be regarded with caution because mussels in each of the two size classes are heterogeneous with respect to age. Accurate testing of hypotheses concerning viability requires good estimates of the age of individual mussels.

Skibinski and Beardmore (1979) used measures of deviation from Hardy-Weinberg equilibrium and di-locus genotypic association as indicators of hybridization and intergradation in mussel samples from British localities. Values of these measures are given in Table I for the different size classes at each locality. The observation that many of the values of *F* and *R* are significant and positive is consistent with the hypothesis that the samples consist of mixtures of populations differing in allele frequencies. The values are not higher in the larger size classes; in fact, there is an indication that *F* may be higher in the smaller size classes. This suggests that the significant positive values are not

Table I. Various measures of genotypic association for Whitsand's Bay 1980–81 and Mevagissey 1981.

Size class (mm)	Whitsand's Bay				Mevagissey			
	n	$F(Est\text{-}D)$	$F(Odh)$	R	n	$F(Est\text{-}D)$	$F(Odh)$	R
10–14	213	0·22[b]	0·15[a]	− 0·03	96	0·43[c]	0·03	0·29[a]
15–19	178	0·23[b]	0·15[a]	0·37[c]	85	0·50[c]	0·29[b]	0·92[c]
20–24	157	0·33[c]	0·22[b]	0·33[b]	63	0·38[b]	0·04	0·63[c]
25–29	170	0·21[b]	0·22[b]	0·62[c]	71	0·20	− 0·03	0·73[c]
30–34	126	0·17	0·15	0·51[c]	108	− 0·01	0·18	0·62[c]
35–39	106	− 0·04	0·23[a]	0·35[a]	117	0·01	0·09	0·32[a]
40–44	46	− 0·02	− 0·00	0·45	40	− 0·07	0·20	0·32
≥45	44	0·16	0·04	0·63[a]				

[a] $P < 0{\cdot}05$. [b] $P < 0{\cdot}01$. [c] $P < 0{\cdot}001$.

If single locus genotype frequencies are $E/E(a)$, $E/G(b)$, and $G/G(c)$, F, the measure of deviation from Hardy-Weinberg, $= (4ac - b^2)/((2a + b)(2c + b))$ (Brown 1970). If the di-locus double homozygote frequencies are $E/E\ E/E\ (r)$, $E/E\ G/G\ (t)$, $G/G\ E/E\ (x)$, and $G/G\ G/G\ (z)$, R, a measure of inter-locus genotypic correlation $= (rz - xt)/((r + t)(x + z)(r + x)(t + z))^{1/2}$ (Skibinski and Beardmore 1979).

generated by any differential viability acting but are more likely to occur as a result of gene flow from predominantly *edulis* or *gallo-provincialis* areas to the east and west, perhaps aided by a difference in reproductive period between the two forms.

The results of the morphometric analysis are shown in Fig. 6 in which scores of shells on the first two factors are plotted. The first factor is most useful and has high positive loadings of adductor scar length and colour, a moderately high negative loading for shell curvature and a positive loading for distance of maximum width from the ventral edge. *Edulis* genotypes should, therefore, have high positive scores on factor 1. This is what is observed. The homozygotes *E/E E/E* and *G/G G/G* cluster in different regions, and there is an indication that a majority of the double heterozygote individuals are intermediate, rather than resembling a mixture of the two homozygote groups. The results, therefore, provide unambiguous quanti-

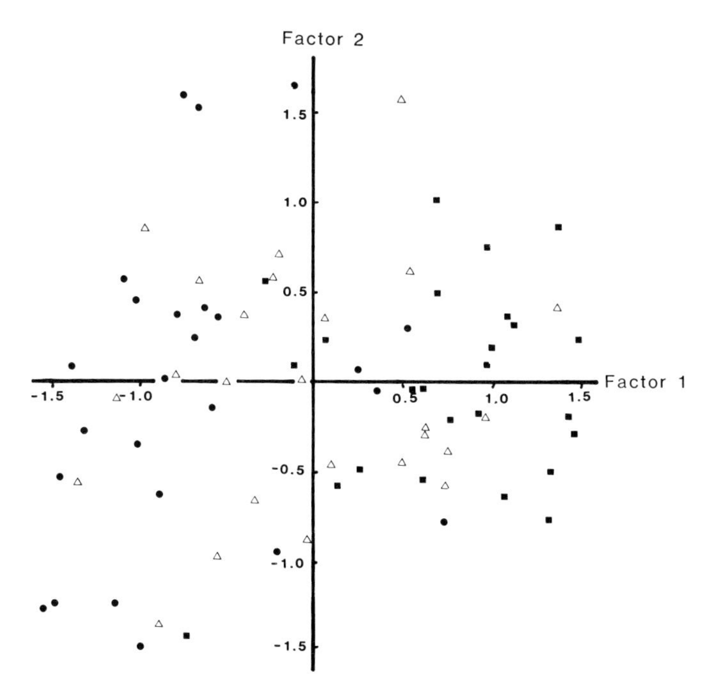

Fig. 6. Morphometric analysis of mussel shells from Whitsand's Bay. The scores on factor 1 are plotted against the scores on factor 2 for each of 75 mussels. ●, *Est-D*^{G}/*Est-D*^{G} *Odh*^{G}/*Odh*^{G}; ■, *Est-D*^{E}/*Est-D*^{E} *Odh*^{E}/*Odh*^{E}; △, *Est-D*^{E}/*Est-D*^{G} *Odh*^{E}/*Odh*^{G}.

tative evidence of an association between the allozyme variants and morphological variation. The hypothesis that any differential viability is acting only or primarily through the allozyme variants must be viewed with scepticism.

In conclusion, if the genetic structure is more or less stable over time, differential viability in favour of the *galloprovincialis* form must be balanced by some advantage of *edulis*. This could be a higher fecundity or preferential immigration and settlement of *edulis* zygotes. Also, the higher viability of *E*/*G* heterozygotes relative to *E*/*E* suggests that, at least with respect to this component of fitness, heterozygote inferiority plays no part in the maintenance of the *edulis*-*galloprovincialis* cline in south-west England.

ACKNOWLEDGEMENTS

I wish to thank R. Armitage, M. Power and particularly E. Roderick for technical assistance. This work was supported in part by grant GR3/3988 from the Natural Environment Research Council to the author.

REFERENCES

Ahmad, M. and Beardmore, J. A. (1976). Genetic evidence that the "Padstow Mussel" is *Mytilus galloprovincialis*. *Marine Biol.* **9**, 139–147.

Ahmad, M., Skibinski, D. O. F. and Beardmore, J. A. (1977). An estimate of the amount of genetic variation in the common mussel *Mytilus edulis*. *Biochem. Genet.* **15**, 833–846.

Avise, J. C. and Smith, M. H. (1974). Biochemical genetics of sunfish I. Geographic variation and subspecific integradation in the bluegill. *Lepomis macrochirus*. *Evolution, Lancaster, Pa.* **28**, 42–56.

Barton, N. H. (1979). Gene flow past a cline. *Heredity* **43**, 333–339.

Beaumont, A. R., Day, T. R. and Gäde, G. (1980). Genetic variation at the octopine dehydrogenase locus in the adductor muscle of *Cerastoderma edule*. (L) and six other bivalve species. *Marine Biol. Letters* **1**, 137–148.

Blackwell, J. M. and Bull, C. M. (1978). A narrow hybrid zone between two western Australian frog species *Ranidella insignifera* and *R. pseudinsignifera*: the extent of introgression. *Heredity* **40**, 13–25.

Brown, A. H. D. (1970). The estimation of Wright's fixation index from genotypic frequencies. *Genetica* **41**, 399–406.

Gosling, E. M. and Wilkins, N. P. (1981). Ecological genetics of the mussels *Mytilus edulis* and *M. galloprovincialis* on Irish Coasts. *Marine Ecol.* **4**, 221–227.

Hunt, W. G. and Selander, R. K. (1973). Biochemical genetics of hybridisation in European house mice (*Mus musculus*). *Heredity* **31**, 11–33.
Koehn, R. K., Newell, R. I. E. and Immermann, F. (1980). Maintenance of an aminopeptidase allele frequency cline by natural selection. *Proc. Nat. Acad. Sci., U.S.A.* **77**, 5385–5389.
Koehn, R. K., Milkman, R. and Mitton, J. B. (1976). Population genetics of marine pelecypods. IV. Selection, migration and genetic differentiation in the blue mussel *Mytilus edulis*. *Evolution, Lancaster, Pa.* **30**, 2–32.
Levinton, J. S. and Suchanek, T. H. (1978). Geographic variation, niche breadth and genetic differentiation at different geographic scales in the mussels *Mytilus californianus* and *M. edulis*. *Marine Biol.* **49**, 363–375.
Lewis, J. R. and Seed, R. (1969). Morphological variations in *Mytilus* from south-west England in relation to the occurrence of *M. galloprovincialis* (Lmk). *Cah. Biol. mar.* **10**, 231–253.
Mitton, J. B. and Koehn, R. K. (1973). Population genetics of marine pelecypods. III. Epistasis between functionally related isoenzymes of *Mytilus edulis*. *Genetics* **73**, 487–496.
Seed, R. (1971). A physiological and biochemical approach to the taxonomy of *Mytilus edulis* L. and *M. galloprovincialis* Lmk from south-west England. *Cah. Biol. mar.* **12**, 291–322.
Seed, R. (1972). Morphological variations in *Mytilus* from the French coasts in relation to the occurrence and distribution of *M. galloprovincialis* Lmk. *Cah. Biol. mar.* **13**, 357–384.
Seed, R. (1978). The systematics and evolution of *Mytilus galloprovincialis* Lmk. *In* "Marine Organisms: Genetics, Ecology and Evolution" (B. Battaglia and J. A. Beardmore, eds), pp. 447–468. Plenum, New York and London.
Skibinski, D. O. F. and Beardmore, J. A. (1979). A genetic study of intergradation between *Mytilus edulis* and *Mytilus galloprovincialis*. *Experientia* **35**, 1442–1444.
Skibinski, D. O. F., Ahmad, M. and Beardmore, J. A. (1978a). Genetic evidence of naturally occurring hybrids between *Mytilus edulis* and *Mytilus galloprovincialis*. *Evolution, Lancaster, Pa.* **32**, 354–364.
Skibinski, D. O. F., Beardmore, J. A. and Ahmed, M. (1978b). Genetic aids to the study of closely related taxa of the genus *Mytilus*. *In* "Marine Organisms: Genetics, Ecology and Evolution" (B. Battaglia and J. A. Beardmore, eds), pp. 469–485. Plenum, New York and London.
Skibinski, D. O. F., Cross, T. F. and Ahmad, M. (1980). Electrophoretic investigation of systematic relationships in the marine mussels *Modiolus modiolus* L., *Mytilus edulis* L., and *Mytilus galloprovincialis* Lmk. (Mytilidae; Mollusca). *Biol. J. Linn. Soc.* **13**, 65–73.
Skibinski, D. O. F., Beardmore, J. A. and Cross, T. F. (1982). Aspects of the population genetics of *Mytilus* (Mytilidae; Mollusca) in the British Isles. *Biol. J. Linn. Soc.*, in press.
Thiesen, B. F. (1975). Growth parameters of *Mytilus edulis* L. (Bivalvia) estimated from tagging data. *Meddr Danm. Fisk. Havunders*, n.s. **7**, 99–109.

Thiesen, B. F. (1978). Allozyme clines and evidence of strong selection in three loci in *Mytilus edulis* L. (Bivalvia) from Danish waters. *Ophelia* **17**, 135–142.

Yonge, C. M. and Campbell, J. I. (1968). On the heteromyarian condition in the Bivalvia with special reference to *Dreissena polymorpha* and certain Mytiliacea. *Trans. R. Soc. Edinb.* **68**, 21–43.

Part IV | Population Structure, Breeding Systems, and Molecular Taxonomy

16 Population Structure, Breeding Systems and Molecular Taxonomy

A. SAURA

Department of Genetics, University of Helsinki,
P. Rautatiek. 13, SF-00100 Helsinki 10, Finland

Abstract: Plants possess very variable breeding systems from self-incompatibility (i.e. obligatory outbreeding) to complete inbreeding. Many species are apomictic, i.e. reproduce vegetatively or through agamospermy. Different populations of a single species may exhibit different modes of reproduction. Protein variants have provided useful markers that have been applied to study these evolutionary problems. In contrast to this, the breeding system of animal populations is rather straightforward. Also here parthenogenetic forms (without recombination) are useful in understanding population structure of sexually reproducing forms. From the standpoint of evolution, small and isolated populations are most interesting.

INTRODUCTION

The genetic structure of a group of populations follows from patterns of mating within single populations and from the amount of gene exchange between populations. This can be conveniently expressed as a deviation from Hardy-Weinberg proportions within single populations and as a variance in allele frequencies among populations. Inbreeding within a population causes a proportionate decline of the average expected number of heterozygous loci in an individual, compared to the average number in its ancestors. This may also be expressed as an average decline in population heterozygosity at all loci or, conversely, as an increase in homozygosity. The measure of the

Systematics Association Special Volume No. 24, "Protein Polymorphism: Adaptive and Taxonomic Significance", edited by G. S. Oxford and D. Rollinson, 1983, Academic Press, London and New York.

process is the coefficient of inbreeding, F. For a discussion of various F-statistics, see e.g. Wright (1965). Subdivision of a large population into a set of small ones causes a similar result (Wahlund, 1928).

High rates of gene flow among populations and tendencies toward random mating within populations decrease the amount of genetic structuring. Obstructions of movement, restraints on random mating, and mere physical distances in the absence of a barrier will all promote genetic differentiation and structuring.

According to the shifting balance theory of Wright (1931), conditions most favourable for progressive evolution are found in species subdivided into a set of local populations, at least some of which are of small or medium size. In Wright's terms, the interaction of selection, drift and weak gene flow operating on several local populations give rise to adaptive transitions more effectively than only selection and drift in a single population.

The founder effect hypothesis of Mayr stresses the evolutionary significance of the latter mode. In the course of migration and dispersal, new populations of a species may be founded by a small number of colonists. The genetic variability of such a population will be limited to those alleles introduced by the founders. These alleles may not be representative of the species as a whole. As such, this bottleneck effect is a special case of Wright's shifting balance hypothesis.

In the absence of selection, genetic structure should reflect parameters of population structure, such as population size, amount of gene flow and breeding system. The difficulty here is to ascertain the non-selective nature of the genetic variation used. In the case of electrophoretic variation, non-selectivity cannot be simply assumed. This is reflected in most of the literature reviewed here: when the structure of a population has been defined through electrophoretically detectable variation, the result is used to test the hypothesis of neutrality.

There are several comprehensive reviews on the patterns of electrophoretic variation in animal and plant species, e.g. Powell (1975), Nevo (1978), Brown (1979) and Gottlieb (1981). In general, early animal studies were concerned with measuring the amount of polymorphism, expressed either as P, the proportion of polymorphic loci or as H, the average heterozygosity per locus per individual. The breeding systems of many animals, in particular the insects, widely used as models in population genetics, are often difficult to measure.

On the other hand, a measure such as H, the average heterozygosity, is rather meaningless for most plants.

Plants are, in general, characterized by a wide variety of breeding systems, ranging from outcrossing to selfpollination. These factors are influenced by variables such as population density, the relationships between male and female floral organs (whether in the same plant or in separate male and female plants); the time of function of male and female organs; mechanism of pollination and pollen competition. Asexual and parthenogenetic reproduction, collectively called apomixis in plants, is a mode of reproduction characteristic of plants but rare in higher animals. Finally, all variables mentioned above may be subject to both ecological and/or genetic control. Conversely, the breeding system imposes constraints on the floral structure that are repeated over a wide range of angiosperms (Ornduff, 1969).

Accordingly, the breeding system has a profound influence on the amount and structuring of genetic variation in populations and in sets of populations. The problem has been treated theoretically by Allard *et al.* (1968). In theory, inbreeders are expected to be less heterozygous, contain less genetic diversity and their populations should be more differentiated from each other than populations of outbreeders. The fixation index of Wright, F, is an appropriate measure for expressing the genotypic structure at a single locus,

$$F = 1 - \frac{\text{Observed heterozygosity}}{\text{Heterozygosity predicted under panmixia}}$$

The sampling variance of F is highest between ($0 < F < 0{\cdot}3$), i.e. typical for outbreeding populations. Predominant inbreeders in which $F > 0{\cdot}9$, have sampling variances for F of about a quarter their equivalent in outbreeding populations with identical allele frequencies (Brown, 1979).

PRIMARILY OUTBREEDING ORGANISMS

1. Drosophila *and conifers*

Historically, electrophoretic techniques were first applied to taxonomic problems within the genus *Drosophila* (Hubby and Throckmorton, 1965). It was apparent that the process of genetic differentiation could be followed with the new method. This taxonomic breakthrough was immediately shadowed by the application of the method to the study of even more fundamental problems of evolutionary genetics.

The early studies on natural *Drosophila* species (e.g. Lewontin and Hubby, 1966; Prakash *et al.*, 1969; Lakovaara and Saura, 1971; Ayala *et al.*, 1972) all indicated rather uniform levels of heterozygosity over geographically discontinuous areas. Protein polymorphism in *Drosophila* species is, in general, geographically undifferentiated. This contrasts strongly with the manifest differentiation in chromosomal inversion polymorphism. Concepts of the genetic structure of *Drosophila* populations were based on models based on inversion polymorphisms (Carson, 1959). We may, of course, assume that much of the electrophoretic variation is selectively neutral and, further, that the species as a whole behaves as a panmictic unit. This would lead to the evenly distributed electrophoretic allozyme frequencies over both central and marginal populations of *Drosophila*. This situation is probably the most commonly held view on electrophoretic protein variation, as it has entered the college textbook level.

Now the situation is not that simple. There is evidence for geographic clines as well as deviations from Hardy–Weinberg proportions in marginal populations (Saura *et al.*, 1973b). There have been numerous attempts to measure the population structure of natural *Drosophila* populations. Actually, the effective size, N_e, of a population should be estimated before population genetics theories are applied to natural populations. In practice, population geneticists assume that N_e for outbreeding species is sufficiently large to be ignored. Many should be relieved by the fact that, at least geographically central *Drosophila* populations, are extremely large and effectively infinite (e.g. Begon *et al.*, 1980).

Conifers provide another example of predominant outbreeders. The pollen flow between continuous populations is not entirely unrestricted, but enzyme markers can be conveniently used to estimate the rates of inbreeding (Müller, 1976). Species such as the Norway spruce (*Picea abies*) have very little allozyme differentiation between central and marginal populations (Tigerstedt, 1973; Lundkvist, 1979). There is also an ample literature on Douglas fir (*Pseudotsuga menziesii*) with data on the uniformity of allele frequencies (e.g. Yang *et al.*, 1977) and estimates of the rates of outcrossing derived by the use of enzyme markers (Shaw and Allard, 1982).

The inbreeding coefficients of many forest trees range from about zero in species such as the pitch pine (*Pinus rigida*) (Guries and Ledig, 1982) to moderate inbreeding, in cases where populations are differentiated over relatively short distances. The Scots pine (*Pinus silvestris*) is an example of a population structure in which the offspring of a single maternal parent falls close to the maternal tree (Rudin, 1975). This pattern of shedding seeds gives rise to clusters of relatives, a family structure, in natural stands. This pattern of inbreeding is a part of the mating system. A close look at *Drosophila* populations shows similar situations (e.g. Saura *et al.*, 1973b; Taylor and Powell, 1977). In general, however, genetic and geographic distances are weakly correlated in the above cases.

2. *Marine organisms*

The effects of migration, environmental heterogeneity and selection on genetic markers are difficult to measure in marine organisms. There are cases in which the population structure should be simple. Fishes like the eel provide such an example. Both American and European eels spawn in the Atlantic in a fashion that strongly suggests that each species behaves as a panmictic unit. Williams *et al.* (1973) and Koehn and Williams (1978) argue that the allozyme frequency changes observed for incoming juveniles along the east coast of North America must reflect natural selection rather than the consequences of population structuring.

Mussels have free-swimming larvae that are able to disperse freely. This should create a uniform population structure. Contrary to expectation, studies on mussels indicate a deficit of heterozygotes in comparison with Hardy-Weinberg expectations (e.g. Zouros *et al.*, 1980; Colgan, 1981). The deficit varies from locus to locus, but locus-specific deficits do not necessarily change from cohort to cohort. In general, the heaviest mussels are the most heterozygous ones. This could be explained easily as a consequence of overdominance in growth rate – an explanation that does not account for the deficit in the number of heterozygotes. It might be due to the breeding system (Wahlund effect) but this appears improbable in species with freely dispersing larvae. An age-structuring of populations may give rise to such effects. There is, indeed, evidence of age-dependent protein variation in diverse organisms (Fujino and Kang, 1968; Tinkle and Selander, 1973). The simplest explanation is that there are errors in scoring the genotypes in the enzyme system with the most pronounced deficit in heterozygosity, namely leucine aminopeptidase (LAP).

The genetic structures of marine organisms respond to environmental changes. Information on the structure can be used to monitor the effect of marine pollution. Nevo *et al.* (1978, 1980) have shown that genotype rather than allozyme frequencies of the barnacle *Balanus amphitrite* and the shrimp *Palaemon elegans* respond to different degrees of pollution.

3. *Migration effects*

Salmonid fishes have been intensely studied (for a review, see e.g. Allendorf and Utter, 1979). Birds have been virtually neglected. This is most probably due to the fact that electrophoresis does not yield much information on bird populations. Most systems are monomorphic. Isoelectric focusing would offer a tool that could be applied in bird populations.

The monarch butterfly (*Danaus plexippus*), studied by Eanes and Koehn (1978) illustrates the effects of migration versus stationary populations. The inbreeding coefficient is the same in migratory and non-migratory collections of the monarch. In the migratory collections, the mixing of subpopulations of different origins gives clear evidence for the Wahlund effect (F_{ST} is decreased while F_{IS} is increased).

Water striders (*Gerris*) have proven to constitute a good testing situation for the study of the effects of migration on population structure. There are differing degrees of winglessness both within and between species (Varvio-Aho, 1981; Zera, 1981). Different species also occur in habitats differing in size and permanence. Different degrees of winglessness result in very different genetic structures in these species (Varvio-Aho, this volume).

PARTHENOGENETIC AND INBREEDING FORMS

1. *Parthenogenesis and selfing*

In the preceding examples, species of non-social animals such as many insects and marine organisms were supposed – and often proven – to behave as a single panmictic unit. There are also phenomena associated with geographical isolation in such forms. From the assumption of panmixis, one may turn to the other extreme, population structure in parthenogenetic, apomictic or self-fertilizing organisms. Parthenogenesis is a term applied to animals. In automictic parthenogenesis a meiotic mechanism exists, but the reduction in chromosome number is compensated for by a process that restores the original number. In apomictic parthenogenesis, meiosis is replaced by mitosis. Apomixis in plants covers agamospermy and vegetative reproduction. Both are common in plants.

In animals with apomictic parthenogenesis (or automixis accompanied by a mechanism that does not enforce homozygosity), the population structure reflects the pattern of environment. This is best seen in a monophyletic parthenogenetic lineage, in which new genotypes are produced through mutations. The distribution of these types depends on their powers of dispersal as well as on their genetic constitution. Apomictic insects are often wingless forms. The population structure typical for these sessile parthenogenetic insects (Saura *et al.*, 1976), earthworms (Christensen *et al.*, 1977; Jaenike *et al.*, 1980), fishes (Vrijenhoek, 1978) and lizards (Parker and Selander, 1976) consists of a set of genotypes, each with slightly differing adaptive properties. The distributions of these genotypes may overlap widely. They are distributed in an orderly succession along an ecological gradient (Saura *et al.*, 1976). The situation is illustrated in the accompanying diagram (Fig. 1).

The population structure of forms without recombination is, accordingly, in stark contrast to the evenly distributed allozyme frequencies in outbreeding organisms. An assessment of these differences appears problematic. There has been a general adherence to the principle of competitive exclusion (cf. White, 1978) and a consequent reluctance to admit that different parthenogenetic genotypes might coexist. The exclusion principle does not, however, hold in these

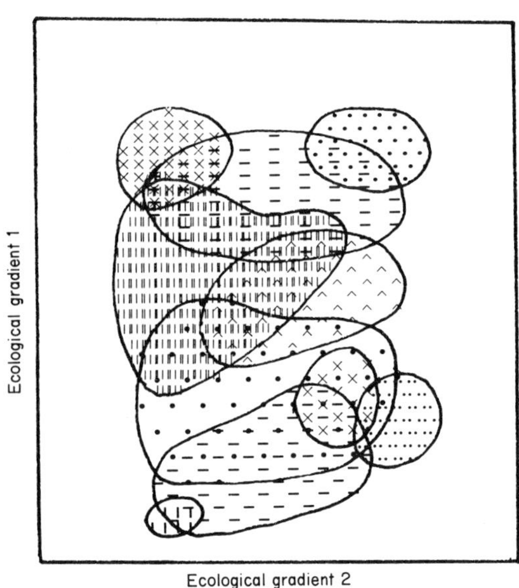

Fig. 1. The population structure of a parthenogenetic animal consists of separate genotypes with overlapping distributions. The distributions of genotypes are indicated with different shading. Each genotype is adapted to a slightly different environment.

forms. An interaction of genetic and ecological factors affecting the population structure appears here to be simple. Evidently, we need more experimental data on comparable situations. The relative advantages of different breeding systems must be weighed against a poor theoretical background for understanding the effects of genetic recombination on evolution (see the reviews by Felsenstein, 1974 and Maynard Smith, 1978). Parthenogenetic or asexual species should have higher levels of heterozygosity than outcrossing species, both within populations and over the entire species, as they can retain mutant alleles by sheltering them from segregation (Lokki, 1976).

Selfing species should be less heterozygous and polymorphic than outbreeding species (Jain, 1976). An animal example of self-fertilization or close inbreeding is the land snail *Rumina decollata* (Selander and Kaufman, 1973; Selander and Hudson, 1976). The American (introduced) populations of this snail originating from the Mediterranean region lack electrophoretically detectable variation. In the Mediterranean region, the species is a complex of strains maintained by facultative self-fertilization. These monogenic strains remain distinct in nature, even though they possess the potential for introgression. The distinctness implies that the strains possess "general purpose genotypes". Likewise, parthenogenetic insects with an efficient dispersal ability have widely distributed genotypes (Lokki *et al.*, 1976).

2. *Inbreeding*

The simple dichotomy between inbreeders and outbreeders is by no means a complete description of the breeding system. Breeding systems vary in space and time, and this difference certainly has very profound effects on the genetic structure of populations. This variation is subject to genetic control.

An important consequence of inbreeding is the maintenance of coadapted gene complexes. The concept of coadaptation is, in general, understood in terms of chromosomal inversions in *Drosophila* (Dobzhansky, 1970). Inversions prevent recombination in heterozygotes and give rise to coadapted blocks of genes. Coadaptation was shown to be associated with inbreeding in the wild oats species *Avena barbata* in California. *A. barbata* is an introduced species originating from the Mediterranean area. Clegg and Allard (1972) showed that five enzyme loci were monomorphic in the semi-arid grasslands region of California. An opposite allele set was prevalent in ecologically more variable cooler regions of the state. Populations in intermediate habitats were polymorphic for these loci.

In the topographically diverse northern and cooler region, changes from mesic to xeric habitats occur over short distances. Such situations provide a testing ground for the study of between locus associations on a microgeographic scale (Hamrick and Allard, 1972; Allard *et al.*, 1972). The results show a consistent pattern of allele frequencies that are correlated with ecological factors. The alleles characteristic of mesic locations steadily decrease in frequency from the mesic environment at the bottom to the xeric environment at the top of a hillside transect. Correlation coefficients between alleles at pairs of loci are of the order of 0·90, that is, statistically significant. As any one allele changes in one locus, this is accompanied by a change in other loci as well. These changes reflect the changes in the degree of xerism with remarkable precision.

A study of the gametic types (Allard *et al.*, 1972) showed that out of very many possible combinations, two types predominate. These two are the 12221 and 21112 types characteristic for xeric and mesic habitats. They are present in great excess over expectation based on single-locus allele frequencies. The loci in question are scattered over the chromosomes of *A. barbata*, and, accordingly, inbreeding has been the factor binding the entire genotype together as a coadapted complex that may be compared to a single locus with two alleles. The average amount of outcrossing in *A. barbata* is about 2%. This gives the species enough recombination potential to produce all possible combinations in large populations.

Barley and its wild progenitor, *Hordeum spontaneum*, have also been extensively studied (Brown *et al.*, 1978 and the review by Brown, 1979). Here a correlation is found between the outcrossing rate and environment. Outcrossing is significantly more common in populations growing in the mesic regions of Israel (2·1%) than in the more xeric regions (0·4%). A feature associated with heterozygosity in this predominantly inbreeding species is multilocus heterozygosity. This is a direct consequence of occasional outcrossing in populations composed of several homozygous lines differentiated from each other at many loci. The effect may be magnified by an excess of certain favoured gametic types. A concentration of heterozygosity in outcrossing cases gives the occasional outcrossing in an otherwise inbreeding species a greater evolutionary significance than the small proportion of heterozygotes otherwise would indicate.

There are allozyme studies on closely related plants with contrasting breeding systems (e.g. Solbrig, 1972; Ellstrand and Levin, 1980). Within species studies are probably more valuable as they allow comparisons to be made between closely related populations. There have been relatively few studies on species containing both selfing and outcrossing populations. The studies on wild tomato species (*Lycopersicon*) (Rick *et al.*, 1977, 1979) illustrate many phenomena associated with populations of plants. In the Canadian wild rye (*Elymus canadensis*), the outcrossing rates vary widely from population to population (Sanders and Hamrick, 1980). An interesting observation in the genetic structure of many plant species is that variations in the breeding system in combination with ecological changes produce clear marginal effects. In the wild barley species *Hordeum jubatum*, the interaction between selection and drift has promoted a differentiation of populations along each type of ecological margin (Shumaker and Babbel, 1980). The marginal populations are less variable than central populations. This is in remarkable contrast to the genetic structure of such model organisms as wild *Drosophila* species.

Summing up, the breeding system has a profound effect on the genetic structure of plant populations. Inbred species have little pollen flow between populations and, as a consequence of this, they are capable of maintaining adaptations to micro-environmental variation more readily than outbred species.

There are certain general results that have been observed by electrophoretic studies in plant populations (Brown, 1979). In outbreeding species, there is a general deficit of heterozygotes in comparison with panmictic expectations, while, in inbreeding species, there is commonly an excess of heterozygotes. Short-lived or herbaceous outbreeders tend to show a large deficit, while trees often conform to Hardy-Weinberg expectations. The deficit is less pronounced in the wind pollinated versus animal pollinated species.

The two results superficially contradict the assumption that outbreeding should favour heterozygosity. Likewise, if heterozygosity is favoured under inbreeding, why should a species adhere to inbreeding and not produce the favoured types through outbreeding? These observations are in contrast to the models of evolution of selfing and outcrossing (Maynard Smith, 1978). Brown (1979) terms these observations the heterozygosity paradox.

The paradox can be solved, in part, in terms of the examples discussed above. Plant populations are capable of forming very small subdivisions and they may respond to very minute environmental pressures. There will always be some Wahlund effect. Likewise, there will often be some inbreeding present in plant populations. These factors together decrease heterozygosity in populations. Increased levels of inbreeding give rise to coadapted multilocus associations and overdominance with an increase in heterozygosity.

PATTERNS OF POPULATION STRUCTURES

1. *Area effects*

Snails have contributed much to our knowledge on the structure of natural populations. They are animals that do not move far within the lifetime of a single individual. Studies on their shell characters have shown that extensive changes in allele frequencies may be found over very short distances. Typical examples of such steep clines are the area effects of *Cepaea nemoralis* in southern England. Populations are characterized by uniformity of morphological polymorphism over large areas – in fact, far larger than the area of a single panmictic unit (Murray, 1972).

White (1978) has interpreted the concept of an area effect in terms of a sympatric speciation model, the area effect speciation. Area effects should represent geographic races with widely divergent allele frequencies across the entire genome. Johnson (1976) studied allozyme polymorphism in association with area effects on the Berkshire Downs. He noticed a correlation at two enzyme loci with a shell banding pattern polymorphism (midbands). Studies on a larger scale, both in Wales and in the Pyrenees (Jones *et al.*, 1980; Caugant *et al.*, 1982) show that while some enzyme alleles are associated with the geographic pattern of morphological variation, most are not; there is, in fact, little concordance of geographic pattern of morphological and enzyme loci.

Selander and Kaufman (1975) have studied the population structuring of the brown snail (*Helix aspersa*). Local populations are much smaller than in *Cepaea*, populations are subdivided on a "microgeographic" scale. There is evidence of random drift over short distances at enzyme loci. McCracken and Brussard (1980) showed that the white-lipped land snail (*Triodopsis albolabris*) is, in fact, a complex of three separate species, distinguishable through electrophoretic variants. The population structures are, in general, similar to those in other snails.

2. *Host races*

Phytophagous insects are considered to be capable of sympatric speciation through the formation of host races. The idea goes back to Fisher's notion on the adaptive divergence of ecological races in situations where hybrids are at a disadvantage. Genetic variation in habitat preference or mating preference could lead to the evolution of a decreased frequency of hybridization. Sufficient conditions for this postulated mode of sympatric speciation have been formulated by Maynard Smith (1966). A stable polymorphism can be established in a heterogeneous environment, in which alternative alleles are selected in different niches. If this polymorphism is coupled with a strong habitat selection and mating within this niche, sympatric speciation can occur. These concepts have attained much prominence as a mechanism of sympatric speciation (Bush, 1975; White, 1978).

There has been rather little testing of these hypotheses by electrophoretic techniques. Guttman *et al.* (1981) have shown that the sympatric "host races" of the Membracid *Enchenopa binotata* represent separate gene pools and are biological species reproductively isolated from each other. Jaenike and Selander (1979) did not find genetic evidence for host race formation in *Drosophila falleni*. This species lives in several species of mushrooms. Jaenike and Selander attributed their results to the lack of sufficient resource predictability, which would allow for the formation of host races in this species.

Parthenogenetic species also show, in this case, a straightforward adaptive response to the environment. Mitter *et al.* (1979) have studied the population structure of the parthenogenetic and flightless moth *Alsophila pometaria*. Different genotypes of this moth occur sympatrically in mixed forests. The genotypes are adapted to hatch at a time of foliation of their particular tree species. This structuring of populations present in parthenogenetic species is absent in related, sexually reproducing species (Mitter and Futyuma, 1979).

3. *Social structures*

The social structuring of mammal populations has been attributed an important role in the rapid evolution that our class of animals has undergone (Bush *et al.*, 1977). The social structure of mammal populations should, indeed, impose constraints on the genetic structure of populations. This was clearly shown by Selander (1970), who studied the mouse (*Mus musculus*) by a grid trapping pattern in a single barn. The barn proved to be occupied by many semi-isolated mouse families. Allele distributions within the barn show a patchy pattern differentiated over very short distances.

Such situations may be analysed further by electrophoretic techniques. Miller Baker (1981) studied the mice in a situation comparable to Selander's with the aid of a different electrophoretic allele introduced to the population. The result showed that the social structuring due to behavioural isolation is not strong enough to prevent gene flow through the breeding of adjacent neighbours. The rate of gene flow suffices to prevent the effects of random drift. In general, small mammals are characterized by a multitude of mating systems, so that results obtained on a single species are not necessarily applicable to others (see e.g. Folz, 1981).

The social structuring and distribution of genetic variation among mammal colonies may, in fact, retard the fixation of genetic variation (Schwartz and Armitage, 1980). Such a result supports the traditional theory of gradual evolution in mammals, rather than the model of Bush *et al.* (1977), which envisages accelerated evolution in social mammals.

Very much attention has been paid to the social structure of eusocial Hymenoptera. Hymenopteran males are haploid and the females diploid. This gives rise to an interesting system of relatedness. There are very many questions that may be answered by electrophoretic methods, such as the origin of the reproductive castes, the evolution of several reproductive females (polygyny) and the evolution of a system with several nests (polydomy) (Ward, 1980; Pamilo, 1982).

The swarming behaviour of certain insect groups, most notably Mayflies and Chironomids, might be assumed to give rise to rapid changes in mating preferences. The Mayfly, *Leptophlebia marginata*, provides a case where sympatric and morphologically similar populations swarm synchronously in different fashions (Saura *et al.*, 1979 and unpublished). The two populations orient to different swarm markers, which gives rise to an efficient ethological isolation. An electrophoretic analysis showed that there were two reproductively isolated forms, i.e. biological species. There is, however, the possibility that gene flow may occur between them through northern populations.

4. *Cyclic populations*

The regular herbivore-predator cycles in population density belong to central problems in ecology. Certain insects may have population size varying 10 000-fold over several years. Voles vary about 100-fold. The first genetic explanation was that of Chitty (1960). He hypothesized that selection operates with unequal intensity upon different phases of the cycle. This changes the genetic constitution of the population. Pimentel (1968) proposed a genetic feedback mechanism, in which the level of herbivore numbers is determined by an adaptation involving both the herbivore and its food plants. Expressed in ecological terms, there should be genotypes with *r*- and *K*-strategies in the population. The social structure of the population contributes also to the different intensities of selection and leads to different modes of dispersal.

Chitty (1960) visualized a polymorphism involving an aggressive, efficiently reproducing morph and a docile, slowly reproducing morph. In the decline phase of the cycle, the aggressive morph decreases in frequency, while the frequency of the other morph remains more or less constant. A genetic change is here accompanying a demographic change; both changes are controlled by selection.

This hypothesis may, of course, be tested. It should be demonstrated that allele frequencies change in different phases of the cycle. Semeonoff and Robertson (1968) demonstrated allozyme frequency changes during a population decline of the field vole (*Microtus agrestis*) in Scotland. Changes in allele frequencies at a transferrin locus (Tamarin and Krebs, 1969) and at both transferrin and aminopeptidase loci (Gaines and Krebs, 1971) were shown to be correlated with changes in the density of the populations. The cause and effect relationship remains largely unsolved, i.e. do the allele frequency changes cause the demographic change or is the genetic change only a consequence of the change in demography. These concepts have been tested in very many experimental designs. Tests involve parallel observations on behaviour and allozyme frequencies (e.g. Rasmuson *et al.*, 1977; Kohn and Tamarin, 1978; Nygren, 1980). Electrophoretically detectable changes need not be responsible for the cyclicity but they indicate how selection acts. The loci relevant in the regulation of demography may, in part, be enzyme loci. Behavioural selection is probably the most effective driving force here. It can certainly be studied by electrophoretic markers.

Another form of regular population cyclicity is the cyclic parthenogenesis in aphids and Cladocerans. Both groups have been studied electrophoretically (e.g. Hebert *et al.*, 1971; Young, 1979; Suomalainen *et al.*, 1980). Parthenogenesis fixes certain allele combinations and allows the effects of selection to be studied in different experimental situations.

5. *Island populations*

Small island populations have characteristics that make them particularly useful in population genetic studies. They are characterized by reproductive isolation from other populations of a species. The number of individuals is small and can often be quite easily estimated. The habitat area is well defined and reduced. The population structure is simple in comparison with mainland populations. Comparisons can be made between island and mainland populations that allow inferences on the structure of the latter. The evolutionary factors associated with island populations (in particular random drift) decrease genetic variability and promote variability between populations.

Electrophoretic studies made on small island populations often clearly show the effects of drift. Saura *et al.* (1973a) studied small island populations of the spittlebug (*Philaenus spumarius*). The spittle masses produced by nymphs of these insects can be counted and the total population size is obtained accurately on a small island. Due to land uplift on the northern coast of the Baltic Sea, the maximum age of these populations can also be estimated. The results show clearly the effects of drift and selection in small island populations. Soulé and Yang (1973), Avise *et al.* (1974) and Schmitt (1978) have also found evidence for reduced variability in island populations.

There are also studies on island populations that do not show effects of reduced variability. Ayala *et al.* (1971) found similar amounts of electrophoretic variation in the Caribbean island populations of *Drosophila willistoni*. This was an interesting observation, as these – quite large – populations lacked the chromosomal inversion polymorphism typical of the mainland populations. Other studies on island populations that have not shown reduced amounts of enzyme polymorphism concern island populations of mice (Berry and Murphy, 1970; Berry and Peters, 1975).

6. *Cave populations and bottleneck effects*

Cave animals (troglodytes or obligatory cavernicoles), in typical cases, inhabit limestone caverns. Many of these forms are aquatic. They have probably evolved from surface populations of troglophiles that have become isolated from each other in the formation of cave habitats. The Pleistocene glaciations have certainly affected the evolution of these forms, and often mark the starting point in their evolution.

Cave populations show typically reduced amounts of electrophoretic variability. The differences in heterozygosity are easily explained as a consequence of the age of the cave population since the last established population bottleneck (cf. Sbordoni *et al.*, 1981). Typical effects of population bottlenecks are evident in the studies by Avise and Selander (1972) of cave fishes, Cockley *et al.* (1977) on cave crickets and Gooch and Hetrick (1979) on *Gammarus*.

An extreme case of a historically established bottleneck is provided by the northern elephant seals, which were very nearly exterminated about 1880. The uniform homozygosity noted in them (Bonnell and Selander, 1974) is probably due to the fixation of alleles at the extreme population bottleneck.

Introductions of animals and plants into new habitats or continents may also give rise to bottleneck effects, as the number of introduced specimens is small; examples range from plants (Schwaegerle and Schaal, 1979) to fishes (Avise and Felley, 1979) and flies (Bryant *et al.*, 1981).

SYNOPSIS

The evolutionary effect of breeding systems was clearly seen by Charles Darwin. He demonstrated an association between hermaphroditism and a sessile habit in animals and plants (Darwin, 1876). The interaction between breeding systems and population structure has had a very prominent role in the development of population genetics theory. The evolution of species consists of tendencies towards change in small isolated populations. The main range of a species is characterized by relative uniformity. Continuous interbreeding between populations holds the species as a biological entity.

As pointed out by Lande (1980), gene flow is not an effective homogenizing agent against spatially varying selection. Now we have very large amounts of electrophoretic variation with unknown function. Genetic homeostasis and stabilizing selection would provide at least one solution. There is some controversial evidence for and against developmental homeostasis at the allozyme level. This is hardly surprising, as electrophoresis detects only degrees of heterozygosity in a fraction of the total genome.

An interaction of selection, drift and weak gene flow operating more or less independently in several variously isolated populations is the agent of adaptive transition allowing evolutionary experiments in a diverse set of environments. A new phenotype with superior fitness may arise and displace pre-existing phenotypes first in one population. It would then spread to other populations. There are numerous studies that illustrate certain aspects of such a development, e.g. the family structure of a viviparous fish (Christiansen, 1977), studies on the isolation-by-distance model in newts (Hedgecock, 1978) and evidence for sympatric speciation under intense selection (Vuorinen *et al.*, 1981).

The populations of parthenogenetic species are highly structured. These species are characteristically sessile forms. Selfing species possess, likewise, a structuring manifest in intense allele correlations over loci. By inbreeding, species can track microgeographic changes. Selection can modify the breeding system with apparent ease.

REFERENCES

Allard, R. W., Jain, S. K. and Workman, P. L. (1968). The genetics of inbreeding populations. *Adv. Genet.* **14**, 55–131.

Allard, R. W., Babbel, G. R., Clegg, M. T. and Kahler, A. L. (1972). Evidence for coadaptation in *Avena barbata*. *Proc. Nat. Acad. Sci., U.S.A.* **69**, 3043–3048.

Allendorf, F. W. and Utter, F. M. (1979). Population genetics. *In* "Fish Physiology, Vol. VIII" (W. S. Hoar and D. J. Randall, eds), pp. 407–454. Academic Press, New York.

Avise, J. C. and Felley, J. (1979). Population structure of freshwater fishes I. Genetic variation of bluegill (*Lepomis machrochirus*) populations in man-made reservoirs. *Evolution, Lancaster, Pa.* **33**, 15–26.

Avise, J. C. and Selander, R. K. (1972). Evolutionary genetics of cave-dwelling fishes of the genus *Astyanax*. *Evolution, Lancaster, Pa.* **26**, 1–19.

Avise, J. C., Smith, M. H., Selander, R. K., Lawlor, T. E. and Ramsay, P. R. (1974). Biochemical polymorphism and systematics in the genus *Peromyscus*. V. Insular and mainland species of the subgenus *Haplomylomus*. *Syst. Zool.* **23**, 226–238.

Ayala, F. J., Powell, J. R. and Dobzhansky, T. (1971). Polymorphisms in continental and island populations of *Drosophila willistoni*. *Proc. Nat. Acad. Sci., U.S.A.* **68**, 2480–2483.

Ayala, F. J., Powell, J. R., Tracey, M. L., Maurão, C. A. and Pérez-Salas, S. (1972). Enzyme variability in the *Drosophila willistoni* group. IV. Genic variation in natural populations of *Drosophila willistoni*. *Genetics* **70**, 113–139.

Begon, M., Krimbas, C. B. and Loukas, M. (1980). The genetics of *Drosophila subobscura* populations. XV. The effective size of a natural population estimated by three independent methods. *Heredity* **45**, 335–350.

Berry, R. J. and Murphy, H. M. (1970). The biochemical genetics of an island population of the house mouse. *Proc. R. Soc. Lond.,* B **176**, 87–103.
Berry, R. J. and Peters, J. (1975). Macquarie Island house mice: A genetical isolate on a sub-Antarctic island. *J. Zool.* **176**, 375–389.
Bonnell, M. L. and Selander, R. K. (1974). Elephant seals: genetic variation and near extinction. *Science, N.Y.* **184**, 908–909.
Brown, A. H. D. (1979). Enzyme polymorphism in plant populations. *Theor. Pop. Biol.* **15**, 1–42.
Brown, A. H. D., Zohary, D. and Nevo, E. (1978). Outcrossing rates and heterozygosity in natural populations of *Hordeum spontaneum* Koch in Israel. *Heredity* **41**, 49–62.
Bryant, E. H., van Dijk, H. and van Delden, W. (1981). Genetic variability of the face fly, *Musca autumnalis* De Geer. *Evolution, Lancaster, Pa.* **35**, 872–881.
Bush, G. L. (1975). Modes of animal speciation. *A. Rev. Ecol. Syst.* **6**, 339–364.
Bush, G. L., Case, S. M., Wilson, A. C. and Patton, J. L. (1977). Rapid speciation and chromosomal evolution in mammals. *Proc. Nat. Acad. Sci., U.S.A.* **74**, 3942–3946.
Carson, H. L. (1959). Genetic conditions which promote or retard the formation of a species. *Cold Spring Harbor Symp. quant. Biol.* **24**, 87–105.
Caugant, D., Selander, R. K. and Jones, J. S. (1982). Geographic structuring of molecular and morphological polymorphism in Pyrenean populations of the snail *Cepaea nemoralis. Genetica* **57**, 177–191.
Chitty, D. (1960). Population processes in the vole and their relevance to general theory. *Can. J. Zool.* **38**, 99–113.
Christensen, B., Jelnes, J. and Berg. U. (1977). A comparative study on enzyme polymorphism in sympatric diploid and polyploid populations of *Lumbricillus lineatus* (O.F.M.) Enchytraeidae, Oligochaeta. *In* "Lecture Notes in Biomathematics 19" (S. Levin, ed.), pp. 365–379. Springer-Verlag, Berlin.
Christiansen, F. B. (1977). Population genetics of *Zoarces viviparus* (L.): A review. *In* "Lecture Notes in Biomathematics 19" (S. Levin, ed.), pp. 21–47. Springer-Verlag, Berlin.
Clegg, M. T. and Allard, R. W. (1972). Patterns of genetic differentiation in the slender wild oats species, *Avena barbata. Proc. Nat. Acad. Sci., U.S.A.* **69**, 1820–1824.
Cockley, D. E., Gooch, J. L. and Weston, D. P. (1977). Genetic diversity in cave dwelling crickets (*Ceuthophilus gracilipes*). *Evolution, Lancaster, Pa.* **31**, 313–318.
Colgan, D. J. (1981). Spatial and temporal variation in the genotypic frequencies of the mussel *Brachidontes rostratus. Heredity* **46**, 197–208.
Darwin, C. (1876). "The Effects of Cross and Self Fertilisation in the Vegetable Kingdom." John Murray, London.
Dobzhansky, T. (1970). "Genetics of the Evolutionary Process". Columbia University Press, New York.
Eanes, W. F. and Koehn, R. K. (1978). An analysis of the genetic structure in

the monarch butterfly, *Danaus plexippus* L. *Evolution, Lancaster, Pa.* **32**, 784–797.

Ellstrand, N. C. and Levin, D. A. (1980). Recombination system and population structure in *Oenothera*. *Evolution, Lancaster, Pa.* **34**, 923–933.

Felsenstein, J. (1974). The evolutionary advantage of recombination. *Genetics* **78**, 737–756.

Folz, D. W. (1981). Genetic evidence for long-term monogamy in a small rodent, *Peromyscus polionotus*. *Am. Nat.* **117**, 665–675.

Fujino, K. and Kang, T. (1968). Transferrin groups of tunas. *Genetics* **59**, 79–91.

Gaines, M. S. and Krebs, C. J. (1971). Genetic changes in fluctuating vole populations. *Evolution, Lancaster, Pa.* **24**, 702–723.

Gooch, J. L. and Hetrick, S. W. (1979). The relation of genetic structure to environmental structure: *Gammarus minus* in a karst area. *Evolution, Lancaster, Pa.* **33**, 192–206.

Gottlieb, L. D. (1981). Electrophoretic evidence and plant populations. *Progr. Phytochem.* **7**, 1–46.

Guries, R. P. and Ledig, F. T. (1982). Genetic diversity and population structure in pitch pine (*Pinus rigida* Mill.). *Evolution, Lancaster, Pa.* **36**, 387–402.

Guttman, S., Wood, T. K. and Karlin, A. A. (1981). Genetic differentiation along host plant lines in the sympatric *Enchenopa binotata* Say complex (Homoptera: Membracidae). *Evolution, Lancaster, Pa.* **35**, 205–217.

Hamrick, J. L. and Allard, R. W. (1972). Microgeographical variation in allozyme frequencies in *Avena barbata*. *Proc. Nat. Acad. Sci., U.S.A.* **69**, 2100–2104.

Hebert, P. D., Ward, R. D. and Gibson, J. B. (1972). Natural selection for enzyme variants among parthenogenetic *Daphnia magna*. *Genet. Res., Camb.* **19**, 173–176.

Hedgecock, D. (1978). Population subdivision and genetic divergence in the red-bellied newt, *Taricha rivularis*. *Evolution, Lancaster, Pa.* **32**, 271–286.

Hubby, J. L. and Throckmorton, L. H. (1965). Protein differences in *Drosophila*. II. Comparative species genetics and evolutionary problems. *Genetics* **52**, 203–215.

Jaenike, J. and Selander, R. K. (1979). Ecological generalism in *Drosophila falleni*: genetic evidence. *Evolution, Lancaster, Pa.* **33**, 741–748.

Jaenike, J., Parker, E. D. and Selander, R. K. (1980). Clonal niche structure in the parthenogenetic earthworm *Octolasion tyrtaeum*. *Am. Nat.* **116**, 196–205.

Jain, S. K. (1976). The evolution of inbreeding in plants. *A. Rev. Ecol. Syst.* **7**, 469–495.

Johnson, M. S. (1976). Allozymes and area effects in *Cepaea nemoralis* on the western Berkshire Downs. *Heredity* **36**, 105–121.

Jones, J. S., Selander, R. K. and Schnell, G. D. (1980). Patterns of morphological and molecular polymorphism in the land snail *Cepaea nemoralis*. *Biol. J. Linn. Soc.* **14**, 359–387.

Koehn, R. K. and Williams, G. C. (1978). Genetic differentiation without isolation in the American eel, *Anguilla rostrata* II. Temporal stability of geographic patterns. *Evolution, Lancaster, Pa.* **32**, 624–637.
Kohn, P. H. and Tamarin, R. H. (1978). Selection at electrophoretic loci for reproduction parameters in island and mainland voles. *Evolution, Lancaster, Pa.* **32**, 15–28.
Lakovaara, S. and Saura, A. (1971). Genic variation in natural populations of *Drosophila obscura*. *Genetics* **69**, 377–384.
Lande, R. (1980). Genetic variation and phenotypic evolution during allopatric speciation. *Am. Nat.* **116**, 463–479.
Lewontin, R. C. and Hubby, J. L. (1966). A molecular approach to the study of genic heterozygosity in natural populations. II. Amounts of variation and degree of heterozygosity in natural populations of *Drosophila pseudoobscura*. *Genetics* **54**, 595–609.
Lokki, J. (1976). Genetic polymorphism and evolution in parthenogenetic animals. VII. The amount of heterozygosity in diploid populations. *Hereditas* **83**, 57–64.
Lokki, J., Saura, A., Lankinen, P. and Suomalainen, E. (1976). Genetic polymorphism and evolution in parthenogenetic animals. V. Triploid *Adoxus obscurus* (Coleoptera: Chrysomelidae). *Genet. Res., Camb.* **28**, 27–36.
Lundkvist, K. (1979). Allozyme frequency distributions in four Swedish populations of Norway spruce (*Picea abies* K.) I. Estimations of genetic variation within and among populations, genetic linkage and a mating system parameter. *Heredity* **90**, 127–143.
Maynard Smith, J. (1966). Sympatric speciation. *Am. Nat.* **100**, 637–650.
Maynard Smith, J. (1978). "The Evolution of Sex." Cambridge University Press, Cambridge.
McCracken, G. F. and Brussard, P. F. (1980). The population biology of the white-lipped land snail, *Triodopsis albolabris*: genetic variability. *Evolution, Lancaster, Pa.* **34**, 92–104.
Miller Baker, A. E. (1981). Gene flow in house mice: introduction of a new allele into freeliving populations. *Evolution, Lancaster, Pa.* **35**, 243–258.
Mitter, C. and Futyuma, D. J. (1979). Population genetic consequences of feeding habits in some forest Lepidoptera. *Genetics* **92**, 1005–1021.
Mitter, C., Futyuma, D. J., Schneider, J. C. and Hare, J. D. (1979). Genetic variation and host-plant relations in a parthenogenetic moth. *Evolution, Lancaster, Pa.* **33**, 777–790.
Müller, G. (1976). A simple method of estimating rates of self-fertilization by analysing isozymes in tree seeds. *Silvae Genet.* **25**, 15–17.
Murray, J. (1972). "Genetic Diversity and Natural Selection". Oliver & Boyd, Edinburgh.
Nevo, E. (1978). Genetic variation in natural populations: patterns and theory. *Theor. Pop. Biol.* **13**, 121–177.
Nevo, E., Shimony, T. and Libni, M. (1978). Pollution selection of allozymes in barnacles. *Experientia* **34**, 1562–1564.

Nevo, E., Perl, T., Beiles, A., Wool, D. and Zoller, U. (1980). Genetic structure as a potential monitor of marine pollution. *V^es Journees Étud. Pollutions*, pp. 61–68. Cagliari, C.I.E.S.M.

Nygren, J. (1980). Allozyme variation in natural populations of field vole, (*Microtus agrestis* L.) III. Survey of a cyclically density-varying population. *Hereditas* **93**, 125–136.

Ornduff, R. (1969). Reproduction biology in relation to systematics. *Taxon* **18**, 121–133.

Pamilo, P. (1982). Genetic population structure in polygynous *Formica* ants. *Heredity* **48**, 95–106.

Parker, E. D. and Selander, R. K. (1976). The organization of genetic diversity in the parthenogenetic lizard *Cnemidophorus tesselatus*. *Genetics* **84**, 791–805.

Pimentel, D. (1968). Population regulation and genetic feedback. *Science, N.Y.* **159**, 1432–1437.

Powell, J. R. (1975). Protein variation in natural populations of animals. *Evol. Biol.* **8**, 79–119.

Prakash, S., Lewontin, R. C. and Hubby, J. L. (1969). A molecular approach to the study of genic heterozygosity in natural populations. IV. Patterns of genic variation in central, marginal and isolated populations of *Drosophila pseudoobscura*. *Genetics* **61**, 841–858.

Rasmuson, B., Rasmuson, M. and Nygren, J. (1977) Genetically controlled differences in behaviour between cycling and non-cycling populations of field vole (*Microtus agrestis*). *Hereditas* **87**, 33–42.

Rick, C. M., Fobes, J. F. and Holle, M. (1977). Genetic variation in *Lycopersicon pimpinellifolium*: evidence of evolutionary change in the mating systems. *Plant Syst. Evol.* **127**, 139–170.

Rick, C. M., Fobes, J. F. and Tanksley, S. D. (1979). Evolution of mating systems in *Lycopersicon hirsutum* as deduced from genetic variation in electrophoretic and morphological characters. *Plant Syst. Evol.* **132**, 279–298.

Rudin, D. (1975). Inheritance of glutamate-oxaloacetate transaminases (GOT) from needles and endosperms of *Pinus sylvestris*. *Hereditas* **80**, 296–300.

Sanders, T. B. and Hamrick, J. L. (1980). Variation in the breeding system of *Elymus canadensis*. *Evolution, Lancaster, Pa.* **34**, 117–122.

Saura, A., Halkka, O. and Lokki, J. (1973a). Enzyme gene heterozygosity in small island populations of *Philaenus spumarius* (L.) (Homoptera). *Genetica* **44**, 459–473.

Saura, A., Lakovaara, S., Lokki, J. and Lankinen, P. (1973b). Genic variation in central and marginal populations of *Drosophila subobscura*. *Hereditas* **75**, 459–473.

Saura, A., Lokki, J., Lankinen, P. and Suomalainen, E. (1976). Genetic polymorphism and evolution in parthenogenetic animals. III. Tetraploid *Otiorrhynchus scaber* (Coleoptera: Curculionidae). *Hereditas* **82**, 79–100.

Saura, S., Lokki, J. and Savolainen, E. (1979). Ethological isolation and genetic diversity. *Aquilo, ser. zool.* **20**, 13–16.

Sbordoni, V., Allegrucci, G., Caccone, A., Cesaroni, D., Cobolli Sbordoni, M. and de Matthaeis, E. (1981). Genetic variability and divergence in cave populations of *Troglophilus cavicola* and *T. andreinii* (Orthoptera, Rhaphidophoridae). *Evolution, Lancaster, Pa.* **35**, 226–233.

Schmitt, L. H. (1978). Genetic variation in isolated populations of the Australian bush-rat, *Rattus fuscipes*. *Evolution, Lancaster, Pa.* **32**, 1–14.

Schwaegerle, K. E. and Schaal, B. A. (1979). Genetic variability and founder effect in the pitcher plant *Sarracenia purpurea* L. *Evolution, Lancaster, Pa.* **33**, 1210–1218.

Schwartz, O. A. and Armitage, K. B. (1980). Genetic variation in a social mammal: the marmot model. *Science, N. Y.* **207**, 665–666.

Selander, R. K. (1970). Behaviour and genetic variation in natural populations. *Am. Zool.* **10**, 53–66.

Selander, R. K. and Hudson, R. O. (1976). Animal population structure under close inbreeding: the land snail *Rumina* in southern France. *Am. Nat.* **110**, 695–718.

Selander, R. K. and Kaufman, D. W. (1973). Self-fertilization and population structure in a colonizing land snail. *Proc. Nat. Acad. Sci., U.S.A.* **70**, 1186–1190.

Selander, R. K. and Kaufman, D. W. (1975). Genetic structure of populations of the brown snail (*Helix aspersa*). I. Microgeographic variation. *Evolution, Lancaster, Pa.* **29**, 385–401.

Semeonoff, R. and Robertson, F. W. (1968). A biochemical and ecological study on plasma esterase in natural populations of the field vole, *Miçrotus agrestis*. *Biochem. Genet.* **1**, 205–227.

Shaw, D. V. and Allard, R. W. (1982). Estimation of outcrossing rates in Douglas-fir using isozyme markers. *Theor. Appl. Genet.* **62**, 113–120.

Shumaker, K. M. and Babbel, G. R. (1980). Patterns of allozyme similarity in ecologically central and marginal populations of *Hordeum jubatum* in Utah. *Evolution, Lancaster, Pa.* **34**, 110–116.

Solbrig, O. T. (1972). Breeding systems and genetic variation in *Leavenworthia*. *Evolution, Lancaster, Pa.* **26**, 155–160.

Soulé, M. and Yang, S. Y. (1973). Genetic variation in side-blotched lizards in islands on the Gulf of California. *Evolution, Lancaster, Pa.* **27**, 593–600.

Suomalainen, E., Saura, A., Lokki, J. and Teeri, T. (1980). Genetic polymorphism and evolution in parthenogenetic animals. Part 9. Absence of variation within parthenogenetic aphid clones. *Theor. Appl. Genet.* **57**, 129–132.

Tamarin, R. H. and Krebs, C. J. (1969). *Microtus* population biology. II. Genetic changes at the transferrin locus in fluctuating populations of two vole species. *Evolution, Lancaster, Pa.* **23**, 183–211.

Taylor, C. E. and Powell, J. R. (1977). Microgeographical differentiation of chromosomal and enzyme polymorphisms in *Drosophila persimilis*. *Genetics* **85**, 681–695.

Tigerstedt, P. M. A. (1973). Studies on isozyme variation in marginal and central populations of *Picea abies*. *Hereditas* **75**, 47–60.

Tinkel, D. W. and Selander, R. K. (1973). Age-dependent allozymic variation in a natural population in lizards. *Biochem. Genet.* **8**, 231–237.

Varvio-Aho, S. (1981). The effects of ecological differences on the amount of enzyme gene variation in Finnish water-striders (*Gerris*) species. *Hereditas* **94**, 35–39.

Vrijenhoek, R. C. (1978). Coexistence of clones in a heterogeneous environment. *Science, N.Y.* **199**, 549–553.

Vuorinen, J., Himberg, M. K.-J. and Lankinen, P. (1981). Genetic differentiation in *Coregonus albula* (L.) (Salmonidae) populations in Finland. *Hereditas* **94**, 113–121.

Wahlund, S. (1928). Zusammensetzung von Populationen und Korrelationsescheinungen vom Standtpunkt der Vererbungslehre aus betrachtet. *Hereditas* **11**, 65–106.

Ward, P. S. (1980). Genetic variation and population differentiation in the *Rhytidoponera impressa* group, a species complex of ponerine ants. *Evolution, Lancaster, Pa.* **34**, 1060–1076.

White, M. J. D. (1978). "Modes of Speciation". Freeman, San Francisco.

Williams, G. C., Koehn, R. K. and Mitton, J. B. (1973). Genetic differentiation without isolation in the American eel, *Anguilla rostrata. Evolution, Lancaster, Pa.* **27**, 197–204.

Wright, S. (1931). Evolution in Mendelian populations. *Genetics* **16**, 97–159.

Wright, S. (1965). The interpretation of population structure by F-statistics with special regard to systems of mating. *Evolution, Lancaster, Pa.* **19**, 395–420.

Yang, J. C., Ching, T. M. and Ching, K. K. (1977). Isoenzyme variation in coastal Douglas-fir. I. A study of geographical variation in three enzyme systems. *Silvae Genet.* **26**, 10–18.

Young, J. P. W. (1979). Enzyme polymorphism and cyclic parthenogenesis in *Daphnia magna*. II. Heterosis following sexual reproduction. *Genetics* **92**, 971–982.

Zera, A. J. (1981). Genetic structure of two species of waterstriders (Gerridae, Hemiptera) with differing degrees of winglessness. *Evolution, Lancaster, Pa.* **35**, 218–225.

Zouros, E., Singh, S. M. and Miles, H. E. (1980). Growth rate in oysters: an overdominant phenotype and its possible explanations. *Evolution, Lancaster, Pa.* **34**, 856–867.

17 | Genetic Variation in Relation to Dispersal Efficiency

S.-L. VARVIO-AHO

Department of Genetics, University of Helsinki,
P. Rautatiek. 13, SF-00100 Helsinki 10, Finland

Abstract: Enzyme gene variation of eight Finnish water-strider (*Gerris*) species was studied, and sufficient causes of within-species and between-species variation patterns were sought. The existence of wing-length polymorphism and its differing patterns among the species enabled the comparison of genic differentiation with and without intervening barriers; short-winged individuals do not have dispersion ability and populations consisting of such individuals are isolated. Within populations, significant gene frequency changes can occur. These changes are most plausibly explained by the finite size of populations, i.e. random drift. Genic differentiation within a species and between species show clear dependence on dispersion ability. Despite differentiation at specific loci, the gene frequency variation, when analysed simultaneously at many loci, was homogeneous over the loci. This indicates that non-selective forces are sufficient explanations for population differentiation. On a micro scale, the variation pattern resembles that observed on a geographic scale. The level of genic variation in different species correlates with effective population sizes, which could be roughly estimated on the basis of dispersion efficiencies, abundances and habitat stabilities of the species. The two most common and abundant species are much more variable than the others. One species completely lacks enzyme gene variation; this is the rarest species studied and also the only one which totally lacks flight ability.

Systematics Association Special Volume No. 24, "Protein Polymorphism: Adaptive and Taxonomic Significance", edited by G. S. Oxford and D. Rollinson, 1983, Academic Press, London and New York.

INTRODUCTION

Patterns of genic variation within and between natural populations are produced by the interaction of selection and non-selective forces in such a complicated way that it is very difficult to separate their roles. Most empirical studies on the maintenance of enzyme gene variation have been interpreted as providing evidence about the adaptive significance of this variation. Statistical tests of various properties of the random drift–neutral mutation hypothesis have shown, however, that the majority of the data on enzyme gene variation is in agreement with this hypothesis (e.g. Fuerst *et al.*, 1977; Latter, 1981).

The relative importance of genetic drift in natural populations is the subject of controversy, "remarkable for its temporal stability and its lack of resolution" (Endler, 1977). The assumption of genetic drift operating on the gene frequencies has often been made when one does not know, or cannot measure any selective differences, and cannot conceive of any reason for them. Considering enzyme gene variation, the measurement of selective differences is virtually an impossible task, as it is not known what kind of selection pressures ought to be looked for. Hence, when interpreting the patterns of enzyme gene variation, it is very reasonable first to include the basic factors which are known to affect genic variation, i.e. genetic drift (all populations are finite in size) and gene flow. When a possible selective factor can be assumed, it is, of course, important to make testable predictions based on it. Selection, balancing selection and environmental heterogeneity are all too broad phenomena and generate so immense a number of possibilities that one could predict any possible relationship between environmental and genetic variation. On the other hand, it may be, as stated by Christiansen and Frydenberg (1974) that any pattern in a polymorphism can also be explained by the neutrality hypothesis, if one is free to postulate a suitable population structure.

The present paper concerns the causes of various patterns of enzyme gene variation within and between water-strider (*Gerris*) species. These are species, for which one is not free to postulate any population structure as much is actually known about their ecological characteristics (see Vepsäläinen, 1973, 1978). The population sites, various kinds of water bodies, are clearly distinguishable entities, and the distances between the sites can be used for estimating the isolation between populations. Relative isolation can also be estimated from the proportion of individuals without flying ability, as water-striders have wing length polymorphism; populations which are monomorphically short-winged are completely isolated. Further, there are several species which differ from each other in the patterns of wing length variation, in the amount of dispersion efficiency and in abundance, depending on the habitat and distribution range. The effects of population size on genic variation can be examined, as well as the effects of gene flow on population differentiation; various wing length patterns enable the comparison of the degree of differentiation with and without intervening barriers. Finally, temporal gene frequency changes from one generation to another can be easily followed, as it is possible to distinguish the individuals belonging to different generations.

In the following, I shall search for those evolutionary forces, which are *sufficient* explanations for patterns of genic variation within a single water-strider population, between populations on a micro and macro scale, and between species in the degree of differentiation and the level of variation.

MATERIALS AND METHODS

Eight of the nine Finnish *Gerris* Fabr. species (Heteroptera, Gerridae) were studied. The species are *G. argentatus* Schumm. = *AR, G. lacustris* (L.) = *LAC, G. lateralis* Schumm. = *LAT, G. najas* (De G.) = *NAJ, G. odontogaster* (Zett.) = *OD, G. paludum* Fabr. = *PAL, G. rufoscutellatus* Lt. = *RU* and *G. thoracicus* Schumm = *THO*.

AR and *PAL* are southern and rare species. *RU* is fairly common, with a southern trend. *THO* is a coastal and archipelagic species. *NAJ* is a very rare species, perhaps only occasional. The most common and abundant species in South and Central Finland are *OD* and *LAC* and in North Finland *LAT* and *OD* (Vepsäläinen, 1973).

The life cycle of the Finnish populations is as follows (Vepsäläinen, 1971, 1974a, b). The imagos, having overwintered on land, colonize the breeding sites in May. Eggs are laid during a period of one to two months. The offspring eclose by mid-July and later reproduce after overwintering. In southern Finland, *OD* and *AR* are partially bivoltine.

LAC and *LAT* are permanently wing-polymorphic species. Diapause generations of *OD, AR, PAL* and *THO* are long-winged, but if a summer generation (non-diapause generation) exists, it is short-winged. *RU* is monomorphically long-winged. All known *NAJ* populations are monomorphically short-winged.

The number of populations studied of each species were as follows. Thirty-five *LAC* populations from southernmost Finland to the species distribution limit (Polar Circle) and two from Sweden, 29 *OD* populations from southernmost to northernmost Finland and two from Sweden, two *LAT* populations from South Finland and 15 from Central and North Finland, four *AR*, two *THO*, three *RU*, three *PAL* and five *NAJ* populations. In each species, except in *THO*, the populations sampled represent quite well the distribution area of the species in Finland; the two *THO* populations are from southernmost Finland and separated only by three kilometres. Three *OD*, one *AR* and 13 *LAC* populations were sampled more than once, i.e. from successive generations and/or different years. One *LAC* and one *OD* population consisted of several subpopulations, which were studied in two successive generations.

Horizontal starch gel electrophoresis was used as explained in e.g. Varvio-Aho and Pamilo (1979, 1980a). The following enzyme systems were studied: acid phosphatases, aldehyde oxidase, esterases, glucose-6-phosphate dehydrogenase, isocitrate dehydrogenase, leucine aminopeptidase, malate dehydrogenase, malic enzyme, phosphoglucomutase, phosphoglucose isomerase, 6-phosphogluconate dehydrogenase and xanthine dehydrogenase.

WITHIN-POPULATION PHENOMENA: GENE FREQUENCY CHANGES

Detecting the operation of even strong selection in natural populations without studying the temporal patterns of gene and genotype frequencies is unlikely. The primary consequence of selection is change in frequencies, except at equilibrium.

No indications for long-term systematic changes of gene frequencies in water-strider populations were found. The general finding was that statistically even highly significant frequency changes take place from overwintered generation to offspring generation at many loci in the three species, *AR*, *OD* and *LAC* (Varvio-Aho *et al.*, 1979). At one locus, *esterase-3* of *LAC*, the changes appeared systematic, but only in one year. In that year, the most frequent electromorph was *100* in the overwintered populations, but the frequency of *96* was higher, or at least as high as that of *100* in the offspring populations. In all offspring populations, there was also a heterozygote excess at this locus. These results indicate that some form of selection was operating on the locus studied or some associated locus or loci. Further studies revealed, however, quite opposite changes at this locus even in nearby populations (Varvio-Aho, 1981a).

One way to estimate the relative roles of selective and non-selective forces in temporal changes is the use of Wright's expression, which relates the standardized gene frequency variance (F_{ST}) to effective population size, $F_{ST} = 1/N_e$. The homogeneity of F_{ST} values over loci, can be tested by the variance ratio of $s^2_{F_{ST}}/2\bar{F}^2_{ST}$, where the denominator gives the approximate expected variance of F_{ST} values (Lewontin and Krakauer, 1973).

F_{ST} values were calculated for alleles within the frequency limits 0·05–0·95 in both generations. Testing the observed and expected variances revealed no inhomogenities. This indicates that the changes could be due to the finite size of populations, i.e. due to genetic drift (Varvio-Aho, 1981a).

The role of drift in populations is that it changes gene frequencies by an amount which depends on the effective population size. This has led some scientists to estimate population size on the basis of observed gene frequency changes by applying Wright's expression (e.g. Krimbas and Tsakas, 1971; Loukas *et al.*, 1979). Unfortunately, the method used seems to be inadequate for that purpose. Computer simulations of Varvio-Aho (1981a) and Pamilo and Varvio-Aho (1981) showed that values of $N_e = 1/F_{ST}$ approximately equal the number of individuals sampled, not N_e (40, 20 or 10 individuals sampled from populations of size 200 or 100). This indicates that absolute values of $1/F_{ST}$, calculated on the basis of observed gene frequency changes, should be compared with the sample size used. If the values obtained are significantly lower than sample sizes, the agent causing gene frequency changes in populations is not mere binomial sampling of offspring from the parental gene pool.

The study of absolute values of standardized variances revealed that the gene frequency changes were in most cases too great to be explained by binomial sampling of offspring. So, homogeneity of F_{ST} values indicate random processes, but magnitudes of $1/F_{ST}$ values indicate some other mechanism than simple binomial sampling. The resolution of this problem can be found by considering the sampling processes of genes in *Gerris* populations. If the variance of offspring number per female is greater than expected on the basis of chance, the effective population size is reduced (e.g. Crow and Kimura, 1970). In that case, relatively great changes in gene frequencies could be accounted for by random processes. *Gerris* females lay eggs in batches, and there is also considerable variation in the length of egg-laying periods among females. Egg batches may increase the variance in offspring number and, because the eggs laid simultaneously develop simultaneously, favourable and unfavourable food and weather conditions during the larval period may also have a similar effect on all the siblings of a batch, but other siblings may succeed much better or much worse. In consequence, random sampling of sibling groups may change gene frequencies quite considerably.

In conclusion, the gene pools within local populations of water-striders are not stable. The most probable evolutionary factor causing gene frequency changes at many loci is genetic drift. This deduction is based on statistical inference and also on common sense: invoking selection is more unwarranted than invoking various random effects, as many simultaneous and strong selection pressures ought to be assumed. The instability of local gene pools indicates between-population differences.

BETWEEN-POPULATION PHENOMENA: DIFFERENTIATION

The consequences of drift are impoverishment of genetic variation within local populations and differentiation of populations. A well-known rule of thumb, dictated by the theoreticians, is that genetic differentiation should become appreciable only when the rate of gene flow is very low. This prediction refers, of course, to an equilibrium situation. At a single point in time, we can observe e.g. differentiation against high gene flow etc., depending on the recent history of the actual populations studied.

In the following, I shall describe the differentiation patterns of three species, *LAC, LAT* and *OD*, as expected on the basis of their population characteristics. On the basis of flight ability, *OD* is the most efficient disperser. As *LAC* populations often have a high proportion of short-winged individuals, we should expect more pronounced differentiation. *LAT* populations are usually almost completely wingless and thus isolated to a high degree.

Gene flow between *Gerris* populations is mostly dependent on dispersal flights in spring from overwintering sites; in the early summer, the flight muscles degenerate. In *OD*, the overwintering population is monomorphic for long-wingedness but, in *LAC* and *LAT*, a certain part of a population, the short-winged one, is fixed throughout the year to the locality in which it arose.

1. *Differentiation on a geographic scale*

In all three species, the general pattern of variation at most loci studied was in accordance with results from many other animal populations: across quite large geographical areas, one major allele predominates, and various degrees of differentiation are found, depending on the degree of isolation.

Sampling sites of *LAC* were selected so that they form sets of closely located populations in different areas. The populations near the species distribution limit are relatively more isolated from each other than the southern populations: available population sites are more isolated and more permanent, and the populations have a higher frequency of short-winged individuals without flying ability (Järvinen and Vepsäläinen, 1976). These populations showed considerable amounts of differentiation, and they clearly differed from southern Finnish populations, which were significantly more homogeneous in their genetic constitution (Fig. 1) (Varvio-Aho and Pamilo, 1979). On the basis of genetic distances (Nei, 1972), populations near each other tended to be more similar than populations which were far apart.

Genetic distance between *OD* were rather low, and the populations were much more similar to each other than were *LAC* populations, the mean distance between population pairs being 0·009. This is of the same order of magnitude as in many other animal populations, whereas in *LAC* the mean distances are of the same order as are often found

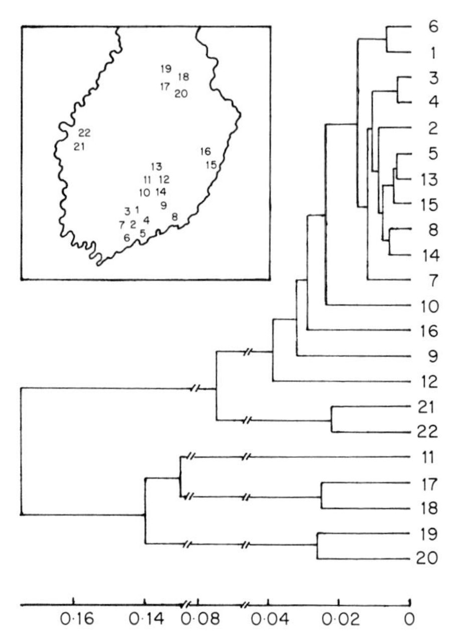

Fig. 1. Dendrogram of genetic distances, based on 11 polymorphic loci, between Finnish *LAC* populations.

between subspecies (Varvio-Aho, 1979; reference data from Nei, 1975).

Differentiation among *LAT* populations was higher than among *OD* populations but lower than among *LAC* populations. This is not fully in accordance with the predictions, but it can be connected with conspicuously lower levels of total genetic variation in this species (Varvio-Aho and Pamilo, 1980b). A well-known phenomenon without obvious explanation is that within-population variation correlates positively with differentiation among populations (Kluge and Kerfoot, 1973; Sokal, 1976).

The results show that differences in dispersion capacity and, consequently, in the amount of gene flow, affect the degree of genetic differentiation. This holds for a single species in different areas (*LAC* populations forming isolated populations and not so isolated populations) and for species having differing capacities for population mixing. Similar results have been described by Zera (1981). Two North American *Gerris* species, which differ in their dispersal capacities, also differ in their genetic structure.

The results do not, as such, throw light on the problem of selection *versus* drift. This problem can be studied by the standardized variances of gene frequencies: if they are not inhomogeneous over the loci among the populations studied, it can be deduced that the variation in allele frequencies may be entirely the result of the breeding structure of the species (see Lewontin and Krakauer, 1973; but also Nei and Maruyama, 1975; Robertson, 1975). In *Gerris* populations, the observed and expected variances of standardized allele frequencies were not significantly different. This implies that mere stochastic factors are sufficient to explain the between-population differences within each species.

2. *Differentiation on a micro-scale*

If we want to determine the effects of stochastic processes on genetic variation, the best demonstration can be obtained by studying micro-geographic patterns (e.g. Cavalli-Sforza and Bodmer, 1971). Such a study avoids the problems of population histories and possible geographic differences in selection. This approach was applied in a study of the genetic structure of populations in *LAC* and *OD* (Varvio-Aho and Pamilo, 1981). In both species, a population subdivided into a number of subpopulations inhabiting a set of closely located ponds was studied.

OD population consisted of subpopulations inhabiting ponds on an open bog within an area of 200 × 400 m. *LAC* populations inhabited ponds in an open sandy field, separated by distances of a few metres. The degree of migration between ponds was determined by mark-recapture studies. About 8% of *OD* individuals visited at least two ponds in the study period. In *LAC*, only one individual changed pond during the study. Subpopulation sizes ranged from some tens to some hundreds of individuals.

At many loci, significant gene frequency differences between near-by ponds were observed. The situation resembles that in house mice studied by Selander (1970), in which the mosaic gene frequency distribution within a single barn looked like a variation pattern in large geographic areas. The relative genetic differentiation was measured by standardized allele frequency variances (F_{ST}) and Nei's (1973) G_{ST} statistics. The G_{ST}'s for the overwintered and offspring generations were 0·041 and 0·036 in *OD* and 0·025 and 0·061 in *LAC*. The corresponding F_{ST} values, when calculated for the commonest alleles and averaged over all loci, were 0·036 and 0·031 in *OD* and 0·026 and 0·056 in *LAC*. The values of F_{ST} varied somewhat from locus to locus, but the variance ratios revealed no significant heterogeneities, which suggests that the loci tested were under the influence of similar gene frequency-changing agents.

In general, the degree differentiation observed on a micro scale, was of the same order, or even higher at some loci, than among geographically distant populations. The studies were undertaken in so restricted and apparently homogeneous areas, that it is difficult to imagine that differing selection pressures were present. As the only reasonable explanation for the patterns is genetic sampling drift within the subpopulations, we need not invoke other explanations for similar differences observed on a much wider scale.

The estimated migration rate between *OD* populations should be large enough to smooth out any differentiation *in the long run*. But, apparently, there has not been enough time. What we observed, was the result of recent history of the subpopulations.

3. *A special search for selection*

I have suggested above that random processes are sufficient explanations for observations on population differentiation at many levels. Yet any searches for selection could be valuable, bearing in mind that it is not possible to relate enzyme phenotypes with specific selection pressures with the present data, and in general. According to Clarke (1975), one of the most powerful methods of detecting selection as an agent influencing the gene frequencies at specific loci in natural populations is the following. A correlation between gene frequencies at homologous loci in closely related sympatric species can only be due to selection.

Water-striders are suitable for such an analysis, as two or three closely related species often coexist at a given site. The analysis of 12 sympatric *OD*/*LAC*, five of *OD*/*LAT* and four *LAC*/*LAT* populations revealed no indications of parallel selection among sympatric populations (Varvio-Aho, 1980; for the method, see Varvio-Aho and Pamilo, 1981).

BETWEEN-SPECIES PATTERN: THE DEGREE OF HETEROZYGOSITY

If genetic variation is neutral, the amount of it should depend on the effective population size. Although effective population size is a most important variable in population genetics theory, it is least amenable to quantification in the real world, and, as a concept, it is somewhat vague (Ewens, 1982). There is the total number of individuals in the species, and a variety of effective sizes depending on population structure, the amount of inbreeding, the history of populations and the gene flow between them.

In Finnish water-striders, the largest populations, with the highest densities, are found in *NAJ, LAC, OD* and *AR*. Population sizes and heterozygosities do not correlate, as the only evident patterns are the following. *LAC* and *OD* are clearly more heterozygous than the other species, and *NAJ* differs from all the other species by being completely monomorphic at all loci studied. Average within-population heterozygosities, based on 16 loci , are (Varvio-Aho *et al.*, 1978; Varvio-Aho and Pamilo, 1979, 1981; Varvio-Aho, 1979, 1981a)

LAC	0·165	*PAL*	0·080
OD	0·146	*RU*	0·044
AR	0·053	*THO*	0·059
LAT	0·055	*NAJ*	0

The effective population sizes in *Gerris* species can be estimated, at least roughly, on the species commonness and dispersion abilities, and on habitat types. Species favouring temporary water bodies often suffer from extinction episodes, which should imply bottlenecks in population sizes and, in consequence, substantial drift. In general, the effects of high rates of dispersion should be to increase the local population size.

Populations of *NAJ* are always monomorphically short-winged, which means that they are completely isolated. This ecological feature clearly distinguishes it from other gerrids. It is also distinguished from the other species by a complete lack of enzyme-gene variation. The most plausible explanation for the monomorphism at enzyme loci is that though census populations sizes are locally among the greatest found in water-striders, the effective population size remains very low, as it is not increased by the mixing of local populations, as in other species. Further, the species is very rare, so there are only a few populations.

The species *OD* and *LAC* are distinguished from all the others in that they are clearly more heterozygous, *and* that they are the only Finnish gerrids that have been classified by Vepsäläinen (1973) as very common and abundant. Commonness and abundancy of a species naturally imply large effective population sizes, when the species is treated as a whole.

In addition to rarity as a factor reducing the level of genetic variation in the other species, the following are plausible: the almost complete winglessness in *LAT* for the same reason as in *NAJ*, and the habitat preferences of *THO* and *RU*. These species reproduce in very small and temporary habitats, which allows local drift; impoverishment of genetic variation apparently cannot be compensated for by the high vagility of the species. *RU* also reproduces in large water bodies, but its populations usually consist of only a few individuals.

Additional indirect evidence for effective population size being an important factor, allowing larger amounts of genic variation to be maintained in large populations, comes from the study of the amount of hidden heterogeneity among electromorphs (Varvio-Aho, 1980). The amount of hidden heterogeneity should depend on effective population size; in large populations considerable amounts of hidden heterogeneity exist, in smaller ones less is expected to be found (Nei and Chakraborty, 1976). By using buffer systems with varying pH values, considerable increase in the number of esterase electromorphs was found. An important feature was that the amount of hidden heterogeneity revealed did not change the order of the species when ranked according to the level of variation. *NAJ* remained completely monomorphic and the most conspicuous increase was found in *OD* (*LAC* was not studied).

A fundamental question for the synthetic theory of evolution is, to what extent can genetic variability among species and populations be predicted as a function of intrapopulation variability (e.g. Sokal, 1978) The classical view is that evolution at the level of species consists of the accumulation of microevolutionary changes. The results from a group of Finnish water-strider species are in accordance with this view. A sufficient explanation for within population phenomena (temporal allele frequency changes), beween-population phenomena (patterns of genetic differentiation within and between species) and between-species patterns (differences in the level of genetic variation) is the same evolutionary force.

ACKNOWLEDGEMENT

I am grateful to Risto Väinölä for invaluable criticism concerning an earlier version of the paper.

REFERENCES

Cavalli-Sforza, L. L. and Bodmer, W. F. (1971). "The Genetics of Human Populations". Freeman, San Francisco.

Christiansen, F. B. and Frydenberg, O. (1974). Geographical patterns of four polymorphisms in *Zoarces viviparus* as evidence of selection. *Genetics* **77**, 765–770.

Clarke, B. (1975). The contribution of ecological genetics to evolutionary theory: detecting the direct effects of natural selection on particular polymorphic loci. *Genetics* **79**, 101–113.

Crow, J. F. and Kimura, M. (1970). "An Introduction to Population Genetics Theory". Harper and Row, New York.

Endler, J. A. (1977). "Geographic variation, Speciation and Clines", p. 33. Princeton University Press, Princeton.

Ewens, W. J. (1982). On the concept of the effective population size. *Theor. Pop. Biol.* **21**, 373–378.

Fuerst. P. A., Chakraborty, R. and Nei, M. (1977). Statistical studies on protein polymorphism in natural populations. I. Distribution of single locus heterozygosity. *Genetics* **86**, 455–723.

Järvinen, O. and Vepsäläinen, K. (1976). Wing polymorphism as an adaptive strategy in water-striders (*Gerris*). *Hereditas* **84**, 61–68.

Kluge, A. G. and Kerfoot, W. C. (1973). The predictability and regularity of character divergence. *Am. Nat.* **107**, 47–56.

Krimbas, C. B. and Tsakas, S. (1971). The genetics of *Dacus oleae*. V. Changes of esterase polymorphism in a natural population following insecticide control – selection or drift. *Evolution, Lancaster, Pa.* **25**, 454–460.

Latter, B. H. D. (1981). The distribution of heterozygosity in temperate and tropical species of *Drosophila*. *Genet. Res.* **38**, 137–156.

Lewontin, R. C. and Krakauer, J. (1973). Distribution of gene frequency as a test of the theory of selective neutrality of polymorphisms. *Genetics* **74**, 175–195.

Loukas, M., Krimbas, C. B. and Vergini, Y. (1979). The genetics of *Drosophila subobscura* populations. IX. Studies on linkage disequilibrium in four natural populations. *Genetics* **93**, 497–523.

Nei, M. (1972). Genetic distance between populations. *Am. Nat.* **106**, 283–292.

Nei, M. (1973). Analysis of gene diversity in subdivided populations. *Proc. Nat. Acad. Sci., U.S.A.,* **70**, 3321–3323.

Nei, M. (1975). "Molecular Population Genetics and Evolution". North-Holland Publishing, Amsterdam.

Nei, M. and Chakraborty, R. (1976). Electrophoretically silent alleles in a finite population. *J. Mol. Evol.* **8**, 381–389.

Nei, M. and Maruyama, T. (1975). Lewontin-Krakauer test for neutral genes. *Genetics* **80**, 395.

Pamilo, P. and Varvio-Aho, S. (1981). On the estimation of population size from allele frequency changes. *Genetics* **95**, 1055–1058.

Robertson, A. (1975). Gene frequency distributions as a test of selective neutrality. *Genetics* **81**, 775–785.

Selander, R. K. (1970). Behaviour and genetic variation in natural populations. *Am. Zool.* **10**, 385–401.

Sokal, R. R. (1976). The Kluge-Kerfoot phenomenon reexamined. *Am. Nat.* **110**, 1077–1091.

Sokal, R. R. (1978). Population differentiation: Something new or more of the same? *In* "Ecological Genetics: The Interface" (P. F. Brussard, ed.), pp. 215–239. Springer-Verlag, New York.

Varvio-Aho, S. (1979). Genic differentiation of *Gerris odontogaster* populations. *Hereditas* **91**, 207–214.

Varvio-Aho, S. (1980). Spatial and temporal enzyme gene variation in water-striders (*Gerris*). Ph.D. thesis, University of Helsinki.

Varvio-Aho, S. (1981a). On the causes of seasonal genetic changes in *Gerris lacustris*. *Hereditas* **94**, 139–142.

Varvio-Aho, S. (1981b). The effects of ecological differences on the amount of enzyme gene variation in Finnish water-strider (*Gerris*) species. *Hereditas* **94**, 35–39.

Varvio-Aho, S. and Pamilo, P. (1979). Genic differentiation of *Gerris lacustris* populations. *Hereditas* **90**, 237–249.

Varvio-Aho, S. and Pamilo, P. (1980a). A new buffer system with wide applicability. *Isozyme Bull.* **13**, 114.

Varvio-Aho, S. and Pamilo, P. (1980b). Genic differentiation of northern Finnish water-strider (*Gerris*) populations. *Hereditas* **92**, 363–371.

Varvio-Aho, S. and Pamilo, P. (1981). Searching for parallel enzyme gene variation among sympatric congeners. *Evolution, Lancaster, Pa.* **36**, 200–203.

Varvio-Aho, S., Järvinen, O. and Vepsäläinen, K. (1978). Enzyme gene variability in three species of water-striders (*Gerris*). *Annls ent. fenn.* **44**, 87–94.

Varvio-Aho, S., Järvinen, O., Vepsäläinen, K. and Pamilo, P. (1979). Seasonal changes of the enzyme gene pool in water-striders (*Gerris*). *Hereditas* **90**, 11–20.
Vepsäläinen, K. (1971). The role of gradually changing day length in determination of wing length, alary dimorphism and diapause in a *Gerris odontogaster* population (Gerridae, Heteroptera) in South Finland. *Ann. Acad. Sci. fenn.*, A, IV **183**, 1–25.
Vepsäläinen, K. (1973). The distribution and habitats of *Gerris* Fabr. species (Heteroptera. Gerridae) in Finland. *Annls. zool. fenn.* **14**, 1–73.
Vepsäläinen, K. (1974a). The life cycles and wing lengths of Finnish *Gerris* Fabr. species (Gerridae, Heteroptera). *Acta Zool. fenn.* **10**, 419–444.
Vepsäläinen, K. (1974b). Determination of wing length and diapause in water-striders (*Gerris* Fabr., Heteroptera). *Hereditas* **77**, 163–176,
Vepsäläinen, K. (1978). Wing dimorphism and diapause in *Gerris*: Determination and adaptive significance. *In* "Evolution of Insect Migration and Diapause" (H. Dingle, ed.), pp. 218–253. Springer-Verlag, New York.
Zera, A. (1981). Genetic structure of two species of water-striders (Gerridae: Hemiptera) with differing degrees of winglessness. *Evolution, Lancaster, Pa.* **35**, 218–225.

18 | Hybrid Zones as Barriers to Gene Flow

N. H. BARTON[1] *and G. M. HEWITT*[2]

[1]*Genetics Department, Downing Street, Cambridge CB2 3EH, U.K.*
[2]*Biological Sciences, University of East Anglia, Norwich NR4 7TJ, U.K.*

Abstract: The hybrid zone which forms when two partially incompatible populations meet acts as a barrier to gene flow. We discuss electrophoretic and theoretical evidence on the strength of such barriers.

Hybrid zones generally involve considerable electrophoretic divergence. The enzyme clines are consistent in position and width; in some cases, they show consistently asymmetric patterns of introgression. This consistency suggests that the clines are maintained primarily by the indirect effects of selection at linked loci, rather than by the effect of each individual locus on fitness.

A cline at a single locus will present some barrier, regardless of the selective mechanism which maintains it. However, unless the locus induces virtually complete assortment or hybrid unfitness, the barrier will be weak. Spreading the same selection over more clines gives a stronger barrier. If the clines are staggered, this barrier is still unlikely to be significant; if they coincide, and if selection is stronger than recombination, then the barrier will be very strong; its strength and asymmetry will be consistent over different loci.

Thus, the taxonomic status of divergent populations cannot be inferred just from the total amount of pre- or post-mating isolation; the number of genetic differences, and the interactions between them are equally important in determining rates of gene flow.

INTRODUCTION

There has been much debate over the taxonomic status of populations

Systematics Association Special Volume No. 24, "Protein Polymorphism: Adaptive and Taxonomic Significance", edited by G. S. Oxford and D. Rollinson, 1983, Academic Press, London and New York.

which hybridize with each other, and yet remain genetically distinct. Some people, including the originators of the biological species concept, have felt that such groups should be regarded as good species (e.g. Bigelow, 1965; Mayr, 1982: 285; Dobzhansky, 1970: 354; Wright, 1978: 5). However, it seems undesirable to regard species as populations which are stable towards introgression of alleles from other such populations; this would transform the definition from one based on gene flow between whole populations, to one based on patterns of selection at particular loci (see Key, 1981: 439). If we keep strictly to the biological species concept, then in order to judge how close populations are to being species, we must measure the rate of gene flow between them. In this paper, our central aims are, first, to summarize electrophoretic observations on the effect of natural hybrid zones on gene flow; and, second, to relate these observations to theoretical models of genetic barriers.

Before discussing the relationship between hybrid zones and speciation, we need to be clear about what we mean by a "hybrid zone". In general, the term is applied to a variety of distinct phenomena, which are correlated in nature. This has led to the proposal of a number of definitions, each with a different basis, but intended to refer to the same set of phenomena. However, these either rely on the unverifiable past history of the zone (Mayr, 1963: 110, 524–5; Short, 1969), or on features of simple clinal variation (e.g. Short, 1969; Endler, 1977: 4; Key, 1981: 439).

We will, therefore, use the term "hybrid zone" to mean the set of clines maintained at the interface between populations resting at different stable equilibria. Alleles which move across the hybrid zone will be selected against, since each parental population is, by definition, stable towards the introduction of alleles from the other. The hybrid zone will, therefore, remain in a balance between dispersal and selection (Bazykin, 1969, 1973).

HYBRID ZONES: OBSERVATIONS

Table I gives a brief summary of electrophoretic observations on putative hybrid zones. Some of these data are contained in more general surveys of differentiation between taxa (e.g. Ayala, 1975; Wright, 1978). We have only considered cases where some samples were collected close to zones of hybridization, and where it is likely that clines are maintained primarily by selection against hybrids.

1. Overall divergence

The most striking feature of these data is the high proportion of enzyme loci which differ between taxa. These hybrid zones were first detected through chromosomal, morphological and behavioural differences which need not have been associated with divergence at more than a few genes; yet, an average of 20% of the enzyme genes have diverged. These figures are only approximate, of course; a limited number of loci and of species were scored, and so sampling errors make it hard to discern the underlying distribution of divergence. It is, in any case, hard to extrapolate from these data up to the rest of the genome; although the abundant, soluble enzymes scored by electrophoresis were initially chosen at random with respect to their pattern of variation, we know that they are not typical of other structural loci (Leigh-Brown and Langley, 1979), and do not correlate closely in variability or diversity with morphological, chromosomal, or phylogenetic characters (Avise, 1977; Douglas and Avise, 1982).

Fortunately, we do have some other evidence that genetic differentiation between distinct hybridizing taxa is generally high. Rare electrophoretic alleles are found at high frequency within four of the hybrid zones listed here; they could be produced by intragenic recombination between cryptic variants at the enzyme loci themselves, or by increased rates of mutation caused by disruption of coadaptation between other loci. Chromosomal aberrations, not due to non-disjunction in chromosomal heterozygotes, are also found within some hybrid zones (e.g. *Podisma, Caledia*). Morphological variability, and in particular, asymmetry, may be greater in hybrids (e.g. *Bombina, Litoria*). Spermiogenesis may be disrupted e.g. in *Caledia* (Moran, 1981) and *Podisma* (Westermann *et al.*, unpublished). Finally, mapping of the region of hybrid inviability in *Podisma* indicates that the inviability is due to differences at a few hundred genetic loci (Barton and Hewitt, 1981a).

We find this surprising, since, in principle, a hybrid zone consisting of a single genetic difference, and subject to only moderate selection, should be perfectly stable. A few zones (e.g. those between Mullerian

Table 1. Electrophoretic data from hybrid zones.

Species	Characters involved	Width[a] (Km)	Preferential mating	Hybrid unfitness	Enzyme[b] divergence	No. of pop'lns	No. of inds	Diseq'bria[c]	Loci[d] with rare alleles	Asymmetry[e]	References
Mammals											
Mus musculus musculus/ domesticus	morphology, behaviour	30–80	–	none noted in F1	41:17:13	44	2696	–	2	consistent at all loci	Selander *et al.* (1969) Hunt and Selander (1973)
Mus musculus (alpine races)	karyotype	20	–	30% sterility in F1	20: ★: 0	9	300	–	–	–	Britton-Davidian *et al.* (1980) Thaler *et al.* (1981)
Spalax ehrenbergi (4 zones)	karyotype, behaviour	3 0·7 (across river) 0·3	yes	Slight non-disjunction, but F1 and Bx in field	17: 5: 1 17: 6: 5 17: 5: 4 17: 3: 1	52	383	–	–	–	Nevo and Shaw (1972) Nevo (1982)
Thomomys bottae ruidosae/actuosus	karyotype, morphology	2	yes	none noted	23: ★: 5	19	175	E/C, but weak E/M	–	–	Patton *et al.* (1979)
Thomomys bottae (many zones)	karyotype	narrow	perhaps	–	27: 13: 4	–	–	–	–	–	Patton *et al.* (1972) Patton and Yang (1977)
Thomomys talpoides (many zones)	karyotype	narrow	–		31: 25: 1	–	–	–	–	–	Nevo *et al.* (1974)
Uroderma bilobatum davisii/convexum	karyotype	50	–	–	22: 11: 1	13	543	C/C	8	1 chrom' cline shows asymm.	Baker (1981) Greenbaum (1981)
Birds											
Icterus galbula/ bullockii	morphology, plumage	450	has evolved in some places	–	19: 2: 2	8	426	Es-1/ plumage in pop'lns with pref. mating	–	–	Corbin *et al.* (1979)

Reptiles and Amphibia										
Sceloporus grammicus	karyotype	0·2	probably not	F1 in HW, but no F2 and few Bx	20: 12: 7	5	270	E/C, E/E	–	–	Hall and Selander (1973)
Ranidella insignifera/ pseudinsignifera	mating call, morphology	2·4	yes	no inviability noted	5: 5: 4 (esterases)	8	184	–	–	–	Bull (1978) Blackwell and Bull (1978)
Pseudophryne bibroni/ semimarmorata	colour pattern	20–30 (may be moving)	no, though breeding dates differ	hybrid embryos less viable	2: 1: 1 (LDH)	10	> 126	–	–	clines in viability and LDH asymm. and shifted	Woodruff (1979, 1981) McDonnell *et al.* (1978)
Litoria ewingi/ paraewingi	mating call, body size	5–11	some	high freq. of eyeless F1	7: 7: 4 (transferrins)	7	178	–	–	–	Gartside (1972a, b)
Bombina bombina/ variegata	morphology, life history	7	in places	none obvious, but hybrids more asymmetric	39: *: 6	55	> 2000	E/E	–	–	Szymura (1976a, b) Syzmura and Farana (1978)
Rana berlandierii/ utricularia	morphology	10–20	possibly	> hybrid larval mortality?	17: 10: 6	13	654	–	5	enzyme clines asymm.	Sage and Selander (1979)
Insects										
Caledia captiva	karyotype	0·2	–	F2 inviable, Bx 50% viable	30: 11: 6	23	1200	C/C, but no E/E	–	chrom. clines consistently asymm.	Daly *et al.* (1981) Shaw *et al.* (1981) Moran *et al.* (1980)
Podisma pedestris	karyotype	0·8	–	F1 and F2 inviability	21: 6: 0	150	2000	–	2	–	Hewitt (1975) Barton and Hewitt (1981b) Halliday *et al.* (1982)
Heliconius erato/ melpomene	Müllerian mimicry races	< 50	no	hybrids viable, but suffer > predation and sexual selection?	17: 9: 0	–	–	–	–	–	Turner *et al.* (1979)

Table I *continued*

Species	Characters involved	Width[a] (Km)	Preferential mating	Hybrid unfitness	Enzyme[b] divergence	No. of pop'lns	No. of inds	Diseq'bria[c]	Loci[d] with rare alleles	Asymmetry[e]	References
Molluscs											
Cepaea nemoralis	("area effects")										
Berks. downs	shell	< 0·1	–	–	*: 6: 3	47	1800	–	–	–	Johnson (1976)
	banding and										
Pyrenees	colour				*: 6: 0	197	–	–	–	–	Jones *et al.* (1982)
Cerion bendalli/ abacoense	shell morphology	< 1	–	–	20: 6: 1	47	1575	–	–	–	Gould and Woodruff (1978) Woodruff (1981)
Partula spp. (many zones)	shell banding, spirality, habitat choice	0·1–3	in places	yes	20: 17: 0	–	–	–	–	–	Johnson *et al.* (1977) Murray and and Clarke (1980)

[a] *Width*: where possible, the width has been taken as the inverse of the maximum gradient in allele frequency.
[b] *Enzyme divergence*: the number of loci scored, the number which were polymorphic or polytypic, and the number which showed a clear difference in allele frequencies between the races.
[c] *Disequilibria*: correlations within populations between different enzyme loci (E/E), between enzymes and morphology (E/M), and between enzymes and chromosomes (E/C).
[d] *Rare alleles*: the number of loci showing a greater frequency of rare alleles within the hybrid zone.
[e] *Asymmetry*: asymmetries in cline position or shape are noted.

mimicry races of *Heliconius*?) may come close to this simple possibility, but several lines of evidence show that hybrid zones, detected through simple characters, usually consist of many coincident clines. This raises three questions: how did all these differences arise, how did they come together, and what now maintains them? Our inferences about the effect of hybrid zones on gene flow depend on the answers to these questions.

First, the differences might have little effect on fitness, and might have arisen through hitch-hiking or random drift. The different alleles would have come together at the hybrid zone, and the sharp cline between them would now be maintained by the reduction in gene flow across the zone. If this is so, then the zone must form a barrier strong enough to prevent the mixing of genes at many loci over the time since the zone was formed.

However, although many of the differences we now observe may be neutral, some of them must be responsible for the selection maintaining the hybrid zone. The selected differences might all have arisen simultaneously, in some sort of "genetic revolution". On the other hand, even if they all arose independently, either through the evolution of coadapted sets of genes, or through random drift in temporarily isolated demes, they might have been brought together by the continual splitting and reunion of the population. Provided that the rate of origin of new incompatibilities is slower than the rate of restructuring, hybrid zones might usually consist of many component zones of independent origin which had been brought together in secondary contact.

If most loci involved in hybrid zones became differentiated as a result of selection, then it is likely that they are still being selected to fit their genetic background. If the enzyme clines are now maintained by the direct effects of coadaptive selection on each individual locus, then it is hard to know the effect of the zone on the diffusion of genes at neutral loci. We would only be able to find how close diverging taxa were to being good biological species by deduction from theoretical models, rather than by direct observation.

There is another possibility; the electrophoretic differences might affect fitness, but the direct effect of each locus might be negligible compared with the indirect effects of selection on all the other loci. If a hybrid zone is a strong barrier to gene flow, then it seems likely that the component clines will be shaped primarily by this barrier, rather than by the particular properties of the separate genes. We may, therefore, be able to judge the effect of hybrid zones as barriers to the flow of neutral genes without making any contentious assumptions about the selection on the individual loci.

2. *Structure of zones: width*

In Fig. 1, we have plotted the shape of the clines in frequency of diagnostic electrophoretic alleles for some of the better studied hybrid zones. In all these cases, the centres of the electrophoretic clines coincide with each other and with clines in other characters. There are a few exceptions to this strict coincidence (e.g. in *Pseudophryne bibroni/ semimarmorata*), but the great majority of hybrid zones consist of a set of tightly clustered clines. This is to be expected, whether the clines are maintained by direct selection on the loci themselves, or by the indirect effect of selection on other loci.

More surprisingly, the widths of different enzyme clines are usually very similar. If the clines are maintained by direct selection on each enzyme locus in response to changes in the genetic background, this would imply an implausibly small variation in the strength of the epistatic interactions between loci. The similarities in width suggest that the main factor maintaining the clines is the general reduction in gene flow induced by selection at other loci.

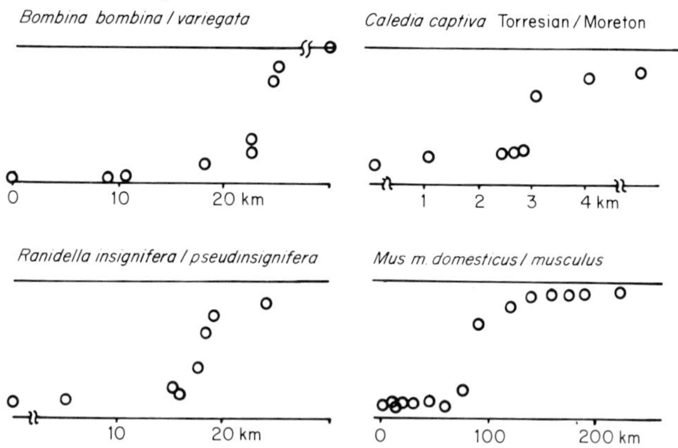

Fig. 1. Average frequency of characteristic alleles across 4 hybrid zones: *Bombina bombina/variegata* (5 loci; Szymura 1976a, b); *Caledia captiva* (6 loci; Moran *et al.*, 1980); *Ranidella insignifera/pseudinsignifera* (5 loci; Blackwell and Bull, 1978); and *Mus m. musculus/domesticus* in E. Jutland (7 loci; Hunt and Selander, 1973).

3. *Structure of zones: introgression and asymmetry*

The most usual way of finding the strength of a barrier to gene flow has been to look for "introgression"; that is, to see whether alleles characteristic of one race are found at low frequency in populations of the other race just outside the hybrid zone. This type of "stepped cline" may be discerned in the zone between *Mus musculus musculus*/*domesticus*, and in the zone in *Bombina*, but in most cases, patterns of introgression are hard to detect. First, samples must be collected from many places, both close to and well away from the zone. Secondly, it is hard to detect introgression between two populations which differ only in frequency of alleles. Even when differences are essentially fixed, it is still hard to know whether a rare allele leaked in from another related race, or whether it was always there. Both these points are illustrated by the differences in interpretation of electrophoretic data from *Caledia captiva* by Moran *et al.*, (1980) and Daly *et al.*, (1981); Moran *et al.* suggested, on the basis of a transect across the zone, that there was introgression, but Daly *et al.* analysed samples from further away, and showed that the apparently introgressing alleles are found throughout the population. Third, the barrier must not be too weak in relation to the age of the zone, since otherwise introgression will have been so extensive as to be undetectable. On the other hand, it must not be too strong in relation to the species range, since otherwise introgressing alleles will spread out to a uniform frequency in the population. Finally, hybrid zones tend to lie at natural barriers, and these will themselves induce a "stepped cline". For example, the chromosomal cline in *Podisma pedestris* has a "stepped" shape in some places, which might suggest that hybrid inviability sets up a strong barrier to the diffusion of the chromosome rearrangements. However, in these places, the zone runs along a steep cliff; in places where dispersal is unimpeded, there is a smooth transition from one karyotype to the other (Barton and Hewitt, 1981b).

Although we have argued that patterns of apparent introgression are hard to interpret, it is clear that several of the better-studied hybrid zones are markedly asymmetric. More surprising, this asymmetry tends to be in the same direction at all loci. In the zone between *Mus m. musculus* and *Mus m. domesticus* in Jutland, there are more *domesticus* alleles in the northern *musculus* populations than vice versa. The degree of asymmetry varies somewhat over the 13 differentiated loci, but is always in the same direction. In the zone between the Moreton and Torresian races of *Caledia captiva*, the enzyme clines are symmetrical, but the chromosomal clines (barring the X) show more introgression of Torresian chromosomes into Moreton populations. In the zone between *Pseudophryne bibroni* and *P. semimarmorata*, the enzyme clines are again consistently asymmetric. In the zone between *Cerion bendalli* and *C. abacoense*, there is little overall electrophoretic differentiation, but the region in which rare alleles are found at higher frequency is displaced to one side of the morphological cline.

HYBRID ZONES: THEORY

The above survey of electrophoretic data showed that hybrid zones are generally composed of many coincident clines. These might each be maintained by their individual coadaptation with the genetic background, or might be the indirect result of the genetic barrier induced by selection on other loci. A realistic theory must, therefore, describe the genetic interactions between large numbers of loci; it must be able to explain the observations that clines at different loci do not vary much in width; that hybrid zones may be flanked by long tails of introgressing alleles; and that this introgression may be consistently asymmetric.

First, we must quantify the strength of a barrier to gene flow. When alleles which are dispersing down a frequency gradient meet a localized barrier, they will pile up behind it, and give a sharp step in frequency. The strength of the barrier (B) may be defined as the ratio between the size of the step and the average gradient on either side (Nagylaki, 1976). B can be thought of as the distance which would present an equivalent obstacle to a neutral allele. However, this analogy is misleading if the allele affects fitness, since a slightly advantageous allele will leak past a local obstacle more rapidly than past an equivalent stretch of open habitat.

These considerations apply to both physical and genetic barriers; indeed, it is not easy to separate the two, since genetic as well as physical factors may cause local reductions in density and dispersal. In particular, the lower fitness of individuals within a cline may reduce the density there, giving a "hybrid sink" into which dispersing alleles will be lost (Hall, 1973). However, simple models suggest that this effect is likely to be weak (Barton, 1980), and there is no evidence that the genetic load in hybrid zones reduces population density any more than a minor physical barrier does. We will, therefore, concentrate on the effect of direct interactions between loci, rather than of indirect changes in population structure.

1. *Single-locus zones*

Suppose that a cline at some locus is maintained by a balance between selection and dispersal. Any gradient in the frequency of alleles at linked loci will interact with this cline to generate linkage disequilibrium (Li and Nei, 1974). The selection on the cline will, therefore, spill over onto the weakly selected locus, and will induce a step in the original gradient. The strength of the resulting barrier is proportional to $\sigma \sqrt{2s/r}$, where σ is the dispersal range, s is the selection maintaining the cline, and r is the rate of recombination between the two loci (Barton, 1979). This formula shows that even with quite strong selection, gene flow is only impeded significantly over a small region of chromosome around the selected locus. For example, a cline maintained by 10% selection will give a barrier equivalent to 100 σ at loci 0·5 centi-Morgan away. Neutral alleles will be delayed for about 10 000 generations by such a barrier, but an allele with an advantage of 1% will only be delayed by 300 generations. The assumptions used in these calculations break down when selection is intense ($s \sim 1$), but Bengtsson (1974) has shown that gene flow is not greatly impeded even under these extreme conditions.

Genetic barriers may be asymmetric if the underlying selection pressure is asymmetric. This is essentially because if one allele is fitter than the other, the cline will tend to move in its favour, and so alleles from the receding genotype will be swept into the fitter genotype. Unless the advancing allele is advantageous even when rare, the cline will be halted when it encounters a sufficiently steep gradient in density or dispersal rate; the static cline will still show asymmetric introgression, but this will now be due to the external gradient, rather than to the movement of the cline itself. Observations of asymmetry may, therefore, indicate an intrinsic tendency of the zone to move (as noted by Shaw and Wilkinson, 1980).

2. Multi-locus zones

A single genetic difference can only significantly delay gene flow when it causes virtually complete assortment or hybrid unfitness. However, since most hybrid zones seem to involve many genetic differences, this result is only useful in showing that no single component (for example, a chromosome rearrangement) is likely to be of primary importance. When selection maintains clines at many loci, calculation of the barrier strength becomes complicated; we therefore begin by considering the cumulative effect of a number of clines which lie in different places, and which therefore do not interact with each other.

Suppose a selection pressure s acts at N loci, giving a total selection $S = Ns$; the loci are spaced r map units apart on the chromosome, occupying a total map distance $R = Nr$. The total barrier strength can be found by summing the effect of each locus, and is given by

$$B \simeq \frac{\sigma \sqrt{8Ns}\ln(N)}{R}.$$

A series of weakly selected clines is a much stronger barrier than a single strongly selected cline, for two reasons. First, the effect of linked loci only decreases in inverse proportion to the recombination rate, and so the cumulative effect of distant loci is comparable to that of nearby loci

$$\sum_{1}^{N} \frac{1}{r_i} \sim \ln(N)$$

Secondly, and more important, the barrier strength depends more strongly on recombination than on selection (as $1/R$ vs $\sqrt{S}$), and so as the number of loci increases, the barrier strength increases with $\sqrt{N}$ (Fig. 2).

Although a sufficiently large number of staggered clines could delay gene flow at most loci, they would also occupy a large area; their barrier effect would be negligible compared with simple physical distance. If the clines coincide, as they do in almost all known hybrid

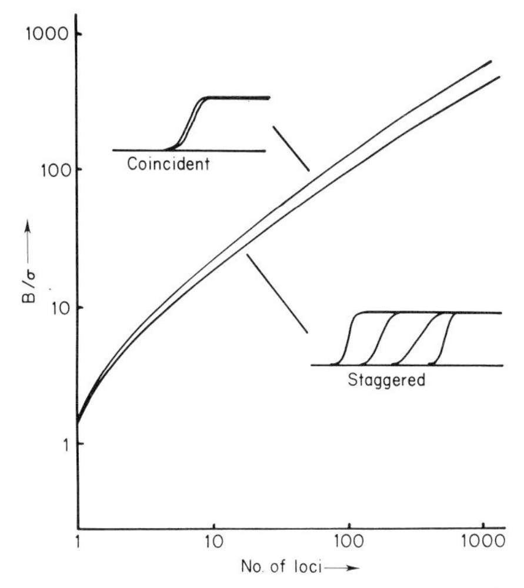

Fig. 2. The strength of the barrier due to a section of chromosome $R = 0{\cdot}5$ map units long, and causing a net selection against heterozygotes $S = 0{\cdot}1$. The barrier strength (measured in dispersal distances; B/σ) is plotted against the number of clines involved; the barrier is slightly stronger when the clines coincide.

zones, then the barrier will be strengthened by interactions between selected loci. Alleles characteristic of the same population will be found together, and so the selection on each locus will be augmented by the selection on correlated loci. The effect of these interactions has been calculated for the simple case where heterozygote disadvantage acts independently on many loci (Barton, 1983). Assortative mating and coadaptation between loci will strengthen the interactions, but since assortment is far from complete in the hybrid zones discussed above, and since the pattern of epistasis between natural populations is hard to assess, we will not consider these processes in detail.

The behaviour of a set of coincident clines depends primarily on the ratio between selection and recombination; the results fall into two distinct classes, according to which of these processes predominates. When selection is weaker than recombination ($S < R$), the barrier is only slightly stronger than when the clines are staggered (Fig. 2). It is still much stronger than if all the selection were concentrated at a single locus, and may give the clines a "stepped" shape (Fig. 3). The centre of each cline will then be steep, and will have much the same width at all loci, whilst the tails will decay more slowly, at a rate determined by the

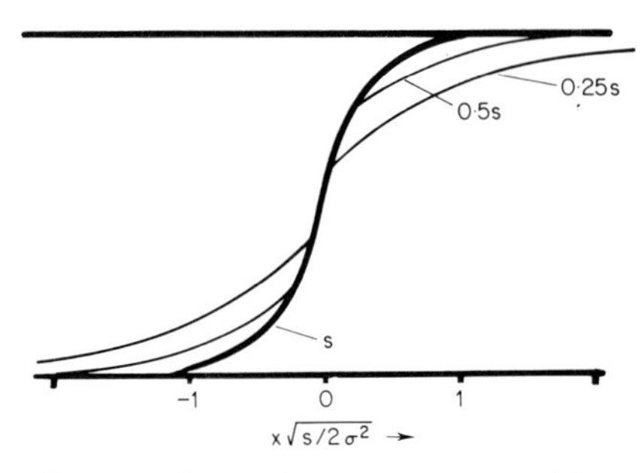

Fig. 3. The shape of a set of 100 clines, maintained by an average selection against heterozygotes *s*, and spaced *2s* map units apart. Different selection pressures may act on each locus; this will not affect the shape of the central part of the clines, but will give tails of varying introgression.

particular selection pressures on each locus (c.f. observations of natural zones; Fig. 1). However, the clines are unlikely to present a substantial barrier to gene flow unless they involve a very large number of genes.

When selection is stronger than recombination, the system behaves quite differently. Blocks of alleles which disperse from one side of the zone to the other are removed by selection faster than they can be broken up by recombination, and so the total selection pressure can act on each locus. The clines become very narrow, and gene flow is greatly reduced; alleles embedded in a block of genes will only rarely escape from the block. This pattern may be illustrated by hybrid zones such as *Sceloporus* and *Caledia*, in which strong disequilibria and intense selection are seen.

Finally, we must consider the consistently asymmetric introgression across the zones in *Caledia* and *Mus m. musculus/domesticus*. If selection is stronger than recombination, then the total selection pressure will be acting on every locus, and so any asymmetry will be in the same direction at each. However, if selection is weaker than recombination, the consistency is puzzling. Although the central parts of the clines are maintained by interactive selection, and so have the same shape for all loci, the tails of introgression should be independent of each other. In *Caledia*, the introgressing characters are homologous chromosome rearrangements, and so each may be subject to similar selection pressures. In *Mus*, where many different enzyme loci are involved, such an explanation is implausible.

CONCLUSIONS

These theoretical models are consistent with many features of natural zones: the similarity in width of different clines, "stepped" patterns of introgression, and asymmetric gene flow. However, more detailed studies of the genetic structure of hybrid zones are needed (for example, of the pattern of disequilibria between their components).

The importance of gene flow in promoting divergence is contentious (Ehrlich and Raven, 1969; Jones *et al.*, 1981); still, the delimitation of biological species by genetic barriers to gene flow seems the most objective procedure. The approach of hybridizing populations to full species, therefore, depends on their ability to exchange genes. Unfortunately, it is not possible to determine rates of gene flow from overall observations of the first few generations of hybridization; the number of genes, and the rates of recombination between them must also be known. Gene flow is not determined by simple dispersal, and genetic isolation is not determined by the frequency of fertile hybrids.

REFERENCES

Avise, J. C. (1977). Genic heterozygosity and the rate of speciation. *Paleobiology* **3**, 422.

Ayala, F. J. (1975). Genetic differentiation during the speciation process. *Evol. Biol.* **8**, 1–78.

Baker, R. J. (1981). Chromosome flow between chromosomally characterized taxa of the volant mammal *Uroderma bilobatum* (Chiroptera: Phyllostomatidae). *Evolution, Lancaster, Pa.* **35**, 296–305.

Barton, N. H. (1979). Gene flow past a cline. *Heredity* **43**, 333–339.

Barton, N. H. (1980). The hybrid sink effect. *Heredity* **44**, 277–278.

Barton, N. H. (1983). Multilocus clines. *Evolution, Lancaster, Pa.* **37**, 454–471.

Barton, N. H. and Hewitt, G. M. (1981a). The genetic basis of hybrid inviability between two chromosomal races of the grasshopper *Podisma pedestris*. *Heredity* **47**,367–383.

Barton, N. H. and Hewitt, G. M. (1981b). A chromosomal cline in the grasshopper *Podisma pedestris*. *Evolution, Lancaster, Pa.* **35**, 1008–1018.

Bazykin, A. D. (1969). Hypothetical mechanism of speciation. *Evolution, Lancaster, Pa.* **23**, 685–687.

Bazykin, A. D. (1973). Population genetic analysis of disrupting and stabilising selection. Part II. Systems of adjacent populations and populations within a continuous area. *Genetica* **9**, 156–166.

Bengtsson, B. O. (1974). "Karyotype evolution *in vivo* and *in vitro*" D. Phil. thesis, University of Oxford.
Bigelow, R. S. (1965). Hybrid zones and reproductive isolation. *Evolution, Lancaster, Pa.* **19**, 449–458.
Blackwell, J. M. and Bull, C. M. (1978). A narrow hybrid zone between the Western Australian frog species, *Ranidella insignifera* and *R. pseudinsignifera*: the extent of introgression. *Heredity* **40**, 13–25.
Britton-Davidian, J., Bonhomme, F., Croset, H., Capanna, E. and Thaler, L. (1980). Variabilite genetique chez les populations de souris (genre *Mus* L.) a nombre chromosomique reduit. *C. r. hebd. Séanc. Acad. Sci., Paris* **290**, 195–198.
Bull, C. M. (1978). The position and stability of a hybrid zone between the Western Australian frogs *Ranidella insignifera* and *R. pseudinsignifera. Aust. J. Zool.* **26**, 305–322.
Corbin, K. W., Sibley, C. G. and Ferguson, A. (1979). Gene changes associated with the establishment of sympatry in orioles of the genus *Icterus. Evolution, Lancaster, Pa.* **33**, 624–633.
Daly, J. C., Wilkinson, P. and Shaw, D. D. (1981). Reproductive isolation in relation to allozymic and chromosomal differentiation in the grasshopper *Caledia captiva. Evolution, Lancaster, Pa.* **35**, 1164–1179.
Dobzhansky, T. (1970). "Genetics of the Evolutionary Process." Columbia University Press, New York.
Douglas, M. E. and Avise, J. C. (1982). Speciation rates and morphological divergence in fishes: test of gradual vs. rectangular modes of evolutionary change. *Evolution, Lancaster, Pa.* **36**, 324–332.
Ehrlich, P. R. and Raven, P. H. (1969). Differentiation of populations. *Science, N.Y.* **165**, 1228–1232.
Endler, J. A. (1977). "Geographic Variation, Speciation and Clines". Princeton University Press, Princeton.
Gartside, D. F. (1972a). The *Litoria ewingi* complex (Anura: Hylidae) in south-eastern Australia. II. Genetic incompatibility and delimitation of a narrow hybrid zone between *L. ewingi* and *L. paraewingi. Aust. J. Zool.* **20**, 423–433.
Gartside, D. F. (1972b). The *Litoria ewingi* complex (Anura: Hylidae) in south-eastern Australia. III. Blood protein variation across a narrow hybrid zone between *L. ewingi* and *L. paraewingi. Aust. J. Zool.* **20**, 435–443.
Gould, S. J. and Woodruff, D. S. (1978). Natural history of *Cerion*. VIII: Little Bahama Bank – a revision based on genetics, morphometrics, and distribution. *Bull. Mus. comp. Zool. Harv.* **148**, 371–415.
Greenbaum, I. F. (1981). Genetic interactions between hybridizing cytotypes of the tent-making bat (*Uroderma bilobatum*). *Evolution, Lancaster, Pa.* **35**, 306–321.
Hall, W. P. (1973). "Comparative population cytogenetics, speciation and evolution of the Iguanid lizard genus *Sceloporus*" Ph.D. thesis, University of Harvard.

Hall, W. P. and Selander, R. K. (1973). Hybridization in karyotypically differentiated populations of the *Sceloporus grammicus* complex (Iguanidae). *Evolution, Lancaster, Pa.* **27**, 226–242.

Halliday, R. B., Barton, N. H. and Hewitt, G. M. (1982). Electrophoretic analysis of a chromosomal hybrid zone in the grasshopper *Podisma pedestris*. *Biol. J. Linn. Soc.* **19**, 51–62.

Hewitt, G. M. (1975). A sex chromosome hybrid zone in the grasshopper *Podisma pedestris* (Orthoptera: Acrididae). *Heredity* **35**, 375–387.

Hunt, W. G. and Selander, R. K. (1973). Biochemical genetics of hybridization in European house mice. *Heredity* **31**, 11–33.

Johnson, M. S. (1976). Allozymes and area effects in *Cepaea nemoralis* on the Berkshire downs. *Heredity* **36**, 105–121.

Johnson, M. S., Clarke, B. C. and Murray, J. (1977). Genetic variation and reproductive isolation in *Partula*. *Evolution, Lancaster, Pa.* **31**, 116–126.

Jones, J. S., Bryant, S. H., Lewontin, R. C., Moore, J. A. and Prout, T. (1981). Gene flow and the geographical distribution of a molecular polymorphism in *Drosophila pseudoobscura*. *Genetics* **98**, 157–178.

Jones, J. S., Selander, R. K. and Schnell, G. D. (1982). Patterns of morphological and molecular polymorphism in the land snail *Cepaea nemoralis*. *Biol. J. Linn. Soc.* **14**, 359–387.

Key, K. H. L. (1981). Species, parapatry, and the morabine grasshoppers. *Syst. Zool.* **30**, 425–458.

Leigh-Brown, A. J. and Langley, C. H. (1979). Re-evaluation of the level of genic heterozygosity in natural populations of *Drosophila melanogaster* by two-dimensional electrophoresis. *Proc. Nat. Acad. Sci., U.S.A.* **76**, 2381–2384.

Li, W. H. and Nei, M. (1974). Stable linkage disequilibrium without epistasis in subdivided populations. *Theor. Pop. Biol.* **6**, 173–183.

McDonnell, L. J., Gartside, D. F. and Littlejohn, M. J. (1978). Analysis of a narrow hybrid zone between two species of *Pseudophryne* (Anura: Leptodactylidae) in south-eastern Australia. *Evolution, Lancaster, Pa.* **32**, 602–612.

Mayr, E. (1963). "Animal Species and Evolution". Belknap Press, Cambridge, Mass.

Mayr, E. (1982). "The Growth of Biological Thought: Diversity, Evolution and Inheritance" Belknap Press, Cambridge, Mass.

Moran, C. (1981). Spermatogenesis in natural and experimental hybrids between chromosomally differentiated taxa of *Caledia captiva*. *Chromosoma* **81**, 579–591.

Moran, C., Wilkinson, P. and Shaw, D. D. (1980). Allozyme variation across a narrow hybrid zone in the grasshopper *Caledia captiva*. *Heredity* **44**, 69–81.

Murray, J. and Clarke, B. C. (1980). The genus *Partula*: speciation in progress. *Proc. R. Soc. Lond.* B **211**, 83–117.

Nagylaki, T. (1976). Clines with variable migration. *Genetics* **83**, 867–886.

Nevo, E. (1982). Speciation in subterranean mammals. *In* "Mechanisms of Speciation" (C. Barigozzi, ed.), pp. 191–218. Liss, New York.

Nevo, E. and Shaw, C. R. (1972). Genetic variation in a subterranean mammal, *Spalax ehrenbergi*. *Biochem. Genet.* **7**, 235–241.

Nevo, E., Kim, Y. J., Shaw, C. R. and Thaeler, C. S. (1974). Genetic variation, selection and speciation in *Thomomys talpoides* pocket gophers. *Evolution, Lancaster, Pa.* **28**, 1–23.

Patton, J. L. and Yang, S. Y. (1977). Genetic variation in the *Thomomys bottae* pocket gophers: macrogeographic patterns. *Evolution, Lancaster, Pa.* **31**, 697–720.

Patton, J. L., Hafner, M. S., Hafner, M. S. and Smith, M. F. (1979). Hybrid zones in *Thomomys bottae* pocket gophers: genetic, phenetic and ecologic concordance patterns. *Evolution, Lancaster, Pa.* **33**, 860–876.

Patton, J. L., Selander, R. K. and Smith, M. H. (1972). Genic variation in hybridizing populations of gophers (genus *Thomomys*). *Syst. Zool.* **21**, 263–270.

Sage, R. D. and Selander, R. K. (1979). Hybridization between species of the *Rana pipiens* complex in central Texas. *Evolution, Lancaster, Pa.* **33**, 1069–1088.

Selander, R. K., Hunt, W. G. and Yang, S. Y. (1969). Protein polymorphism and genic heterozygosity in two European subspecies of house mouse. *Evolution, Lancaster, Pa.* **23**, 379–390.

Shaw, D. D. and Wilkinson, P. (1980). Chromosome differentiation, hybrid breakdown, and the maintenance of a narrow hybrid zone in *Caledia*. *Chromosoma* **80**, 1–31.

Shaw, D. D., Moran, C. and Wilkinson, P. (1981). Chromosomal reorganisation, geographic differentiation, and the mechanism of speciation in the genus *Caledia*. *In* "Insect Cytogenetics" (R. L. Blackman, G. M. Hewitt and M. Ashburner, eds), pp. 171–194. Blackwell Scientific Publications, Oxford.

Short, L. L. (1969). Taxonomic aspects of avian hybridisation. *Auk* **86**, 84–105.

Szymura, J. M. (1976a). New data on the hybrid zone between *Bombina bombina* and *B. variegata* (Anura: Discoglossidae). *Bull. Acad. Polon. Sci.*, III **24**, 355–363.

Szymura, J. M. (1976b). Hybridization between discoglossid toads *Bombina bombina* and *Bombina variegata* in southern Poland as revealed by the electrophoretic technique. *Z. Zool. Syst. EvolForsch.* **14**, 227–236.

Szymura, J. M. and Farana, I. (1978). Inheritance and linkage analysis of five enzyme loci in interspecific hybrids of toadlets, genus *Bombina*. *Biochem. Genet.* **16**, 307–319.

Thaler, L., Bonhomme, F. and Britton-Davidian, J. (1981). Processes of speciation and semi-speciation in the house mouse. *Symp. Zool. Soc. Lond.* **47**, 27–41.

Turner, J. R. G., Johnson, M. S. and Eanes, W. F. (1979). Contrasted modes of evolution in the same genome: allozymes and adaptive change in *Heliconius*. *Proc. Nat. Acad. Sci., U.S.A.* **76**, 1924–1928.

Woodruff, D. S. (1979). Postmating reproductive isolation in *Pseudophryne* and the evolutionary significance of hybrid zones. *Science, N.Y.* **203**, 561–563.
Woodruff, D. S. (1981). Towards a genodynamics of hybrid zones: studies of Australian frogs and West Indian land snails. *In* "Evolution and Speciation" (W. Atchley and D. S. Woodruff, eds), pp. 171–200. Cambridge University Press, Cambridge.
Wright, S. (1978). "Evolution and the Genetics of Populations. IV. Variability Within and Among Natural Populations." University of Chicago Press, Chicago.

19 | The Population Structure of Cyclic Parthenogens

J. P. W. YOUNG

*John Innes Institute, Colney Lane,
Norwich NR4 7UH, U.K.*

Abstract: Cyclic parthenogenesis is an interesting reproductive strategy intermediate between conventional sexuality and obligate parthenogenesis; it is restricted to three animal groups: Cladocera (crustaceans), Monogononta (rotifers) and Aphididae (insects). In the past decade, enzyme electrophoresis has played a major role in advancing our knowledge of the population genetics and taxonomy of these organisms, but to a large extent a separate tradition has developed around each of the three taxa. The aim of this review is to bring together all the relevant literature and to compare progress and results. Topics discussed include the mode of inheritance in parthenogenesis and the sexual phase, pseudosexual reproduction (cryptoparthenogenesis) and anholocycly, taxonomic findings, predation and parasitism, the level of genetic polymorphism, deviations from Hardy–Weinberg equilibrium, spatial and temporal variation, evidence for selective differences, and the maintenance of clonal diversity.

INTRODUCTION

What are cyclic parthenogens? They are animals which can reproduce both asexually and sexually. Usually, several or many generations of parthenogenesis intervene between each bout of sexual reproduction, which is triggered by environmental cues and may have an annual periodicity. The sexually-produced offspring are morphologically

Systematics Association Special Volume No. 24, "Protein Polymorphism: Adaptive and Taxonomic Significance", edited by G. S. Oxford and D. Rollinson, 1983, Academic Press, London and New York.

distinctive and are usually diapausing and resistant to unfavourable environments. This type of life history, common amongst plants, is in animals confined to three groups: the rotifer class Monogononta, the crustacean order Cladocera, and the insect (hemipteran) family Aphididae. Within each of these groups, cyclic parthenogenesis (heterogony) is the rule, although there are exceptions which will be discussed later.

The aim of this review is to discuss the diverse contributions which the technique of enzyme electrophoresis has made in recent years to our understanding of these organisms. In the past ten years, there have been about 100 publications reporting electrophoretic studies: I will not cover most of those concerned solely with organophosphate resistance in aphids, which form a large, and largely separate, literature, but I have included nearly all the others. Cladocerans, rotifers and aphids each have a long, separate history of population and genetic studies, which I shall not attempt to cover here, but I hope that my review of the electrophoretic work will emphasize the interesting parallels and differences between them, and perhaps encourage some fruitful intellectual mixis. A good deal of the work on cladocerans has been reviewed previously by Hebert (1978, 1980), and on rotifers by King (1980).

ESTABLISHING THE MODE OF INHERITANCE

1. Parthenogenesis

There are a number of different genetic mechanisms by which unfertilized females can produce fertile female offspring, ranging from effectively mitotic mechanisms which preserve the mother's genotype intact, to processes which lead to complete homozygosis (White, 1970). The genetic nature of the parthenogenesis has been the subject of considerable debate in all cyclic parthenogens, the central issue being whether genetic variation can be generated within a single clone. The evidence was inconclusive because the available morphological variants were of unknown genetic basis, and the chromosomes were hard to study cytologically. Electrophoresis has largely resolved this debate.

Early workers studying cladocerans (Banta and Wood, 1927) found some heritable changes occurring during parthenogenesis, but recognized that the cause could have been mutation rather than segregation. Cytological studies of parthenogenetic egg production (reviewed by Hebert, 1978, 1980) have been conflicting, with some claiming to find chromosome association that suggest the possibility of recombination ("endomeiosis", e.g. Bacci *et al.*, 1961), while others, including the most recent and reliable study (Zaffagnini and Sabelli, 1972), reported only normal mitosis. Studies of the inheritance of electrophoretic variants in cladoceran parthenogenesis have invariably supported the latter view: the mother's genotype is inherited intact in all reported cases, which cover the following species: *Daphnia magna* (Hebert and Ward, 1972), *D. carinata* (Hebert and Moran, 1980), *D. pulex* (Berger and Sutherland, 1978; Hebert and Crease, 1980). *D. middendorffiana* (Hebert and McWalter, 1981), and *Bosmina longirostris* (Manning *et al.*, 1978).

In rotifers, the absence of any visible synapsis of the chromosomes during parthenogenetic oogenesis is corroborated by the failure to observe any intraclonal changes in isozyme patterns (King, 1977).

The parthenogenesis of aphids seems to be basically similar to that of cladocerans, though the evidence is less clear-cut. Blackman (1979) gives a critical review of the relevant literature. As in cladocerans, cytological reports include claims of endomeiosis in several species (e.g. Cognetti, 1961); there are numerous reports of effective selection within clones for quantitative characters, and a number of counterclaims that selection is ineffective. As Blackman (1979) points out, some of the reported changes are almost certainly the result of contamination, the remainder could be ascribed to a variety of genetic effects besides "endomeiosis". Even electrophoretic data have proved controversial in aphids. Reports of electrophoretic changes during parthenogenesis are of two kinds. The first involves the carboxylesterase E4 in *Myzus persicae*. This single enzyme has attracted the lion's share of attention in aphid electrophoresis because of its association with organophosphate resistance (Needham and Sawicki, 1971). Several reports (e.g. Beranek, 1974) have described sudden, heritable changes in the activity of this enzyme, but these changes (which involve band intensity and not mobility, cf. Devonshire's reply to Baker, 1980) are plausibly explained by locus-specific effects such as gene reduplication (Devonshire and Sawicki, 1979) or chromosomal rearrangement (Blackman *et al.*, 1978). Rather different is the claim by

Beranek and Berry (1974) of substantial changes in the band pattern of cholinesterase in clones of *Aphis fabae*. Amongst other changes, they report that three isolated clones changed simultaneously to the same new heritable pattern. The only really plausible explanation is contamination by alien aphids (Blackman, 1979) but, in any case, the phenomenon does not represent segregation, since new bands appeared. In contrast, several authors have reported stable inheritance of heterozygous enzymes during parthenogenesis in *Myzus persicae* (May and Holbrook, 1978; Blackman, 1979) and *Macrosiphum rosae* (Tomiuk and Wöhrmann, 1982). Suomalainen *et al.* (1980) also observed stable inheritance in *Acyrthosiphon pisum*, but were unable to test for segregation since nearly all the enzymes appeared to be homozygous. In conclusion, then, segregation has not been detected in aphid parthenogenesis; other genetic changes, if real, are specialized or infrequent.

2. *Sexual reproduction*

The males of cyclic parthenogens are, of course, produced parthenogenetically. Those of cladocerans are cytologically diploid, and electrophoresis of three enzymes in *Daphnia magna* has confirmed that they inherit their mother's genotype at these loci (Hebert and Ward, 1972). The males of rotifers are cytologically haploid, but have proved too small for electrophoresis (King and Snell, 1977b), while those of aphids have, so far as I know, not been examined.

Female cladocerans are normally amphoteric, that is, each female can, given the right cues, reproduce both sexually and asexually. Murthy and Fraser's (1971) finding that males, parthenogenetic and sexual females each had distinctive LDH patterns was probably an artefact of nutritional status (Hebert, 1973), and does not imply genetic differences. Mictic and amictic reproduction in rotifers commonly appear to be confined to separate, specialized types of female, but King and Snell (1977b) identified amphoteric individuals of *Asplanchna girodi* and used electrophoresis to show that their parthenogenetic progeny were produced by the normal apomictic process.

Electrophoretic markers have been valuable in demonstrating that normal genetic segregation occurs during the sexual reproduction of cyclic parthenogens. The expected Mendelian inheritance has been observed in the cladocerans *Daphnia magna* (Hebert and Ward, 1972;

Ferrari and Hebert, 1982) and *D. carinata* (Hebert and Moran, 1980), and, with some reservations, in the aphid *Myzus persicae* (Blackman, 1979; Takada, 1979a) and the rotifer *Asplanchna girodi* (King, 1977; King and Snell, 1977a).

These results confirm that the normal sexual process does occur; however, a number of species are known in which it is replaced by pseudosexual reproduction (cryptoparthenogenesis), in which females produce, without fertilization, eggs with the morphology typical of sexually-produced eggs (diapausing, resistant etc.). Electrophoresis has confirmed that in *Daphnia pulex, D. middendorffiana* and *D. cephalata* such reproduction is, like normal parthenogenesis, effectively mitotic (Hebert and Crease, 1980; Hebert, 1981; D. B. McWalter and P. D. N. Hebert, unpublished). Analogous pseudo-sexual eggs have been reported in the rotifer *Keratella quadrata* (Ruttner-Kolisko, 1946), but the mode of inheritance has not yet been confirmed genetically.

Species in which pseudosexual replaces sexual reproduction are, of course, obligate rather than cyclic parthenogens, but I shall include them in my brief. Hebert (1978) has pointed out that genes for pseudosexuality might spread into a range of genetic backgrounds provided they did not suppress male meiosis. Incidentally, the widespread occurrence of pseudosexual reproduction poses the question: why is sexual reproduction retained at all? It has been suggested that sib competition may confer a short-term advantage on sexual lineages (Williams and Mitton, 1973; Maynard Smith, 1976); the persistent coexistence of clones (see below) suggests that any such effect may act through the Elbow-Room Model of ecological divergence rather than the Aphid-Rotifer or Lottery Model of competitive exclusion (Young, 1981).

IDENTIFYING SPECIES

1. *Contributions to taxonomy*

Electrophoresis promises to be a useful tool in aphid taxonomy. Closely related species are often difficult to differentiate morphologically, the major differences being in host range which may be difficult to establish. Tomiuk *et al.* (1979) examined species from five genera, and found that, in some cases, electrophoresis could provide a rapid diagnosis. More samples of each species will, however, need to be examined before this method can become routine. An interesting aspect of this work is that species in several groups had identical patterns at all loci examined: further electrophoretic work may help to clarify the true species limits. Furk (1979) examined a number of races of *Aphis fabae s.l.* by enzyme electrophoresis and isoelectric focusing of proteins, and found a good correlation with host range: these techniques may allow genetically-isolated, host range-specific species to be distinguished in this species complex. May and Holbrook (1978) found that samples of *Myzus persicae* were completely invariant, with the exception of four individuals that were homozygous for a novel allele at each of three loci. The authors suggest, quite reasonably, that these represent a hitherto-unrecognized sibling species, but it must be pointed out that they also found a fifth variant individual which appeared to be a hybrid between the two types. It is instructive that some of the variants were found as "contaminants" in a laboratory colony.

Hebert (1977) has used electrophoresis to supplement conventional taxonomy in a revision of the Australian cladocerans formerly grouped as *Daphnia carinata*. He identified nine species on the basis of enzyme differences, with no evidence of introgression, and subsequently found morphological differences to substantiate his classification. Hebert and McWalter (1982) used enzyme data to construct a numerical taxonomy of the arctic species *Daphnia middendorffiana* and *D. pulex*. Pseudosexual parthenogenesis is universal in *D. middendorffiana* and common in *D. pulex*, so these species are not strictly cyclic parthenogens, but are truly clonal. Many workers have had difficulties with the conventional taxonomy, finding many intermediate forms. Hebert and McWalter found that there were indeed two quite distinct taxonomic groups, but that cuticular pigmentation, classically the hallmark of *D. middendorffiana*, occurred in some forms from each group. In a recent study, as yet unpublished, B. J. Hann and P. D. N Hebert examined three populations of *Simocephalus* in Ohio. They recognized four species on the basis of morphological criteria, and found little polymorphism within each, but extensive allelic substitution between species. The interesting background to this work is that a few years earlier these same populations had been studied by other workers (Smith, 1974; Smith and Fraser, 1976) who ascribed all their material to a single species, "*S. serrulatus*", which they found to be very polymorphic, with extreme heterozygote deficits and disequilibria between loci. The reexamination by Hann and Hebert seems to confirm earlier suspicions that these populations were a mixture of reproductively isolated forms (Young, 1979a; Hebert, 1980; Hebert and Moran, 1980). Finally, mention should be made of the evidence by Manning *et al.* (1978) that two species are involved in the seasonal morphological changes observed in *Bosmina longirostris* (see below).

The examples I have cited show that electrophoresis has been valuable both in corroborating conventional taxonomy and in detecting unsuspected sibling species. There are some very promising areas for future study, such as the relationship between host range, reproductive isolation and genetic divergence in aphids, and the taxonomic status of cladoceran species that include two or more distinct morphs, such as *Daphnia longiremis* (Riessen and O'Brien, 1980).

2. *Predators and parasites*

Parasitism by hymenopterans and predation by other invertebrates exert an important influence on aphid populations, and aphid ecologists have developed ingenious electrophoretic methods for their study. Electrophoresis of a parasitized aphid reveals the parasite's isozymes superimposed on those of the host, and this allows parasitized individuals to be recognized before any abnormality is visible externally, and may even allow the species of the parasite to be determined (Wool *et al.*, 1978b; Tomiuk and Wöhrmann, 1980b, 1981). Secrets can be wrenched from predators too: Murray and Solomon (1978) were able to identify various prey species, including aphids, by their esterases which persisted in the guts of predatory bugs or mites for many hours. In view of their long-standing interest in predator–prey interactions, it is surprising that zooplankton biologists have not found more uses for techniques of this kind.

GENETIC DIVERSITY

Many electrophoretic studies, especially in cladocerans, have been motivated by the general problem of discovering the level of genetic polymorphism that is maintained within biological species and the mechanisms that maintain it. From an evolutionist's point of view, the term "cyclic parthenogenesis" is slightly misleading because these creatures are essentially sexual organisms, but the intervening parthenogenesis means that the intensity of selection is much magnified relative to genetic segregation and recombination. An additional complication is that many populations of cladocerans, rotifers and aphids are not sexual at all, having either adopted pseudosexual production of the diapausing stage or colonized an environment which is sufficiently equable that diapause is neither needed nor induced (a phenomenon known as "anholocycly" in aphids, but also frequent in cladocerans). These variations need to be taken into account in any assessment of genetic variation but, unfortunately, there is, in many studies, no clear evidence for the extent of sexual reproduction. This has inevitably made comparisons difficult and, in some cases, confusing.

1. *The level of polymorphism*

Many populations have been studied at only a handful of loci, so only a rough estimate of genetic diversity is possible, but many studies have found polymorphism comparable with that in other organisms, while a substantial minority have reported very little variation. Considerable diversity has been found in the cladocerans *Daphnia magna* (e.g. Hebert, 1974b, c), *D. pulex* (Berger and Sutherland, 1978; Hebert and Crease, 1980), *D. carinata* (Hebert and Moran, 1980), *Bosmina longirostris* (Manning *et al.*, 1978), and the aphids *Macrosiphum euphorbiae* (May and Holbrook, 1978) and *M. rosae* (S. Joseph and R. S. Singh, unpublished). However, very low levels of variation have been reported in *Myzus persicae* (May and Holbrook, 1978; Wool *et al.*, 1978a) and various other aphids (Tomiuk and Wöhrmann, 1980a), and in a number of populations of the cladocerans *Daphnia pulex, D. laevis* and *D. rosea* (S. R. Krepp, unpublished), *D. cephalata* (Hebert, 1981) and several species of *Simocephalus* (B. J. Hann and P. D. N. Hebert, unpublished). The distinction between cyclic and obligate parthenogenesis does not seem to be critical; low and high diversities are found in both groups. It may be that the examples of low polymorphism reflect the colonizing success, perhaps temporary, of single clones.

2. *Heterozygote excesses*

Parthenogenesis frees a polymorphic locus from the constraints of Hardy-Weinberg equilibrium (H-W), and studies of *Daphnia magna* have revealed a consistent pattern of deviations. In populations which are frequently reestablished through the sexual phase, most loci approximate to H-W expectations (Hebert, 1974c), but in populations with long histories of continuous parthenogenesis, extreme deviations, especially heterozygote excesses, are common, together with disequilibria between loci, but many clones continue to coexist (Hebert *et al.*, 1972; Hebert, 1974a, b; Hebert and Ward, 1976; Young, 1979a, b). Berger (1976) suggested that the heterozygote excesses might reflect single-locus heterosis, but there are good reasons for believing the observed selection is a multi-locus effect (Hebert, 1974b; Angus, 1978; Young, 1979b). Indeed, the accumulation of recessive deleterious alleles ("synthetic heterosis") is formally sufficient without the need to postulate heterozygote superiority at any locus (Young, 1979b; *pace* Hebert, 1983). I have emphasized *D. magna* as the best-studied species, but other cladocerans are generally concordant, with H-W approximated in holocyclic *D. carinata* (Hebert and Moran, 1980), and deviations, predominantly heterozygote excesses, in populations that are probably anholocyclic or pseudosexual of *D. pulex* (Hebert and Crease, 1980; Berger and Sutherland, 1978), *D. galeata mendotae* (Mort and Jacobs, 1981) and *Bosmina longirostris* (Manning *et al.*, 1978; reanalysed by Hebert 1980). On the other hand, Berger and Sutherland's (1978) Lake Saratoga population of *D. pulex*, apparently permanent, did not show any consistent pattern of H-W deviations. Heterozygote deficits reported in *Simocephalus* (Smith and Fraser, 1976) probably reflect a mixture of species (see above).

In anholocyclic populations of the aphid *Myzus persicae*, Baker (1978, 1979) did not find marked heterozygote excesses, but he did observe extreme disequilibrium between loci, suggesting a small number of clones. Tomiuk and Wöhrmann (1981) found that MDH frequencies remained near H-W in the holocyclic *Macrosiphum rosae*. In anholocyclic *Sitobion avenae*, H. D. Loxdale (pers. comm.) reports a high level of polymorphism at many loci, but a complete absence of heterozygotes at most: reminiscent of the *Simocephalus* species mixture discussed earlier.

3. *Variation in space*

Migration of cladocerans between ponds is relatively rare, as evidenced by the large differences in genotype composition which are commonly found even between neighbouring populations (Hebert, 1974b, c, 1975; Hebert and Crease, 1980; Hebert and Moran, 1980, Manning *et al.*, 1978). Spatial heterogene of frequencies within a population is often negligible and always much less than the differences between populations (Hebert, 1974a; Hebert and Ward, 1976; Manning *et al.*, 1978; Young, 1979a).

Aphids have the opposite pattern of migration, with alates covering large distances but less thorough mixing of apterous colonies. A population is hard to define in these circumstances, but moderate amounts of spatial heterogeneity have been reported on both a large and a small scale in *Myzus persicae* (Baker, 1977a, b, 1978, 1979; Takada 1979b) and *Macrosiphum rosae* (Tomiuk and Wöhrmann, 1981; S. Joseph and R. S. Singh, unpublished). In order to avoid sampling a single family repeatedly, aphid workers are usually careful to take only one individual from each colony, but there is some interesting evidence that clumps of aphids may not be as clonal as they look. Morphological variation in the gall-forming aphid *Pemphigus populitransversus* has been studied very extensively (e.g. Sokal, 1952, 1962; Sokal *et al.*, 1980), and the supposed genetic identity of individuals within a gall allowed an estimate of non-genetic variation. However, recent electrophoretic work by Setzer (1980) has revealed that at least a quarter of the aphids within a gall may be immigrants from other clones.

4. *Variation in time*

Holocyclic populations of *Daphnia magna* have stable genotype frequencies, but when sex is absent or infrequent, rapid changes have often been observed (review by Hebert, 1978). Temporal changes have also been reported in other cladocerans (Smith and Fraser, 1976; Manning *et al.*, 1978; Berger and Sutherland, 1978; Hebert and Moran, 1980; Hebert and Crease, 1980; M. L. Pace, pers. comm.). None of these studies has been sufficiently long-term to establish whether the changes form part of a regular seasonal cycle, but the available evidence suggests that some are actually unique events.

In some cladocerans, there is a striking seasonal cycle of morphological variation known as "cyclomorphosis". Some of the changes are known to be phenotypic responses to the environment, but electrophoretic study of *Bosmina longirostris* (Manning *et al.*, 1978) has revealed that a major component is the relative abundance of two genetically distinct species. This finding has allowed a reevaluation of the extensive morphometric and ecological studies on this species (Kerfoot, 1980; Kerfoot and Peterson, 1979, 1980; Black, 1980a, b). Brock (1980a, b) has subsequently published a very similar finding in a distinct population of the same species.

An extreme case of temporal change has been recorded in the rotifer *Asplanchna girodi* (King, 1977). At first, only a single genotype was detected in the population, and this persisted for several weeks, but suddenly (within a week) it was almost completely displaced by a new genotype (species?) that differed from it at several loci.

Perhaps the best known example of temporal genetic change in any parthenogen is the development and spread of resistance to organophosphate insecticides in aphids, especially *Myzus persicae*, due to changes in the activity of an esterase. This has generated a large literature in which electrophoresis has played an important role, but I cannot do it justice here (see, e.g. Needham and Sawicki, 1971; Devonshire, 1977; Devonshire and Sawicki, 1979; Beck and Büchi, 1980). Suffice it to say that this drastic recent selection pressure on a largely anholocyclic species may have reduced clonal diversity. Temporal monitoring of other loci has revealed changes in *Myzus persicae* (Baker, 1977b, 1978, 1979) and *Macrosiphum rosae* (Tomiuk and Wöhrmann, 1981; S. Joseph and R. S. Singh, unpublished).

5. *Selective differences amongst clones*

There is considerable evidence that clones, even if coexisting, are not ecologically equivalent. In natural populations of *Daphnia magna*, electrophoretically-distinct genotypes commonly have different parthenogenetic fecundities or frequencies of sexual reproduction (reviewed by Hebert, 1978; Young, 1979a, b). Laboratory comparisons or artificial populations of clones from this and other cladocerans have revealed differences in various components of fitness (Hebert and Crease, 1980; Brookfield, 1981; Mort and Jacobs, 1981; Loaring and Hebert, 1981; Hebert *et al.*, 1982; Ferrari and Hebert, 1982; S. R.

Krepp, unpublished). Similar studies on the rotifer *Asplanchna* (Snell, 1979, 1980) present the same conclusion.

6. *Why do clones coexist?*

On the whole, parthenogen populations that are truly cyclic have a pattern of genetic diversity not dissimilar to that of fully sexual species; it is those with rare or absent sex that are enigmatic. The problem is a familiar one in parthenogens: competitive exclusion should eliminate all but the fittest clone and yet there is good evidence, especially in cladocerans, that many clones coexist. Aphid population structure is probably complex, and certainly poorly understood, so the issue does not present itself so clearly in their case. Indeed, the low level of polymorphism found in some aphids has been a more frequent source of concern (Tomiuk and Wöhrmann, 1980a). It is tempting to invoke competitive exclusion as a cause of this, but such an argument ought to apply *a fortiori* to obligate parthenogens, which are commonly polymorphic (Suomalainen *et al.*, 1976), and low diversity is, in any case, found in both holocyclic and anholocyclic aphids. King (1977, 1980) has suggested that competition leads to the predominance of a single clone in rotifer populations, but that one clone is rapidly replaced by another as the environment changes. However, his evidence is slender.

In cladocerans, though, we know that clones coexist over periods of many months. Heterosis cannot maintain genotypic diversity in a parthenogen, and there is abundant evidence that clones are not selectively neutral, so the only plausible source of stability is ecological specialization leading to frequency-dependent selection: nearly all authors have suggested this. Ironically, the best evidence for such selection comes from an experiment that failed to detect any. Brookfield (1981) set up pairs of clones in artificial populations but failed to find any evidence of stabilizing selection. But, he also failed to obtain coexistence, suggesting that the assemblages of clones that coexist naturally cannot be thrown together arbitrarily. My hunch is that some of our problems stem from the search for a stability which may be illusory: a river changes its water hourly and its course every year, yet maintains its identity for millennia. Our evidence is for coexistence, not for permanence; we need models of persistence, not of equilibrium.

ACKNOWLEDGEMENTS

I wrote to nearly all the people whose electrophoretic studies I have cited here, and am extremely grateful for the many enthusiastic replies I received, without which this review would have been less complete and less up-to-date.

REFERENCES

Angus, R. A. (1978). *Daphnia* and the search for heterosis. *Am. Nat.* **112**, 955–956.

Bacci, G., Cognetti, G. and Vaccari, A. M. (1961). Endomeiosis and sex determination in *Daphnia pulex*. *Experientia* **17**, 505–506.

Baker, J. P. (1977a). Assessment of the potential for and development of organophosphorus resistance in field populations of *Myzus persicae*. *Ann. appl. Biol.* **86**, 1–9.

Baker, J. P. (1977b). Changes in composition in populations of the peach-potato aphid, *Myzus persicae*, overwintering in Scotland 1976–7. *Proc. 1977 British Crop Protection Conference – Pests and Diseases*, 255–261.

Baker, J. P. (1978). Electrophoretic studies on populations of *Myzus persicae* in Scotland from March to July, 1976. *Ann. appl. Biol.* **88**, 1–11.

Baker, J. P. (1979). Electrophoretic studies on populations of *Myzus persicae* in Scotland from October to December, 1976. *Ann. appl. Biol.* **91**, 159–164.

Baker, J. (1980). Evolution by gene duplication in insecticide-resistant *Myzus persicae*. *Nature, Lond.* **284**, 577–578.

Banta, A. M. and Wood, T. R. (1927). Inheritance in parthenogenesis and in sexual reproduction in Cladocera. *Verhandl. V Int. Congr. Vererbungswiss*, 391–396.

Beck, A. K. and Büchi, R. (1980). Esterasetest zum Nachweis der Insektizidresistenz bei der Hopfenblattlaus, *Phorodon humuli* Schrk. *Z. ang. Ent.* **89**, 113–121.

Beranek, A. P. (1974). Stable and non-stable resistance to dimethoate in the peach-potato aphid (*Myzus persicae*) *Entomologia exp. appl.* **17**, 381–390.

Beranek, A. P. and Berry, R. J. (1974). Inherited changes in enzyme patterns within parthenogenetic clones of *Aphis fabae*. *J. Ent.* A **48**, 141–147.

Berger, E. (1976). Heterosis and the maintenance of enzyme polymorphism. *Am. Nat.* **110**, 823–839.

Berger, E. and Sutherland, J. (1978). Allozyme variation in two natural populations of *Daphnia pulex*. *Heredity* **41**, 13–23.

Black, R. W. (1980a). The genetic component of cyclomorphosis in *Bosmina*. *In* "Evolution and Ecology of Zooplankton Communities" (W. C. Kerfoot, ed.), pp. 456–469. University Press of New England, Hanover, New Hampshire.

Black, R. W. (1980b). The nature and causes of cyclomorphosis in a species of the *Bosmina longirostris* complex. *Ecology* **61**, 1122–1132.

Blackman, R. L. (1979). Stability and variation in aphid clonal lineages. *Biol. J. Linn. Soc.* **11**, 259–277.

Blackman, R. L., Takada, H. and Kawakami, K. (1978). Chromosomal rearrangement involved in insecticide resistance of *Myzus persicae*. *Nature, Lond.* **271**, 450–452.

Brock, D. A. (1980a). Genotype structure of *Bosmina longirostris* and *Daphnia parvula* populations. Thesis, University of Texas.

Brock, D. A. (1980b). Genotypic succession in the cyclomorphosis of *Bosmina longirostris* (Cladocera). *Freshwater Biol.* **10**, 239–250.

Brookfield, J. F. Y. (1981). No evidence for frequency-dependent selection acting between clones of the water flea, *Daphnia magna*. *Heredity* **47**, 297–315.

Cognetti, G. (1961). Endomeiosis in parthenogenetic lines of aphids. *Experientia* **17**, 168–169.

Devonshire, A. L. (1977). The properties of a carboxylesterase from the peach-potato aphid *Myzus persicae* (Sulz,), and its role in conferring insecticide resistance. *Biochem. J.* **167**, 675–683.

Devonshire, A. L. (1980). Evolution by gene duplication in insecticide-resistant *Myzus persicae*. *Nature, Lond.* **284**, 577–578.

Devonshire, A. L. and Sawicki, R. M. (1979). Insecticide-resistant *Myzus persicae* as an example of evolution by gene duplication. *Nature, Lond.* **280**, 140–141.

Ferrari, D. C. and Hebert, P. D. N. (1982). The induction of sexual reproduction in *Daphnia magna*: genetic differences between arctic and temperate populations. *Can. J. Zool.* **60**, 2143–2148.

Furk, C. (1979). Field collections of *Aphis fabae* Scopoli *s. lat.* (Homoptera: Aphididae) studied by starch gel electrophoresis and isoelectric focusing. *Comp. Biochem. Physiol.* **62B**, 225–230.

Hebert, P. D. N. (1973). Phenotypic variability of lactate dehydrogenase in *Daphnia magna*. *J. exp. Zool.* **186**, 33–38.

Hebert, P. D. N. (1974a). Ecological differences among genotypes in a natural population of *Daphnia magna*. *Heredity* **33**, 327–337.,

Hebert, P. D. N. (1974b). Enzyme variability in natural populations of *Daphnia magna*. II. Genotypic frequencies in permanent populations. *Genetics* **77**, 323–334.

Hebert, P. D. N. (1974c). Enzyme variability in natural populations of *Daphnia magna*. III. Genotypic frequencies in temporary populations. *Genetics* **77**, 335–341.

Hebert, P. D. N. (1975). Enzyme variability in natural populations of *Daphnia magna*. I. Population structure in East Anglia. *Evolution, Lancaster, Pa.* **28**, 546–556.

Hebert, P. D. N. (1977). A revision of the taxonomy of the genus *Daphnia* in southeastern Australia. *Aust. J. Zool.* **25**, 371–398.
Hebert, P. D. N. (1978). The population biology of *Daphnia*. *Biol. Rev.* **53**, 387–426.
Hebert, P. D. N. (1980). The genetics of Cladocera. *In* "Evolution and Ecology of Zooplankton Communities" (W. C. Kerfoot, ed.), pp. 329–336. University Press of New England, Hanover, New Hampshire.
Hebert, P. D. N. (1981). Obligate asexuality in *Daphnia*. *Am. Nat.* **117**, 784–789.
Hebert, P. D. N. (1983). Clonal diversity in cladoceran populations. *In* "Population Biology: Retrospect and Prospect" (C. E. King and P. S. Dawson, eds), pp. 37–59. Columbia University Press, New York.
Hebert, P. D. N. and Crease, T. J. (1980). Clonal coexistence in *Daphnia pulex* Leydig: another planktonic paradox. *Science, N.Y.* **207**, 1363–1365.
Hebert, P. D. N. and McWalter, D. B. (1982). Cuticular pigmentation in arctic *Daphnia*: adaptive diversification of asexual lineages? *Am. Nat.* (in press).
Hebert, P. D. N. and Moran, C. (1980). Enzyme variability in natural populations of *Daphnia carinata*. *Heredity* **45**, 313–321.
Hebert, P. D. N. and Ward, R. D. (1972). Inheritance during parthenogenesis in *Daphnia magna*. *Genetics* **71**, 639–642.
Hebert, P. D. N. and Ward, R. D. (1976). Enzyme variability in natural populations of *Daphnia magna*. IV. Ecological differentiation and frequency changes of genotypes at Audley End. *Heredity* **36**, 331–341.
Hebert, P. D. N., Ferrari, D. C. and Crease, T. J. (1982). Heterosis in *Daphnia*: a reassessment. *Am. Nat.* **119**, 427–434.
Hebert, P. D. N., Ward, R. D. and Gibson, J. B. (1972). Natural selection for enzyme variants among parthenogenetic *Daphnia magna*. *Genet. Res., Camb.* **19**, 173–176.
Kerfoot, W. C. (1980). Perspectives on cyclomorphosis: separation of phenotypes and genotypes. *In* "Evolution and Ecology of Zooplankton Communities" (W. C. Kerfoot, ed.), pp. 470–496. University Press of New England, Hanover, New Hampshire.
Kerfoot, W. C. and Peterson, C. (1979). Ecological interactions and evolutionary arguments: investigations with predatory copepods and *Bosmina*. *Fortschr. Zool.* **25**, 159–196.
Kerfoot, W. C. and Peterson, C. (1980). Predatory copepods and *Bosmina*: replacement cycles and further influences of predation upon prey reproduction. *Ecology* **61**, 417–431.
King, C. E. (1977). Genetics of reproduction, variation and adaptation in rotifers. *Arch. Hydrobiol., Beih.* **8**, 187–201.
King, C. E. (1980). The genetic structure of zooplankton populations. *In* "Evolution and Ecology of Zooplankton Communities" (W. C. Kerfoot, ed.), pp. 315–328. University Press of New England, Hanover, New Hampshire.

King, C. E. and Snell, T. W. (1977a). Sexual recombination in rotifers. *Heredity* **39**, 357–360.

King, C. E. and Snell, T. W. (1977b). Genetic basis of amphoteric reproduction in rotifers. *Heredity* **39**, 361–364.

Loaring, J. M. and Hebert, P. D. N. (1981). Ecological differences among clones of *Daphnia pulex* Leydig. *Oecologia* **51**, 162–168.

Manning, B. J., Kerfoot, W. C. and Berger, E. M. (1978). Phenotypes and genotypes in cladoceran populations. *Evolution, Lancaster, Pa.* **32**, 365–374.

May, B. and Holbrook, F. R. (1978). Absence of genetic variability in the green peach aphid, *Myzus persicae* (Hemiptera: Aphididae). *Ann. ent. Soc. Am.* **71**, 809–812.

Maynard Smith, J. (1976). A short term advantage for sex and recombination through sib competition. *J. Theoret. Biol.* **63**, 245–258.

Mort, M. A. and Jacobs, J. (1981). Differences among genotypic frequencies of undisturbed and manipulated populations of *Daphnia. Oecologia* **50**, 184–186.

Murray, R. A. and Solomon, M. G. (1978). A rapid technique for analysing the diets of invertebrate predators by electrophoresis. *Ann. appl. Biol.* **90**, 7–10.

Murthy, R. and Fraser, A. (1971). Lactate dehydrogenase (LDH) differences between sexual phases in *Daphnia magna. Genetics* **68**, s46.

Needham, P. H. and Sawicki, R. M. (1971). Diagnosis of resistance to organophosphorus insecticides in *Myzus persicae* (Sulz.). *Nature, Lond.* **230**, 125–126.

Riessen, H. P. and O'Brien, W. J. (1980). Re-evaluation of the taxonomy of *Daphnia longiremis.* Sars 1862 (Crustacea, Cladocera): description of a new morph from Alaska. *Crustaceana* **38**, 1–11.

Ruttner-Kolisko, A. (1946). Über das Auftreten unbefruchteter Dauereier bei *Keratella quadrata. Österr. Zool. Z.* **1**, 179–191.

Setzer, R. W. (1980). Intergall migration in the aphid genus *Pemphigus. Ann. entom. Soc. Am.* **73**, 327–331.

Smith, M. Y. (1974). A study of polymorphism in *Simocephalus serrulatus.* Ph.D. thesis, University of Cincinnati.

Smith, M. Y. and Fraser, A. (1976). Polymorphism in a cyclic parthenogenetic species: *Simocephalus serrulatus. Genetics* **84**, 631–637.

Snell, T. W. (1979). Intraspecific competition and population structure in rotifers. *Ecology* **60**, 494–502.

Snell, T. W. (1980). Blue-green algae and selection in rotifer populations. *Oecologia* **46**, 343–346.

Sokal, R. R. (1952). Variation in a local population of *Pemphigus. Evolution, Lancaster, Pa.* **6**, 296–315.

Sokal, R. R. (1962). Variation and covariation of characters of alate *Pemphigus populitransversus* in eastern North America. *Evolution, Lancaster, Pa.* **16**, 227–245.

Sokal, R. R., Bird, J. and Riska, B. (1980). Geographic variation in *Pemphigus populicaulis* (Insecta: Aphididae) in eastern North America. *Biol. J. Linn. Soc.* **14**, 163–200.

Suomalainen, E., Saura, A. and Lokki, J. (1976). Evolution of parthenogenetic insects. *In* "Evolutionary Biology". Vol. 9 (M. Hecht, W. Steere and B. Wallace, eds), pp. 209–257. Plenum Press, New York.

Suomalainen, E., Saura, A., Lokki, J. and Teeri, T. (1980). Genetic polymorphism and evolution in parthenogenetic animals. *Theor. Appl. Genet.* **57**, 129–132.

Takada, H. (1979a). Esterase variation in Japanese populations of *Myzus persicae* (Sulzer) (Homoptera: Aphididae), with special reference to resistance to organophosphorus insecticides. *Appl. ent. Zool.* **14**, 245–255.

Takada, H. (1979b). Characteristic forms of *Myzus persicae* (Sulzer) (Homoptera: Aphididae) distinguished by colour and esterase differences, and their occurrence in populations on different host plants in Japan. *Appl. ent. Zool.* **14**, 370–375.

Tomiuk, J. and Wöhrmann, K. (1980a). Enzyme variability in populations of aphids. *Theor. Appl. Genet.* **57**, 125–127.

Tomiuk, J. and Wöhrmann, K. (1980b). Population growth and population structure of natural populations of *Macrosiphum rosae* (L.) (Hemiptera, Aphididae). *Z. ang. Ent.* **90**, 464–473.

Tomiuk, J. and Wöhrmann, K. (1981). Changes of the genotype frequencies at the MDH-locus in populations of *Macrosiphum rosae* (L.) (Hemiptera, Aphididae). *Biol. Zbl.* **100**, 631–640.

Tomiuk, J. and Wöhrmann, K. (1982). Comments on the genetic stability of aphid clones. *Experientia* **38**, 320–321.

Tomiuk, J., Wöhrmann, K., and Eggers-Schumacher, H. A. (1979). Enzyme patterns as a characteristic for the identification of aphids. *Z. ang. Ent.* **88**, 440–446.

White, M. J. D. (1970). Heterozygosity and genetic polymorphism in parthenogenetic animals. *In* "Essays in evolution and genetics" (M. K. Hecht and W. C. Steere, eds), pp. 237–262. Appleton-Century-Crofts, New York.

Williams, G. C. and Mitton, J. B. (1973). Why reproduce sexually? *J. Theoret. Biol.* **39**, 545–554.

Wool, D., Bunting, S. and van Emden, H. F. (1978a). Electrophoretic study of genetic variation in British *Myzus persicae* (Sulzer) (Hemiptera, Aphididae). *Biochem. Genet.* **16**, 987–1006.

Wool, D., van Emden, H. F. and Bunting, S. W. (1978b). Electrophoretic detection of the internal parasite, *Aphidius matricariae* in *Myzus persicae*. *Ann. appl. Biol.* **90**, 21–26.

Young, J. P. W. (1979a). Enzyme polymorphisms and cyclical parthenogenesis in *Daphnia magna*. I. Selection and clonal diversity. *Genetics* **92**, 953–970.

Young, J. P. W. (1979b). Enzyme polymorphisms and cyclical parthenogenesis in *Daphnia magna*. II. Evidence of heterosis. *Genetics* **92**, 971–982.

Young, J. P. W. (1981). Sib competition can favour sex in two ways. *J. Theoret. Biol.* **88**, 755–756.
Zaffagnini, F. and Sabelli, B. (1972). Karyologic observations on the maturation of the summer and winter eggs of *Daphnia pulex* and *Daphnia middendorffiana*. *Chromosoma* **36**, 193–203.

20 | Protein Variation in Sexual Diploids and "Conspecific" Parthenogenetic Polyploids

B. CHRISTENSEN

Institute of Population Biology, University of Copenhagen, Universitetsparken 15, DK 2100 Copenhagen, Denmark

Abstract: It is argued that sexual diploids and derived parthenogenetic species represent useful models for studying the mechanisms behind intraspecific variation and the genetic differentiation associated with speciation. In the case of non-hybrid origin of the parthenogenetic species, the two species are initially identical in a genetic sense. If, therefore, a subsequent genetic differentiation is adaptive, it must be strongly dependent on a subsequent ecological divergence.

In the nominal species *Lumbricillus lineatus*, eggs of the triploid parthenogenetic sibling have to be activated by spermatozoa from the diploid sexual species in order to develop normally. Therefore, the two forms invariably practice intimate coexistence and they are supposed to have identical ecologies. Studies on nine polymorphic loci revealed a high degree of genetic identity between sympatric populations, indicating that the intraspecific variation is adaptive.

In the nominal species *Trichoniscus pusillus*, the triploid parthenogenetic sibling is independent of its sexual ancestor and the two units have diverged to some extent both genetically and ecologically. Variation within the *Pgm* locus in a number of sympatric populations revealed parallel trends between the two sibling species, indicating that the intraspecific variation is adaptive in this case also. There is, however, no correlation between the intraspecific variation and the environmental factors underlying the ecological differentiation between the two forms of *T. pusillus*. This is taken to indicate an uncoupling of micro- and macroevolutionary processes.

Systematics Association Special Volume No. 24, "Protein Polymorphism: Adaptive and Taxonomic Significance", edited by G. S. Oxford and D. Rollinson, 1983, Academic Press, London and New York.

INTRODUCTION

It is, of course, a matter of debate whether morphologically identical populations reproducing by parthenogenesis can be said to constitute a species and, consequently, it is also a matter of discussion whether the process through which such forms arise from sexually reproducing ancestors can be called speciation. Even so, it is clear that whenever a parthenogenetic population arises from a sexually reproducing one, an evolutionary event has occurred that involves a splitting into reproductively isolated lineages. And, furthermore, even if the parthenogenetic lineage does not meet the formal requirements of the biological species definition, it behaves in all essential features of its morphology and ecology like a "normal" species. I therefore claim that comparisons between sexually reproducing forms and derived parthenogenetic forms may throw light upon the mechanisms underlying genetical differentiation between "normal" species.

In contrast to speciation in sexual organisms, a new parthenogenetic species may arise in a single generation. This means that the first stages of normal speciation, where genetic differentiation occurs between geographically separated populations and on later contact reproductive isolation emerges, are completely left out. In a sense, a sexual species and a parthenogenetic species arisen from it may be considered two sibling species developed through a single mutational event. The initial genetic difference between such sibling species depends upon the nature of this process. If the parthenogenetic species is of hybrid origin, it differs from its ancestors to an extent that depends upon the genetic difference between these. If, however, it is of non-hybrid origin, the parthenogenetic species and its sexual ancestors are initially identical except for the mutational event that created the sibling pair. Existing genetic differences between sibling species of the latter category are, therefore, likely to have arisen subsequent to the speciation. My second claim is, therefore, that studies on the genetic differentiation between a non-hybrid parthenogenetic species and its sexual ancestor specifically elucidate the nature of the genetic differentiation that occurs in closely related species after speciation.

It is a question of great interest whether the existing genetic differences both within and between species are of a random nature or mainly adaptational. Since there is no initial difference between a non-hybrid parthenogenetic species and its sexual ancestor, and since two such species can occur under various conditions of sympatry, i.e. in more or less similar environments, my third claim is that observations on the genetic differentiation (including protein variation) between them can provide data that enable one to discriminate between the above-mentioned fundamental mechanisms.

OBSERVATIONS

1. *Two different situations*

In principle, the origin of a new parthenogenetic species is a saltational splitting into reproductively isolated lineages which are discrete and independent evolutionary units. However, if the eggs of the parthenogenetic species still have to be activated by spermatozoa from the sexual ancestor in order to develop normally (gynogenesis or pseudofertilization), the independence of the parthenogenetic species is strongly restricted. Several such cases are known, for a review see Christensen (1980a). The sexual diploid and parthenogenetic triploid (auto-) form of the oligochaete worm *Lumbricillus lineatus* represents such a situation. It occurs mainly in wrack beds along the sea coast, and diploids and triploids are found completely intermingled on practically every piece of rotten seaweed. This intimate coexistence shows that the individuals of the two biotypes must be very similar ecologically – probably just as similar as individuals within a single "normal" species. This is illustrated schematically in Fig. 1a. The idea of nearly identical resource utilizations in diploid and triploid

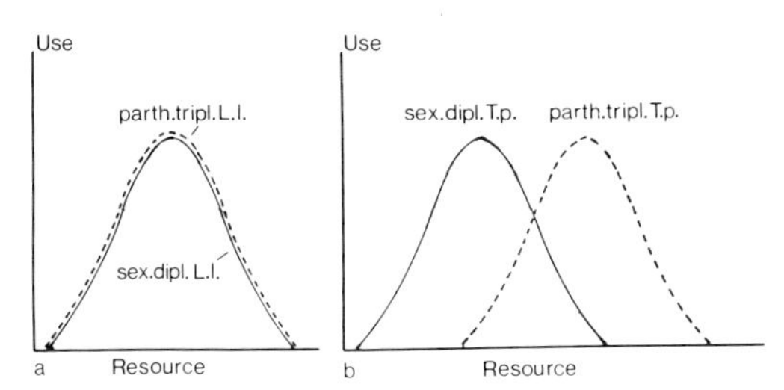

Fig. 1. Schematic illustration of the ecological interactions between sexual diploids and parthenogenetic triploids in (a) *Lumbricillus lineatus*, and (b) *Trichoniscus pusillus*.

L. lineatus is based upon the observed intimate coexistence in the field. In this particular case, where the organisms involved have a low vagility, the strong coexistence may be a consequence of the dependency of the triploid upon the diploid, but among higher animals where similar relationships exist, the partners involved may have different ecologies, see Vrijenhoek (1978).

This kind of reproductive parasitism on the ancestral sexual species must be extremely restrictive to the spread and adaptation of the parthenogenetic species. Natural selection would, therefore, be expected to do away with this dependence and, in the majority of parthenogenetic species, the eggs develop normally without being activated by spermatozoa from the sexual ancestor. Such a situation is known in the woodlouse *Trichoniscus pusillus*, where a sexual diploid and a parthenogenetic triploid biotype also occur, but in this case, the latter can exist independently of its sexual ancestor. In this case as well as in the case of *L. lineatus* mentioned above, the two biotypes are referred to the same nominal species because morphologically they do not differ sufficiently to be given species rank, but genetically speaking they are separated and, as already mentioned above, they are best compared with sibling species.

The independence of triploid parthenogenetic *T. pusillus* is clearly seen in the geographical distribution of the two biotypes. In southern Europe, the diploid form occupies all habitats. Through a zone of sympatry, it is gradually replaced by the triploid until, in northern Europe, the latter occupies all habitats (Vandel, 1940; Sutton, 1980; Fussey and Sutton, 1981). Such a pattern of geographical displacement is a common phenomenon among closely related sexual species. The interactions among such species pairs can be illustrated schematically as shown in Fig. 1b. The niche overlap indicates that the two species co-occur completely intermingled in some habitats, and may here compete for common limited resources.

It should be noted that the locality of the diploid *Trichoniscus pusillus* dealt with below is several hundred kilometres to the north of the continuous range of this form and that it may have been unintentionally introduced by man.

2. *The* Lumbricillus lineatus *case*

If the relationship between diploid and triploid *L. lineatus* indicated in

Fig. 1a is correct, then the two forms have practically identical ecologies, and this must have been the case throughout the entire evolutionary history of the parthenogenetic sibling. The outcome of genetic differentiation within the two independently evolving units is, therefore, expected to be different, depending on whether new allelic variations are neutral or adaptive and, thus, subject to natural selection. In the former case, no correlation in genetic differentiation is expected between sympatric populations of the two siblings sharing the same environment, while in the latter case, a strong correlation is expected between sympatric populations. If the allelic variation is adaptive, the parthenogenetic sibling, due to its dependency upon the sexual diploid, is bound to follow its "reproductive host" more or less like an identical twin in order to persist in the same environments.

These predictions were tested through a comparison of genetic similarity between diploids and triploids occurring in the same locality with that between diploid and triploid populations from two ecologically different environments (Christensen, 1979a). One of the stations, Ho, is on the west coast of Jutland facing the North Sea, the other, Nivå, is located north of Copenhagen on the coast of the far less saline Øresund. The worms were studied electrophoretically for nine polymorphic loci. As an expression of genetic similarity, genotype frequencies were used (Hedrick, 1971) instead of allele frequencies, because it is usually impossible to decide which of the alleles is present twice in heterozygote triploids containing two alleles. The results obtained are presented in Table I. They show a higher degree of genetic similarity between diploids and triploids co-occurring in the same environment than between diploids and triploids from different environments. This agrees with the prediction derived from the hypothesis that protein variation is adaptively relevant.

Table I. Genetic identity between sexual diploid and parthenogenetic triploid *Lumbricillus lineatus* occurring in the same and in different environments.

		Parthenogenetic triploid *L. lineatus*	
		Nivå	Ho
Sexual diploid	Nivå	0·973 ± 0·008	0·573 ± 0·113
L. lineatus	Ho	0·678 ± 0·109	0·905 ± 0·045

Mean and standard error are given.

Another feature, not specifically stated above, also supports the idea of a strong correlation between sympatric populations of diploid and triploid *L. lineatus*. The frequency of triploid genotypes with three different alleles is extremely low in all localities where the triploid coexists with the diploid. This is in sharp contrast to a high frequency of such genotypes in the *Pgi* locus in two unique localities where the diploid form is absent and a tetraploid cytotype of *L. lineatus* contributes the spermatozoa which activate the triploid eggs (see Christensen *et al.* 1976, 1978; Christensen, 1980b). It would seem that the triploid form, when coexisting with the diploid, does not exploit its innate possibilities to develop genotypes that cannot occur in diploids, whereas in the absence of diploids, such constraints upon the triploid do not exist.

3. *The* Trichoniscus pusillus *case*

The relationships between diploid and triploid *T. pusillus* outlined in Fig. 1b resemble the competitive interactions supposed to exist between many closely related sexual species in nature. Observations on the genetic differentiation in this case may, therefore, be of more general interest than the above case of *Lumbricillus lineatus*. Actually, the complete niche overlap between genetically independent units postulated in this latter case may not exist in nature according to conventional ecological thinking.

As mentioned above, the diploid form of *T. pusillus* is probably introduced by man into Denmark and is only known from a very small area with a limited range of environmental conditions. The data presented here are, therefore, based upon observations in a single locality, the bird preserve Rungstedlund, north of Copenhagen, where diploid and triploid *T. pusillus* coexist in 13 out of 16 stations investigated (Fig. 2; Christensen, 1983). Eight different loci were studied. Four of these (*Pgi, Fru 1.6 DP, Amy* and *Got*) are monomorphic and identical in the two forms. *Mdh* shows a single band in all triploids, whereas all the diploids in addition to this also have a slightly slower moving band indicating the presence of a fixed duplication in this form. The three remaining loci (*Pgm, Est* and *Idh*) are polymorphic and they share the following features: (1) both the diploid and the triploid form is polymorphic and (2) no allele is common to the two forms. This can be illustrated by the variation in the *Pgm* locus. In

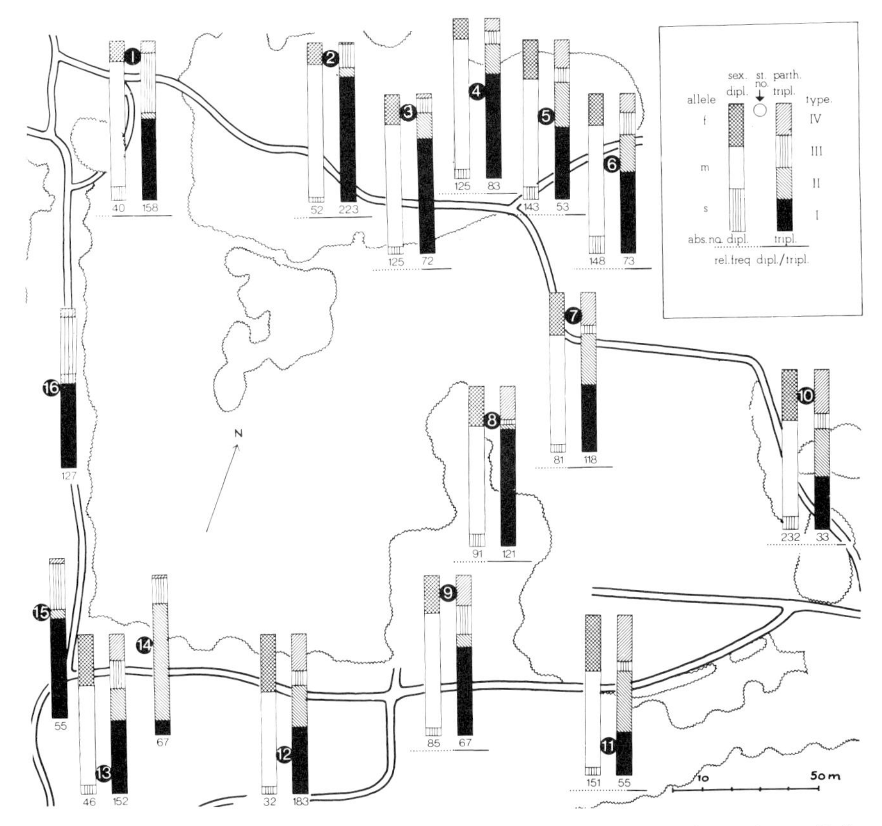

Fig. 2. Map showing positions of sampling sites. For each station, allele frequency in diploids (left column) and frequency of genotypes in triploids (right column) at the *Pgm* locus, sample size and relative frequency of the two forms (below) are given. In sts 14–16 only triploids were found. (From Christensen, 1983).

the diploid, three different alleles are present, *Pgm* 1.51, *Pgm* 1.38 and *Pgm* 1.24. The positions given are relative to that of the most common allele in the triploid, and in the following, the three alleles are termed *f*, *m* and *s*, respectively. In the triploid, four different types have been found, namely *Pgm* 1.00/1.25, *Pgm* 1.00/1.13, *Pgm* 1.00/1.13/1.29 and *Pgm* 0.85/1.00/125, in the following, termed Type I, II, III and IV, respectively (see also Christensen, 1979b).

No quantitative data are available for the *Idh* locus, and with respect to the esterase variation, both diploids and triploids showed only modest differences between stations within the area. But the variable

Pgm locus shows large and constant differences between stations, both in the distribution of alleles among diploids and in the frequency of genotypes among triploids (Christensen, 1983). It appears from Fig. 2 that the frequency of the *f* allele is fairly low among diploids in the northernmost stations, down to 13% in two stations, but between 32 and 36% in the eastern and southern stations. A nearly threefold increase over a distance of approximately 150 m. Similar strong differences occur among triploids. In the western and northern stations, the variants Type II and Type IV are rare, usually below 10%, whereas in the eastern and southern stations, both are of common occurrence, usually between 20 and 30%.

If this protein variation is adaptively neutral, the distribution of genotypes in diploid and triploid *T. pusillus* co-occurring in the same environment (station) is expected to vary at random. If, however, the variation is adaptive, some correlation is expected in such population parameters as genetic diversity and distribution of particular genotypes in co-occurring diploid and triploid populations. The scatter diagrams in Fig. 3 show the correlation between the genetic diversity (a) and the correlation between the frequency of the *f* allele and Type II + IV (b) in co-occurring diploid and triploid populations. A strong positive correlation is revealed in both cases, indicating that the variation is adaptive. Consequently, the diversification that has occurred in

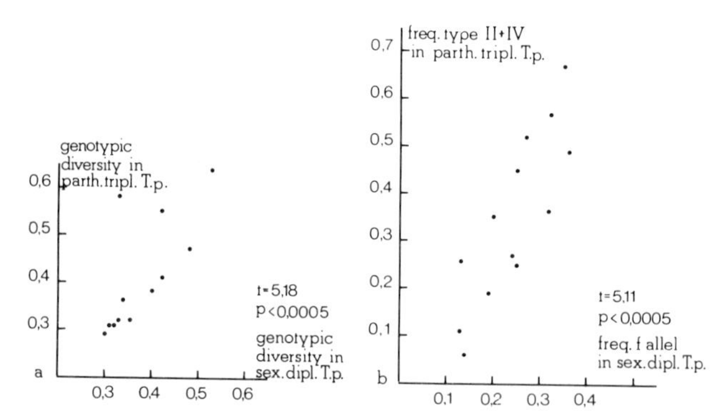

Fig. 3. Correlations between co-occurring diploid and triploid *T. pusillus* with respect to: (a) genotypic diversity (Simpson's index), and (b) frequency of *f* allele (*Pgm*) in diploids and Type II + IV in triploids. Associations measured by Spearman's Rank Correlation for large samples. (From Christensen, 1983).

the *Pgm* locus in these two independently evolving units has resulted in variations that are apparently adaptive *within* both lineages.

DISCUSSION

There is an enormous amount of literature that has discussed the various forces responsible for the maintenance of protein variation in populations. For reviews on the two most important alternative explanations, the neutral model and the adaptive model, see Kimura and Ohta (1974) and Ayala (1975). The arguments presented here in favour of the view that genetic differentiation *within* species is mainly due to natural selection thus agree with the view of Ayala (*op. cit.*). The present contribution indicates that this explanation is also valid within parthenogenetic species (see also Suomalainen *et al.*, 1976). With regard to the theme of this conference, an important question is whether the differentiation *between* species is also due to natural selection. Ayala and Gilpin (1974) and Ayala (1977) are in favour of this view but they have been challenged by Nei and Tateno (1975).

As mentioned above, a new parthenogenetic species arises through a saltational event, and if it is of non-hybrid origin, the ancestral and derived species would be initially identical. If existing genetic differences between two such species are due to natural selection, they have evolved as adaptive responses to subsequent ecological differentiations. Because of the special relationship between the diploid and the triploid form, *Lumbricillus lineatus* does not provide data that can be used in the present context. But the *Trichoniscus pusillus* case is potentially useful. The speciation that gave rise to triploid *T. pusillus* was not only saltational but, by necessity, also sympatric. Initially, therefore, the two species coexisted in the same environment and had practically identical ecologies. Subsequently, a process of ecological exclusion took place, resulting in a partitioning where sexual diploids can be said to inhabit central habitats, whereas the parthenogenetic triploids have taken over the more marginal ones. This is seen both in the geographical distribution of the two forms and in their habitat partitioning when they occur within the same area. The horizontal bars in Fig. 2 show the relative frequency of diploids and triploids within the area studied. It is seen that diploids are common in the eastern stations but are rare or entirely absent in the western where triploids predominate. The former stations are lower lying and with favourable moisture conditions throughout the year, while the latter are somewhat elevated and may experience summer droughts. Considering that woodlice are crustaceans adapted to a terrestrial life, the habitats where diploids predominate may be called central and those where only triploids occur marginal in an evolutionary sense (see also Vandel, 1940; Sutton, 1980). Is then the genetic differentiation that has occurred among diploid and triploid *T. pusillus* and which in some loci has resulted in diagnostically useful differences an adaptive response to the above described ecological differentiation? Our present knowledge is far too restricted to provide an answer. But the question may be rephrased to something less ambitious but still interesting: Can the processes that we have identified within the species account for the observed difference between them? Or, more specifically: Is the adaptive intraspecific variation correlated with the environmental factor(s) that determine the distribution of the two species?

Now an answer can be given. The observed relative frequency of the two forms within a given station is taken as a measure of the environmental factor(s) to which the two species are adapted. They are then plotted against the frequency of the *f* allele in the diploid and the frequency of Type II + IV in the triploid. The resulting diagrams are shown in Fig. 4, and, in none of the cases, is there any correlation between the intraspecific variation and the environmental factors underlying the distribution of the two species. Therefore, in this case at least, the processes occurring within the species are apparently not directly coupled to those occurring between species. I would hesitate

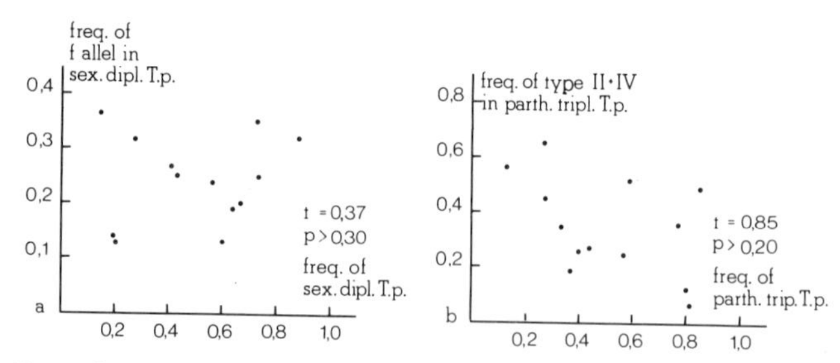

Fig. 4. Correlations between (a) relative frequency of sexual diploid *T. pusillus* and frequency of the *f* allele (*Pgm*), and (b) relative frequency of triploid parthenogenetic *T. pusillus* and frequency of Type II + IV. (From Christensen, 1983).

to conclude that the existing species specific differences are non-adaptive. The resource dimensions to which they might eventually be adaptive responses may well be different from those to which the intraspecific variation is adapted.

In conclusion: the data presented here indicate that intraspecific protein differentiation is mainly adaptive; the same question applied to interspecific differentiation could not be answered, but observations are presented indicating that processes at the two levels are to some extent uncoupled.

ACKNOWLEDGEMENTS

I thank The Danish State Research Foundation for financial support and C. Overgaard Nielsen, H. Enghoff and B Theisen for helpful comments on the manuscript.

REFERENCES

Ayala, F. J. (1975). Genetic differentiation during the speciation process. *Evol. Biol.* **8**, 1–78.

Ayala, F. J. (1977). Protein evolution: Non-random patterns in related species. *In* "Measuring Coefficients of Selection in Natural Populations" (F. B. Christiansen and T. Fenchel, eds), pp. 177–205. Springer-Verlag, Berlin.

Ayala, F. J. and Gilpin, M. E. (1974). Gene frequency comparisons between taxa. Support for the natural selection of protein polymorphisms. *Proc. Nat. Acad. Sci., U.S.A.* **71**, 4847–4849.

Christensen, B. (1979a). Studies on genetic similarity in two obligatorily coexisting forms of *Lumbricillus lineatus* (Müller) (Enchytraeidae). *Zool. Scripta* **8**, 314.

Christensen, B. (1979b). Differential distribution of genetic variants in triploid parthenogenetic *Trichoniscus pusillus* (Isopoda, Crustacea) in a heterogeneous environment. *Hereditas* **91**, 179–182.

Christensen, B. (1980a). Annelida. *In* "Animal Cytogenetics" (B. John, ed.). Borntraeger, Berlin and Stuttgart.

Christensen, B. (1980b). Constant differential distribution of genetic variants in polyploid parthenogenetic forms of *Lumbricillus lineatus* (Enchytraeidae, Oligochaeta). *Hereditas* **92**, 193–198.

Christensen, B. (1983). Genetic variation in coexisting sexual diploid and parthenogenetic triploid *Trichoniscus pusillus* (Isopoda, Crustacea). *Hereditas* **98**, 201–207.
Christensen, B., Berg, U. and Jelnes, J. (1976). A comparative study on enzyme polymorphism in sympatric diploid and triploid forms of *Lumbricillus lineatus* (Enchytraeidae, Oligochaeta). *Hereditas* **84**, 41–48.
Christensen, B., Jelnes, J. and Berg, U. (1978). Long-term isozyme variation in parthenogenetic polyploid forms of *Lumbricillus lineatus* (Enchytraeidae, Oligochaeta) in recently established environments. *Hereditas* **88**, 65–73.
Fussey, G. D. and Sutton, S. L. (1981). The identification and distribution of the bisexual and parthenogenetic forms of *Trichoniscus pusillus* (Isopoda, Oniscoidea) in Ireland. *Ir. Nat. J.* **20**, 196–199.
Hedrick, P. W. (1971). A new approach to measuring genetic similarity. *Evolution, Lancaster, Pa.* **25**, 276–280.
Kimura, M. and Ohta, T. (1974). On some principles governing molecular evolution. *Proc. Nat. Acad. Sci., U.S.A.* **71**, 2848–2852.
Nei, M. and Tateno, Y. (1975). Interlocus variations of genetic distance and neutral mutation theory. *Proc. Nat. Acad. Sci., U.S.A.* **72**, 2758.
Soumalainen, E., Saura, A. and Lokki, J. (1976). Evolution of parthenogenetic insects. *Evol. Biol.* **9**, 209–257.
Sutton, S. L. (1980). "Woodlice". Pergamon Press, Oxford.
Vandel, A. (1940). La parthénogenèse geographique IV. Polyploidie et distribution geographique. *Bull. biol. Fr. Belg.* **74**, 94–100.
Vrijenhoek, R. C. (1978). Coexistence of clones in a heterogeneous environment. *Science, N.Y.* **199**, 549–552.

21 | Concluding Remarks

A. J. CAIN

Department of Zoology, University of Liverpool,
P. O. Box 147, Liverpool L69 3BX, U.K.

One cannot produce a summary of so vast a diversity of topics, enzymes and organisms as have been set before us, but there are a number of concluding reflections that may perhaps be useful. Principally, I am impressed (or rather, depressed) by the very diverse interpretations that have been put on their findings by the speakers, and by workers in these fields generally. The number of explanatory myths that are circulating in those branches of genetics and taxonomy relevant to our themes is really extraordinary, and would do credit to any pre-scientific culture. I call them explanatory myths because, whether or not there is any truth in them, they are used to prop up people's pet theories without the slightest attempt to verify them, even when they are verifiable.

In the first place, of course, there is the myth of the neutrality of enzyme variants. One of those who put it forward has told us that the objectivity of science is a "text-book myth". "Long before there is any direct evidence, scientific workers have brought to the issue deep-seated prejudices" which are ultimately caused by "deep ideological biasses" and so on (Lewontin, 1974). Surely in this case he is right. The dogma was asserted on no direct evidence whatever, supported by a display of mathematical theorizing so simplistic that it could not possibly apply to real organisms, and protected from exposure by an avoidance of any consideration of the probable biology of the variants that has been maintained ever since. The number of workers (especially in the U.S.A. where fashion is a great determinant of what work is

Systematics Association Special Volume No. 24, "Protein Polymorphism: Adaptive and Taxonomic Significance", edited by G. S. Oxford and D. Rollinson, 1983. Academic Press, London and New York.

not done) who happily accepted this explaining away is really alarming. The effect of it was clearly to insinuate that there was no point in looking into the possibility that enzyme variants were selected for different functions. Fortunately the control exercised by this dogma could only found a sect, not a Universal Church, and there have been enough dissidents (some represented in this conference) to do some actual work on the problem, and show the falsity of the dogma as dogma. Even the least reflective followers of the neutralist school must start thinking now. No such *a priori* judgments can be made on enzyme or any other variants; what little positive evidence we have emphasizes the effectiveness of selection on them (Cain, 1977, 1979).

A second myth, arising out of neutrality, is the biological clock. There never was any evidence for a biological clock in the fine structure of proteins. It was a pure inference from the dogma of neutrality and had no more basis (or likelihood) than it. Although the clock still fascinates some workers, quite enough has been done empirically (e.g. Joysey, 1981) to show its worthlessness even to those who had not perceived its theoretical status. Yet why should serious workers have to waste time refuting such ideas? Probably nine-tenths of the work done assuming the neutrality of enzyme variants has been a waste of time, reagents, salaries and scientific print. What accidental benefits may have accrued could have been obtained equally well, if not better, by a more realistic approach.

The myth of genetic revolution in small populations has not affected our themes much, except in the so-called decoupling of macroevolution from microevolution. It is worth recollecting that it was an ingenious hypothesis, deserving investigation not misuse as a myth, produced by Mayr (1954) to account for apparently bizarre features of certain bird populations on small islands off New Guinea. The interrelations, behavioural and ecological, of these forms with other bird species in their small and taxonomically unbalanced faunas, have not been investigated. And although genes at different loci can certainly interact, the *genetic* evidence for coadaptation of the whole genotype is remarkably unconvincing. Indeed, coadaptation of the genotype is itself in a fair way to becoming an explanatory myth (Cain, 1977).

For good measure I will add the myth that the genetic programme of any living organism "is the result of a history that goes back to the origin of life and thus incorporates the "experiences" of all ancestors" – Delbrück as reported approvingly by Mayr (1982: 56, 69). Very little reflection is needed to show that if this were so, evolution would be the most hopelessly inefficient of all natural processes in the Universe. Embryologists, who are well aware of extensive remodellings in development, could not subscribe to it for a moment.

In taxonomy, Hennigian, cladistic and transform cladistic, the number of myths is so great, one can hardly even enumerate them. Some have been introduced as methodologies, but in fact produce the explanations they profess to find. Perhaps the most pervasive is the myth of parsimony, of which we have heard something. Evolution knows nothing whatever of parsimony. Even those workers who insist on using it as a principle in constructing phylogenetic trees (and they include many enzymologists) are surprised by the amount of convergence and parallelism that is detected even by this most inefficient approach (see papers in Joysey and Friday, 1982). In fact, (Cain, 1982) it will underrate seriously the amount of convergence, because it is precisely in those stocks which are already closely allied that convergence is most likely to occur (and most likely to go undetected). Convergence (except by Dr G. M. Davis) has been very carefully skirted around in almost every paper in this conference. Only one has mentioned the importance of discordant characters, the finding of which is the only way we have of getting even a minimum estimate of the amount of convergence in evolution. Even this method has hardly been used, yet as soon as one scans a number of related species, one turns up not only examples of convergence in particular characters, but samples of characters some of which must be convergent, but exactly which we cannot determine (Cain, 1982).

An example is the very common southern and western European juxtalittoral snail *Theba pisana* (Müller). It clearly belongs with the Common Snail *Helix aspersa* and similar forms in the large Western Palaearctic family Helicidae. Since it has a four-bladed dart (used in courtship) it is always put in the subfamily Helicinae. But is this sufficient evidence? The shell pattern agrees far better with those in a different subfamily (Helicellinae), and anatomical features of the reproductive system with a third (Helicigoninae). Now, I have absolutely no means *a priori* of determining which of these three sets of characters is more important phylogenetically than the others. But what can be said is, whatever set is the right one phylogenetically, the other two are convergent. In fact, in the principal taxonomic characters of this form, two thirds are convergent, although no-one knows which two thirds they are.

Extensive convergence implies independent adaptation, and makes neutrality highly unlikely. How do we know, apart from discordant characters, whether similarities are due to convergence or not? This is a difficult problem, irrespective of what type of character we are using. Either (1) one must have genuine dated historical evidence (which we have not for the biological clock), or (2) we must be able to estimate the probability of convergence from the nature of the selection pressures acting upon a given character, or (3) we may have a history of the ecological opportunities open to a particular stock.

In the first case, we need a near-perfect fossil record, in which the affinities of all forms are evident on straight comparative anatomy, without any interpretation from *a priori* principles (especially taxonomic ones) contaminating the classification. This will be available only seldom. In the second, I have suggested (Cain, 1982) that if species-recognition structures and behaviours used in courtship are a genuine arbitrary sign-language, they will be able to change, except for certain species-specific characters, only slowly, and may be the best phyletic indicators. They if anything may give the best approach to a biological clock. In the third case, if we know that a continent or large island (for example) has been isolated for a long time, then a particular group of animals on it, evolving in the face of competition from the rest of the fauna and so confined to a particular set of niches, may be genuinely monophyletic (Cain, 1982), and any resemblances to forms on other landmasses must be convergence. There are classic examples in the fossil history of mammals. Everyone knows of the radiation in South America, but there are also an Australian radiation, a Madagascan one, and two successive ones in Africa. Here, plate tectonics may help, especially when the fossil record is poor. An excellent example is G. M. Davis's study (1979) on the adaptive radiation of pomatiopsid snails in the Mekong River, in which the plate tectonics explains the history of the river system which explains the ecological opportunities that these animals have taken up (an example which I did not believe when I first heard of it).

If we cannot take it for granted that characters are adaptively neutral, then what we need to know is, what is the taxonomic significance of an adaptive character, a question that has never yet been worked out, partly owing to Charles Darwin's influence (Cain, 1964). He took over from Richard Owen the old dichotomy of interpretation, that what characters one can see are adaptive are adaptive, and what one can not are part of God's (or Nature's) plan and exist for edification. Darwin produced a new interpretation for the second category – they must be ancestral. Logically, this is a cross-classification, because what it ancestral can also be adaptive – in fact is most likely to be retained if it is adaptive. The questions of the extent of convergence and taxonomic importance of adaptive characters are the only important ones in evolutionary taxonomy, and are sedulously avoided by those who play museum parlour-games with taxonomic characters.

Fortunately there are positive gains that this conference has shown up well. The most striking is the number of sibling species that have already been detected by enzymic investigation. The idea which afflicted certain drosophilists early in this century, that species must be distinct enough for taxonomists to be able to separate them on museum material, goes back to an old Talmudic tradition. When the animals were brought before the Protoplast (the first-formed Man) to receive their names, each kind bore its signature – its distinguishing characters – because it would be beneath his dignity to have to strain his eyes. The importance of sibling species to us is that they reveal how incompetent we are, without special techniques, both to recognize them, and to appreciate the environment in the way that they do. They can find different energy-pools stable enough to maintain them and accessible in ways different enough to require a different species, where we can see no difference at all. How important it is in biology not to ask the pundits, always to ask the organisms.

In close connection with this is the amount of variation that can be found in parthenogenetic and similar forms, which certainly cannot be dismissed as evolutionary dead ends.

Equally important are the results of surveys and direct experimental investigations which either reveal selection acting on enzyme variants, or reveal such associations with climatic and habitat variables that selection is the most likely explanation. Again, it is absolutely useless for us to look casually at some peculiar variation and proclaim that because we can see no sense in it, it must be due to genetic drift, or founder effects, or chance, or history or what not. Genetic drift is of course a serious scientific hypothesis. Given the right conditions, the process is certain to occur. The question is, how often do those conditions actually occur in the wild? Drift has been used in practice as one of the biggest explanatory myths of all.

I have just written a review of terrestrial molluscan ecogenetics, and what has impressed me most as a result of it is the absolute necessity of temporal as well as spatial surveys. In work on the snail *Cepaea*, we are, of course, lucky in some respects. There are studies of populations that have been going on actively or intermittently for up to fifty years. There are others that can use data (from agricultural and forestry records) that go back for about a thousand years. And some work, on subfossil material, spans several thousand years. Even in those areas, such as the sand-dunes at Berrow mapped by Diver and then by Clarke and Murray, where the morph frequency patterns seem remarkably random, long-term surveys show a general stability accompanied by overall fluctuations both of which are incompatible with anything but selection.

It is of course true, as I have pointed out before (Cain, 1977, 1979) that work on selection in the wild is chancy in the extreme. One cannot apply for a grant in the happy confidence that striking results will be reported in two or three years' time. And if one does not find selection, that does not mean that it was not acting. Nevertheless, if we are to advance in knowledge instead of merely in argumentation and personal publicity, what we need more than anything else is hard fact on what goes on in the wild. What with those who are prejudiced against selection, those that won't undertake the hard work necessary, and those that know it all already, it will be impossible soon to master the so-called literature before getting out into the field. This is indeed the age of the loudmouth.

REFERENCES

Cain, A. J. (1964) The perfection of animals. *Viewpts Biol.* **3**, 36–63.

Cain, A. J. (1977) The efficacy of natural selection in wild populations. *In* "Changing Scenes in Natural Sciences 1776–1976" (C. E. Goulden, ed.), pp. 111–133. Academy of Natural Sciences, Philadelphia.

Cain, A. J. (1979). Introduction to general discussion. *Proc. R. Soc. Lond.*, B **205**, 599–604.

Cain, A. J. (1982). On homology and convergence. *In* "Problems of Phylogenetic Reconstruction" (K. A. Joysey and A. E. Friday, eds), pp. 1–19. Academic Press, London.

Davis, G. M. (1979). The origin and evolution of the gastropod family Pomatiopsidae, with emphasis on the Mekong River Triculinae. *Monogr. Acad. nat. Sci. Philad.* **20**, 1–120.

Joysey, K. A. (1981). Molecular evolution and vertebrate phylogeny in perspective. *Symp. zool. Soc. Lond.* **46**, 189–218.
Joysey, K. A. and Friday, A. E. (eds) (1982). "Problems of Phylogenetic Reconstruction". Academic Press, London.
Lewontin, R. C. (1974). "The Genetic Basis of Evolutionary Change". Columbia University Press, Columbia.
Mayr, E. (1954). Change of genetic environment and evolution. *In* "Evolution as a Process" (J. S. Huxley, A. C. Hardy and E. B. Ford, eds), pp. 157–180. Allen and Unwin, London.
Mayr, E. (1982). "The Growth of Biological Thought. Diversity, Evolution and Inheritance". Belknap Press, Cambridge, Mass.

INDEX

N

O

P

Systematics Association Publications

1. BIBLIOGRAPHY OF KEY WORKS FOR THE IDENTIFICATION OF THE BRITISH FAUNA AND FLORA
   3rd edition (1967) [Out of print]
   Edited by G. J. Kerrich, R. D. Meikle and N. Tebble
2. FUNCTION AND TAXONOMIC IMPORTANCE (1959)
   Edited by A. J. Cain
3. THE SPECIES CONCEPT IN PALAEONTOLOGY (1956)
   Edited by P. S. Sylvester-Bradley
4. TAXONOMY AND GEOGRAPHY (1962)
   Edited by D. Nichols
5. SPECIATION IN THE SEA (1963)
   Edited by J. P. Harding and N. Tebble
6. PHENETIC AND PHYLOGENETIC CLASSIFICATION (1964)
   Edited by V. H. Heywood and J. McNeill [Out of print]
7. ASPECTS OF TETHYAN BIOGEOGRAPHY (1967)
   Edited by C. G. Adams and D. V. Ager
8. THE SOIL ECONOSYSTEM (1969)
   Edited by J. Sheals
9. ORGANISMS AND CONTINENTS THROUGH TIME (1973)†
   Edited by N. F. Hughes

LONDON. Published by the Association

Systematics Association Special Volumes

1. THE NEW SYSTEMATICS (1940)
   Edited by Julian Huxley (Reprinted 1971)
2. CHEMOTAXONOMY AND SEROTAXONOMY (1968)*
   Edited by J. G. Hawkes

* Published by Academic Press for the Systematics Association
† Published by the Palaeontological Association in conjunction with the Systematics Association

3. DATA PROCESSING IN BIOLOGY AND GEOLOGY (1971)*
   Edited by J. L. Cutbill
4. SCANNING ELECTRON MICROSCOPY (1971)*
   Edited by V. H. Heywood
5. TAXONOMY AND ECOLOGY (1973)*
   Edited by V. H. Heywood
6. THE CHANGING FLORA AND FAUNA OF BRITAIN (1974)*
   Edited by D. L. Hawksworth
7. BIOLOGICAL IDENTIFICATION WITH COMPUTERS (1975)*
   Edited by R. J. Pankhurst
8. LICHENOLOGY: PROGRESS AND PROBLEMS (1977)*
   Edited by D. H. Brown, D. L. Hawksworth and R. H. Bailey
9. KEY WORKS (1978)*
   Edited by B. G. Kerrich, D. L. Hawksworth and R. W. Sims
10. MODERN APPROACHES TO THE TAXONOMY OF RED AND BROWN ALGAE (1978)*
    Edited by D. E. G. Irvine and J. H. Price
11. BIOLOGY AND SYSTEMATICS OF COLONIAL ORGANISMS (1970)*
    Edited by G. Larwood and B. R. Rosen
12. THE ORIGIN OF MAJOR INVERTEBRATE GROUPS (1979)*
    Edited by M. R. House
13. ADVANCES IN BRYOZOOLOGY (1979)*
    Edited by G. P. Larwood and M. B. Abbot
14. BRYOPHYTE SYSTEMATICS (1979)*
    Edited by G. S. Clarke and J. G. Duckett
15. THE TERRESTRIAL ENVIRONMENT AND THE ORIGIN OF LAND VERTEBRATES (1980)*
    Edited by A. L. Panchen
16. CHEMOSYSTEMATICS: PRINCIPLES AND PRACTICE (1980)*
    Edited by F. A. Bisby, J. G. Vaughan and C. A. Wright
17. THE SHORE ENVIRONMENT: METHODS AND ECOSYSTEMS (2 Volumes) (1980)*
    Edited by J. H. Price, D. E. G. Irvine and W. F. Farnham

18. THE AMMONOIDEA (1981)*
Edited by M. R. House and J. R. Senior
19. BIOSYSTEMATICS OF SOCIAL INSECTS (1981)*
Edited by P. E. Howse and J.-L. Clément
20. GENOME EVOLUTION (1982)*
Edited by G. A. Dover and R. V. Flavell
21. PROBLEMS OF PHYLOGENETIC RECONSTRUCTION (1982)*
Edited by K. A. Joysey and A. E. Friday
22. CONCEPTS IN NEMATODE SYSTEMATICS (1983)*
Edited by A. R. Stone, H. M. Platt and L. F. Khalil
23. EVOLUTION, TIME AND SPACE (1983)*
Edited by R. W. Sims, J. H. Price and P. E. S. Whalley
24. PROTEIN POLYMORPHISM: ADAPTIVE AND TAXONOMIC SIGNIFICANCE (1983)*
Edited by G. S. Oxford and D. Rollinson
25. CURRENT CONCEPTS IN PLANT TAXONOMY (1983)*
Edited by V. H. Heywood and D. M. Moore
26. DATABASES IN SYSTEMATICS (1984)*
Edited by R. Alkin and F. R. Bisby